# Electronic Techniques

Fifth Edition

# Electronic Techniques

## SHOP PRACTICES AND CONSTRUCTION

Robert S. Villanucci
Alexander W. Avtgis
William F. Megow

*Wentworth Institute of Technology*
*Boston, Massachusetts*

Prentice Hall, Inc.
*Englewood Cliffs, New Jersey*      *Columbus, Ohio*

**Library of Congress Cataloging-in-Publication Data**

Villanucci, Robert S.
    Electronic techniques : shop practices and construction / Robert
S. Villanucci, Alexander W. Avtgis, William F. Megow.– 5th ed.
      p.    cm.
    Includes bibliographical references and index.
    ISBN 0–13–361965–6
    1. Electronic apparatus and appliances–Design and construction.
  2. Electronic packaging.    I. Avtgis, Alexander W.
II. Megow, William F.      III. Title.
TK7836.V55   1996
621.381–dc20                          95–32971
                                            CIP

Cover photo: © K. Harriger/American Stock Photography
Editor: Charles E. Stewart, Jr.
Production Coordination: bookworks
Cover Designer: Patti Ann C. Okuno-Levering
Production Manager: Laura Messerly
Marketing Manager: Debbie Yarnell
Illustrations: Diphrent Strokes, Inc.

This book was set in Times Roman by Pine Tree Composition and was printed and bound
by Semline, Inc., a Quebecor America Book Group Company. The cover was printed by
Phoenix Color Group.

Printed in the United States of America

10 9 8 7 6 5 4 3 2 1

ISBN: 0–13–361965–6

Prentice-Hall International (UK) Limited, *London*
Prentice-Hall of Australia Pty. Limited, *Sydney*
Prentice-Hall of Canada, Inc., *Toronto*
Prentice-Hall Hispanoamericana, S. A., *Mexico*
Prentice-Hall of India Private Limited, *New Delhi*
Prentice-Hall of Japan, Inc., *Tokyo*
Simon & Schuster Asia Pte. Ltd., *Singapore*
Editora Prentice-Hall do Brasil, Ltda., *Rio de Janeiro*

*To*
*Annette, Gloria, and Helen*

# Brief Contents

# Unit FOUR
# Printed Circuit Board Component Assembly and Soldering  295

# Unit FIVE
# Chassis Fabrication and Metal Finishing Techniques  361

# Unit SIX
# Chassis Hardware and Wiring  431

# Unit SEVEN
# Student Projects  487

# Contents

## Unit TWO
# Designing Printed Circuit Boards with CAD Techniques 63

# 5 Printed Circuit Board Materials 64

# 6 Single-Sided Printed Circuit Board Design: The Preliminary Sketch 82

# 7 Drawing PCB Artworks with Computer-Aided Drafting Techniques 105

# 8 Double-Sided Plated-Through-Hole Printed Circuit Artwork Design 132

# Unit THREE
# Printed Circuit Board Fabrication and Inspection 145

# 9 PCB Photographic Processes 146

# 10 Single-Sided PCB Processing: Print-and-Etch Technique 164

# 11 Double-Sided PCB Processing: Electroless Copper Deposition and Copper Flash Plating 185

# 12 Double-Sided PCB Processing: Imaging 210

# 13 Double-Sided PCBs: Final Fabrication Processes 227

# 14 Multilayer PCB Fabrication 247

# 15 Quality Assurance Inspection Procedures for PCBs 268

# Unit FOUR
# Printed Circuit Board Component Assembly and Soldering 295

## 16 PCB Hardware and Component Assembly 296

## 17 PCB Soldering 316

## 18 Surface Mount Technology 339

## Unit FIVE
# Chassis Fabrication and Metal Finishing Techniques 361

# 19 Shearing 362

# 20 Chassis Layout 369

# 21 Drilling, Reaming, and Punching 381

# 22 Metal Cutting 397

# 23 Bending 412

# 24 Metal Finishing and Labeling 425

# Unit SIX
# Chassis Hardware and Wiring 431

# 25 Chassis Hardware and Assembly 432

LEARNING OBJECTIVES 432

# 26 Wire and Interconnection Techniques 453

LEARNING OBJECTIVES 453

# Unit SEVEN
## Student Projects 487

# 27 Projects 488

# Appendices 525

# Bibliography 543

# Index 545

# Preface

The electronic packaging industry continues to undergo rapid changes, and there appears to be no end to the progress in sight. The adoption of new device packages for high-density assemblies in conjunction with more functionally complicated printed circuit board technologies has continued to bring about evermore sophisticated design, construction, and packaging techniques that require today's technician to possess special skills. It is the purpose of this fifth edition to continue to present a practical and realistic approach for developing these new skills in planning, designing, and constructing electronic equipment.

Owing to the wide acceptance of the first four editions of *Electronic Techniques,* much of the material has been retained, but several changes—additions, deletions, and re-arrangements—have been incorporated into this new edition to both strengthen and update the material.

### What's New in This Fifth Edition

1. To better organize the material, the book has been rearranged into seven separate units. Each unit groups common processes, materials, or fabrication techniques, allowing the user access to all material necessary to study and understand a complete topic, in one unit. The new unit designations are given below:

Unit 1: Circuit Design and Project Planning
Unit 2: Designing PCBs with CAD Techniques
Unit 3: PCB Fabrication and Inspection
Unit 4: PCB Component Assembly and Soldering
Unit 5: Chassis Fabrication and Metal Finishing Techniques
Unit 6: Chassis Hardware and Wiring
Unit 7: Student Projects

2. A **new** chapter on electronic circuit design, using a modern 5-V switching regulator as the vehicle, is developed with the aid of National Semiconductor's SWITCHERS MADE SIMPLE design software. Circuit analysis of the unregulated portion of the power supply project is developed with a popular circuit simulation software package called PSpice from MicroSim Corporation.

3. The chapter on **Packaging Design** has been completely rewritten to accommodate the new switching regulator power supply project.

4. The chapter on **Preparing Detailed Drawings** has been revised to reflect the changes associated with the new switching regulator power supply project.

5. The chapter on **Single-Sided Printed Circuit Board Design** has been completely revised and concentrates on single-sided pc board design concepts and sketching the preliminary component and conductor pattern artwork.

6. The chapter on single-sided pc board design is followed by the addition of a **new** chapter that details the techniques necessary to design and layout PCBs using a general-purpose, computer-aided design package—**AutoCAD-12**—to produce component and conductor pattern artwork for single-sided pc boards.

7. A revised chapter on double-sided printed circuit board design, using **EAGLE 2.6** from CadSoft Computer, Inc., highlights a software tool that was specifically developed to design printed circuit boards. The use of a *Schematic Module, Layout Editor,* and *Autorouter* to develop the top and bottom conductor pattern artworks for a sample double-sided PCB is presented.

8. A **new project** using Motorola's M68HC11 EVBU microcontroller development platform has been included. This project reflects current state-of-the-art technology and is intended to generate student interest.

9. A general statement on **safety** has been added to each chapter whose subject matter warrants it, and a more comprehensive discussion on the various aspects of general safety, basic precautions, and available information and regulations has been added after this preface.

## To the Educator

This text has been designed to fulfill the needs of a typical electronic shop course, manufacturing processes program, or package design project. The material is directed toward educational institutions such as technical–vocational schools, technical institutes, and junior colleges, as well as industrial and military training programs.

The training of a skilled craftsman requires not only clear and detailed explanations, but also visual and graphic aids. For this reason, over 500 figures and drawings have been included to illustrate fundamental techniques used in electronic design, construction, and packaging. In addition, many exercises have been included to aid in a thorough understanding of these basic concepts.

Although the chapters are grouped in units of similar material, individual chapters may be studied independently. No prerequisite knowledge of electronic circuits is necessary to understand the package design and fabrication techniques presented, although a deeper insight may be realized if the technician has a fundamental background in electronic circuits and devices. Chapter 1, however, does require some background in electronics for a more complete understanding.

**UNIT 1:** In Chapter 1, the design, with popular software tools, of a 5-volt switching-regulator power supply is introduced and used as a teaching vehicle for all the remaining topics in this first unit. Chapter 2 is a discussion of the general factors that must be considered in packaging any electronic system. Chapter 3 reintroduces the switching-regulator project and its packaging design. In Chapter 4, preliminary considerations of the package are converted to detailed engineering drawings and sketches.

**UNIT 2:** The four chapters in this unit detail the information necessary to design both single- and double-sided printed circuit boards. Chapter 5 supplies pertinent information on available printed circuit board material needed to make an informed selection. Chapter 6 covers the logical design process necessary to generate a "sketch" of the component placement and conductor pattern routing for single-sided PCBs. Then, in Chapter 7, using a general-purpose CAD drafting package, detailed information on keystrokes and procedures necessary to convert the "sketch" into a conductor pattern artwork is covered. AutoCAD, Release 12, was chosen to illustrate this chapter, but the presentation can be adapted to other manufacturers' products. This unit concludes with Chapter 8 on the design of double-sided PCBs. In this chapter a software tool developed specifically to design PCBs, EAGLE 2.6, was selected to illustrate the use of a Schematic Model, Layout Editor, and Autorouter to develop the top and bottom conductor pattern artworks for a simple double-sided PCB. Concepts such as Electrical Rules Check (ERC), Design Rules Check (DRC), and RATSNEST are presented.

**UNIT 3:** Chapters 9 through 15 provide detailed information on fabricating both single- and double-sided printed circuit boards. Chapter 9 introduces the photographic process, includes information on diazo film to make positive phototools, and covers the use of a densitometer. Chapter 10 supplies detailed information on the fabrication of single-sided printed circuit boards using the print-and-etch method. Chapters 11 through 13 cover the plated-through-hole process, the print-plate-and-etch technique for double-sided board fabrication, and the use of dry-film photo resist. Chapter 14 provides basic design and fabrication procedures for multilayer boards (MLBs). Visual inspection and testing procedures for printed circuit boards are presented in Chapter 15. This chapter also includes the use of a microprocessor-based bed-of-nails tester, a Caviderm, and microsectioning techniques.

**UNIT 4:** Chapter 16 covers printed circuit board hardware component assembly and Chapter 17 includes information on PCB soldering. Topics range from basic hand soldering to industrial processes such as wave soldering. This unit concludes with Chapter 18, which details information on surface-mount technology, including component assembly and soldering.

**UNIT 5:** The layout, fabrication, forming, and finishing of sheet metal chassis elements are covered in Chapters 19 through 24.

**UNIT 6:** Chassis hardware assembly and wiring techniques are covered in Chapters 25 and 26 in order to combine the various elements into a finished system.

**UNIT 7:** Chapter 27 contains a broad selection of projects to (1) further develop student interest and (2) provide practice in developing the skills discussed in this book.

### Regarding Software

In keeping with the industrial trend of using more computer software packages to facilitate electronic designs, enhance quality, and improve productivity by reducing the packaging-time cycle, five separate software packages have been included for presentation in the fifth edition of this book. Each application package was chosen as a representative example to illustrate the solution to a particular problem presented. However, many other products are available from a very large group of suppliers, a group that is itself expanding rapidly. Please note also that the selection was made without regard to price. The cost of the software illustrated in this book ranges from free (available from the manufacturer) to industrial-grade products—large programs—that are rather expensive. The following is a list of the computer software products used here and the name and address of the manufacturer or supplier for each.

> AutoCAD—general-purpose drafting software
>> Autodesk, Inc.
>> 2320 Marinship Way
>> Sausalito, CA 94965
>> (415) 332-2344

EAGLE (Version 2.6)—PCB layout software
    CadSoft Computer, Inc.
    801 South Federal Highway
    Suite 201
    Delray Beach, FL 33483-5185
    (407) 274-8355

SWITCHERS MADE SIMPLE—dc-to-dc converter design software
    National Semiconductor Corporation
    2900 Semiconductor Drive
    P.O. Box 58090
    Santa Clara, CA 95052-8090
    1-800-272-9959 or (408) 721-5000

PSpice—circuit analysis software
    MicroSim Corporation
    20 Fairbanks
    Irvine, CA 92718
    1-800-245-3022

LABTECH ACQUIRE—process control software
    Laboratory Technologies Corporation
    400 Research Drive
    Wilmington, MA 01887
    (508) 657-5400

## A Word About Safety

Modern industrial printed circuit board fabrication processes require the use of hazardous chemicals as described in this book. The authors cannot emphasize strongly enough that the handling, use, shipping, storage, and disposal of these chemicals be performed under supervision by people who are trained in this area and who are familiar with local, NFPA, NIOSH, OSHA, and EPA regulations. This is a learning text and, as such, describes in detail the chemical propcesses required. However, in no way does it provide the training and conditions necessary for the proper handling of these chemicals. In this edition, however, we have added a section on safety after the preface. This new section is intended to help those interested find the necessary information on safety before attempting to work with any chemical or processes.

## Acknowledgments

We express our gratitude to Robert F. Coughlin and Frederick F. Driscoll for their many helpful comments and suggestions. We also thank the many manufacturers who have furnished photographs. Our sincere thanks also are extended to Thomas E. Naylor, Jr., for processing the photographs for all the editions and to Phyllis Wolff for her skillful typing of the manuscript. Finally, we are grateful to our families for their continued support and patience throughout the preparation of this fifth edition.

Robert S. Villanucci
Alexander W. Avtgis
William F. Megow

Boston, Massachusetts

# Safety Concerns and Information Sources

Techniques presented in this text involve the use of hand, machine, and power tools, electrical equipment, and printed circuit board processing systems, in addition to hazardous chemicals and materials. The authors, therefore, urge readers intending to perform any of the procedures described to first seek information regarding applicable safety practices. You must be fully aware of safety rules and regulations, precautions to take, and safety equipment to use to prevent putting your health at risk, sustaining personal injury, or causing injury to others working nearby. To assist the reader in becoming safety conscious, concerns for safety appear in the introductory sections of chapters that contain techniques where safety practices must be exercised.

There are many organizations from which to obtain the most current safety practices and procedures or from which approved safety equipment is in compliance with specific conditions, equipment, and materials being used. Three examples of such organizations are the American National Standards Institute (ANSI), the National Institute of Occupational Safety and Health (NIOSH), and the National Fire Protection Association (NFPA). ANSI is a privately funded organization that identifies and coordinates the development of industrial/public, national, consensus standards for safety. Many ANSI standards apply to the safe design and performance of equipment and safe practices and procedures. NIOSH is an agency that tests and certifies respiratory and air sampling devices and recommends exposure limits to the Occupational Safety and Health Administration (OSHA). OSHA is part of the U.S. Department of Labor and is the regulatory and enforcement agency for safety and health in U.S. industrial sectors. NFPA is an organization that promotes fire protection and prevention and safeguards against loss of life and property. NFPA publication 704M is the code for showing hazards of materials on familiar diamond-shaped container labels. The selection and use of equipment such as safety glasses or goggles, full face shields, ear protectors, dust masks, respirators, and chemical-resistant gloves, clothing, and footwear is primarily based on the item being in compliance with and meeting ANSI and NIOSH approval/certification.

Other helpful sources of technical safety information are distributors of safety equipment and materials. They will answer questions regarding the most appropriate equipment for your specific needs. Their catalogs also include ANSI and NIOSH certification and approval numbers.

When working in an environment in which hand, machine, or power tools are used, injuries may be sustained if one is unfamiliar with the correct use of the tool. You must develop respect for even the simplest cutting or forming tools. Very often tool manufacturers include safety information regarding correct operating procedures. It is essential that you put this information to use. For additional safety information, the Government Printing Office may be contacted for listings of available occupational safety literature.

In cases of procedures where hazardous materials or chemicals are used, health risks are of great concern. Inhalation, skin absorption, and ingestion are the principal ways by which hazardous materials and chemicals enter the body. You must be aware of the safeguards required for your protection. Manufacturers are required to supply, on request, comprehensive information about their chemicals and materials that can pose health problems if improperly used. Mandatory requirements for providing Material Safety Data Sheets (MSDS), labeling chemicals, and making safety training available in the workplace are the result of the Occupational Safety and Health Act (OSH ACT), which established the Hazard Communication Rule (HCR). To better understand the type of information provided in an MSDS, a sample form that the manufacturer must prepare for you, on request, is shown in Appendix XII. To further assist you, chemical container labels give the identity of the chemical, who made it, why it is hazardous, what might happen if it is mishandled, and how to protect yourself. It is important that no chemical or material be used before the information provided in the MSDS and on the container label is read, understood, and applied.

When in doubt about any procedure or your safety in the workplace, consult the MSDS provided at the Right-To-Know (RTK) station in your facility (academia or industry). Also know who your designated safety officer is for obtaining assistance and safety training if required. Although safety rules and regulations are in place for your protection, it is also an individual responsibility always to be in compliance.

Addresses and phone numbers of several sources of safety information and equipment are provided in Appendix XIII.

Unit

# ONE

# Circuit Design
## and
# Project Planning

# 1

# Designing a 5-Volt Switching-Regulator Power Supply

## LEARNiNG OBJECTIVES

*Upon completion of this chapter on the design of a 5-volt switching-regulator power supply, the student should be able to*

- Identify the three terminals of a standard household wall receptacle and explain its connection to the service entrance panel, and then to the power company lines.
- Understand the function of the transformer, bridge rectifier, and filter capacitor of an unregulated power supply.
- Use a circuit analysis software package called PSpice to model the unregulated supply, and obtain its performance characteristics.
- Know the differences between an unregulated and a regulated power supply's output.
- Compare and contrast the advantages and disadvantages of linear regulators and switched-mode dc-to-dc converters.
- Use a design software package called SWITCHERS MADE SIMPLE from National Semiconductor first to choose the correct switching regulator IC and then to define the components of a buck-type converter.

## 1.0 INTRODUCTION

In this first chapter we begin the design of a high-quality and reliable piece of electronic equipment. The system chosen to illustrate electronic design and packaging principles is a switched-mode power supply. When completed, this power supply project will output a regulated 5 volts dc with a ripple voltage of less than 100 millivolts, and deliver a load current of up to 1 ampere.

This project has several desirable characteristics. Its relatively straightforward theory is easy to understand, it incorporates the most current integrated circuit technology available for switched-mode power supply design, and it requires the use of a printed circuit board (PCB) for effective operation. Further, computer-assisted techniques are available to both analyze the unregulated supply section and then design the switched-mode dc-to-dc portion of the project.

Before we begin to examine the numerous and common packaging problems involved in the preliminary planning stage of an electronic system, we will consider the electronic design, analysis, and "breadboarding" phases of the switching power supply circuit. Each phase of this design is considered in detail in the sections to follow.

## 1.1 STANDARD HOUSEHOLD/INDUSTRIAL POWER

The most convenient source of electrical power is a standard household wall outlet, or receptacle, like that shown in Fig. 1.1. In the United States, the available voltage is usually a

**Figure 1.1** 115-volt ac wall outlet wired to the service entrance panel. Wiring shows grounding techniques and standard connections to power company lines.

115-volt ac sine wave at a power frequency of 60 hertz, and the receptacle is wired in accordance with the National Electric Code.

The wide slot of the receptacle is called the *neutral,* or *common,* connection. It is routed to the neutral bus at the service entrance panel with a white insulated wire. The neutral bus is then wired to the power company's ac earth ground at the power pole.

The narrow slot of the receptacle is the *hot* connection and is wired to a fuse in the service entrance panel with a black (or red, blue, etc.) insulated wire. This fused hot wire is then connected to the power company's 115-volt ac hot wire. Thus, the hot wire is at 115 volts ac with respect to the neutral wire. The hot line is fused at the service panel with either thermally activated fuses or circuit breakers. If American Wire Gauge (AWG) No. 14 wire is used at the receptacle, the line is fused for 15-A operation, and a 20-ampere fuse is used for AWG No. 12 wire.

Finally, the U-shaped slot of the receptacle is connected to earth ground. The connection at the wall receptacle is made with either a bare copper wire, green insulated wire, or using the metal shield or conduit pipe of the household service. At the service panel the connection to ground is made with a bare copper wire connected to a water pipe buried in the ground. The U-shaped slot of the system is known by many names; *safety, equipment, chassis, utility, third wire,* or *U-ground.*

In a good power distribution system, the U-shaped slot should be at zero volts with respect to earth or ground. However, it is not uncommon to measure a several-volt potential between the neutral wire and this third wire ground. This is due to return currents flowing through the wire resistance of the neutral line. ***Caution:*** It is both dangerous and often unnecessary to measure the voltage between the hot and neutral slots directly. Before attempting this measurement, obtain appropriate instruction from your instructor.

Since most modern electronic equipment runs on low-voltage dc, an ac-to-dc converter is needed to convert the 115-volt ac household power at the wall receptacle into a lower dc voltage to power the equipment. The first stage in this conversion process is usually an unregulated power supply.

## 1.2 THE UNREGULATED SUPPLY

Figure 1.2 illustrates a typical unregulated supply. When plugged into a household wall receptacle this "raw" supply will convert the ac input voltage into a low-level dc output voltage. The main components in Fig. 1.2 are a step-down transformer ($T_1$), four rectifying diodes ($D_1$ to $D_4$), and a filter capacitor ($C_1$). While there are other support components (fuse, on/off switch, indicator lamp, etc.) needed to complete this subsystem, only the main components are discussed in detail here.

**Step-Down Transformer** The step-down transformer consists of a primary winding connected to the household power and a secondary winding connected to the diode rectifier circuit. Since both windings are wound on a laminated steel core with insulated wire, the transformer provides electrical isolation between household power and the output of the unregulated supply. Furthermore, the transformer is manufactured with approximately 10 times more primary winding turns than secondary winding turns. With 115 volts ac applied to the primary, the secondary winding output is therefore reduced to about 12.6 volts ac (12.6 volts rms) hence the name *step-down transformer*.

The transformer chosen for this project is referred to as a filament transformer and has a nameplate rating of 12.6 volts ac at 2 amperes. By these specifications the manufacturer guarantees that the secondary winding output ($V_{sec}$) will not fall below the rated 12.6 volts ac when the secondary is delivering its rated load current of 2 amperes rms. To guarantee this output level and overcome winding resistance at full load, the transformer manufacturer winds less than the theoretical 10:1 turns ratio on the core. Therefore, at no load, when the secondary winding is delivering very low load currents, the secondary output will be above the rated 12.6 volts ac, dropping to its rated value when the rated load current of 2 amperes is drawn from the secondary winding.

**Diode Bridge Rectifier** The four diodes ($D_1$ to $D_4$) shown in Fig. 1.2 are wired in a configuration called a full-wave bridge rectifier. The transformer's secondary winding is connected to opposite corners of the diode bridge. These are referred to as the ac connec-

**Figure 1.2** An unregulated power supply, with 4 bridge rectifier diodes, outputs pulsating dc when filter is removed.

tions. Note the ∽ symbol in Fig. 1.2. The filter capacitor, and load, are connected to the remaining corners of the bridge, labeled the dc connections. Note also the dc symbol in Fig. 1.2. This circuit will convert the transformer's secondary ac voltage, $V_{sec}$, into a pulsating dc output as shown.

To understand how the diode bridge functions as a rectifier, assume that the filter capacitor is removed and replaced with just a load resistor ($R_L$). Then assume that the transformer secondary is going through its positive half-cycle. When the $V_{sec}$ exceeds two diode drops (about 1.2 volts), diodes $D_1$ and $D_3$ begin conducting current through the load. The output polarity is shown and the load voltage raises to about 1.2 volts less than the peak secondary voltage.

During the negative half-cycle, the polarity of the transformer's secondary voltage reverses, diodes $D_2$ and $D_4$ begin conducting, and current is routed through the load in the *same direction* as D1 and D3 during the positive half-cycle. The result is a pulsating dc output voltage. Under light loads, when the load current ($I_L$) is low, the peak output voltage will be maximum, dropping in value as $I_L$ increases until, at a maximum load of 2 amperes, the peak output voltage will drop to about 18 to 20 volts.

**Filter Capacitor**   When a filter capacitor ($C_1$) is connected across the load, the output voltage more closely resembles the desired constant value voltage. During each half-cycle, when the secondary voltage is above the load voltage, the transformer furnishes current to the load and charges the filter capacitor. When the transformer's secondary voltage drops below the load voltage, the capacitor furnishes load current in an attempt to maintain the output voltage constant. The transformer is "disconnected" from the load by the rectifying diodes during this time.

Under no-load conditions, when $I_L$ is 0 amperes, the filter capacitor charges to its maximum level, and the output is shown in Fig. 1.3a as a constant dc voltage. When a load is connected, i.e., when $I_L$ increases, the average or dc output voltage decreases due primarily to transformer-winding resistance losses. In addition, since the capacitor must furnish load current by discharging through the load, a small component of ripple voltage ($V_r$) will be present at the output (Fig. 1.3b). Under maximum-load conditions, when $I_L$ equals 2 amperes for the transformer specified, the dc output will decrease to its minimum value and the ripple voltage will increase to its maximum level (Fig. 1.3c).

With a basic understanding of how the unregulated section of our power supply works, we turn to a circuit analysis software package called PSpice. PSpice will be used to analyze the performance of the unregulated section of our power supply.

## 1.3  CIRCUIT ANALYSIS USING PSPICE

Before we can design the switching regulator portion of this project it will be necessary to determine several parameters of the unregulated section discussed above. Specifically, we must determine the raw supply's dc output voltage at both no load ($I_L$ = 0 amperes) and at full load ($I_L$ = 1 ampere). Also useful will be the worst-case ripple voltage when the load current equals 1 ampere.

Until recently it would have been necessary to breadboard Fig. 1.2 on the bench, connect an adjustable power load resistor across the filter capacitor, and set it to draw the necessary maximum-load current. We would then measure the average output voltage with a dc-reading voltmeter. Ripple voltage could be found with an ac coupled oscilloscope connected at the output. The measurements would be repeated at no-load (or very light load, i.e., about 1% of maximum) conditions.

Today breadboarding a circuit and analyzing its characteristics can be done more conveniently using a personal computer and a circuit analysis software package. To illustrate this new method of "breadboarding" the authors have chosen to use the circuit analy-

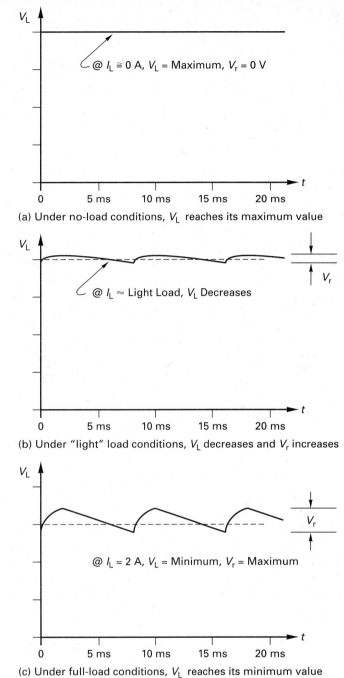

@ $I_L \cong 0$ A, $V_L$ = Maximum, $V_r$ = 0 V

(a) Under no-load conditions, $V_L$ reaches its maximum value

@ $I_L \approx$ Light Load, $V_L$ Decreases

$V_r$

(b) Under "light" load conditions, $V_L$ decreases and $V_r$ increases

@ $I_L$ = 2 A, $V_L$ = Minimum, $V_r$ = Maximum

$V_r$

**Figure 1.3** An unregulated supply exhibits a decrease in dc load voltage as $I_L$ increases. Also ripple voltage increases as $I_L$ increases.

(c) Under full-load conditions, $V_L$ reaches its minimum value

sis software called PSpice from MicroSim Corporation (Irvine, CA) to analyze the performance of the unregulated supply.

## 1.4  PSPICE INPUT FILE TO SIMULATE AN UNREGULATED SUPPLY

Figure 1.4a is a redrawn circuit schematic of the raw supply, with accompanying nodes identified. The step-down transformer is being modeled as two inductors coupled together magnetically. Primary and secondary winding resistance is also shown as part of the

(a) Unregulated supply with nodes identified

UNREGULATED POWER SUPPLY
```
*************************************************************************
*                                                                       *
*                 This PSpice net list evaluates the performance        *
*            of an unregulated power supply under varying load conditions. *
*                                                                       *
*************************************************************************
*INPUT VOLTAGE
VAC 1 2 SIN(0 115 60)                                              ; line   1
*TRANSFORMER MODEL
Lpri 3 2 2H                                                       ; line   2
Rpri 1 3 0.001                                                    ; line   3
Lsec 4 5 0.065H                                                   ; line   4
Rsec 4 6 1.0                                                      ; line   5
Kps Lpri Lsec 0.995                                              ; line   6
Rdc 2 5 1Tohm                                                     ; line   7
*DIODE BRIDGE RECTIFIER
D1 6  7 MOD1                                                      ; line   8
D2 0  6 MOD1                                                      ; line   9
D3 0  5 MOD1                                                      ; line  10
D4 5  7 MOD1                                                      ; line  11
.MODEL MOD1 D(RS=0.1)                                             ; line  12
*FILTER CAPACITOR
C1 7 0 1000U                                                      ; line  13
*LOAD RESISTOR
RL 7 0 10K              ; NO LOAD = 10kohms, FULL LOAD = 15ohms   ; line  14
*TRANSIENT ANALYSIS
.TRAN 1M 34M                                                      ; line  15
*OPTIONS
.OPTION ITL4=50                                                   ; line  16
*OUTPUT
.PROBE                                                            ; line  17
.END                                                              ; line  18
```

(b) PSpice input file

**Figure 1.4**  The transient analysis of the unregulated supply (a) is achieved by running the PSpice input file (b).

transformer model and a completely annotated PSpice input file for Fig. 1.4a is given in Fig. 1.4b.

Note in Fig. 1.4b that line 1 of the input file represents the wall receptacle output voltage. In PSpice, it is modeled as an independent sine wave voltage source set to an amplitude of 115 volts ac and a power frequency of 60 hertz. This voltage source drives the primary winding of transformer $T_1$, shown by dashed lines in Fig. 1.4a and represented in PSpice by lines 2 through 7 of the accompanying input file.

The primary and secondary winding inductances, $L_{pri}$ and $L_{sec}$, are in units of henrys and are obtained by measurement. Their values for the transformer selected are given in lines 2 and 4, respectively. Primary winding resistance (line 3) is included, since PSpice will not allow a voltage source ($V_{AC}$) to be connected directly across an inductor. The 1-milliohm (0.001-ohm) value was selected so that it would have no adverse effect on performance. Secondary winding resistance (line 5) is set to 1 ohm, a typical measured value for the secondary of a good quality filament transformer. The transformer's coefficient of coupling $K$ is modeled by line 6 and set to a value approximately equal to 1, the coupling coefficient of a lossless transformer.

One requirement of PSpice is that each node must have a dc path to ground (zero node). This is necessary so that a small-signal bias solution can be found by the software. Resistor $R_{dc}$ is added to the circuit so that the primary winding circuitry has a path to ground, and satisfies this software requirement. As line 7 of the input file shows, $R_{dc}$ is set to 1 teraohm ($1 \times 10^{12}$ ohms), resulting in negligible effect on circuit performance.

The diode bridge rectifier consists of four silicon power devices represented in PSpice with lines 8 to 11 of the input file. Note that the anode node occurs first in the listing, followed by the cathode node, for each diode. A model card, line 12, is included to indicate the data sheet specification of ohmic resistance for the power diodes chosen for the project.

An electrolytic capacitor $C_1$, with a typical value of 1000 microfarads, is added for output filtering and modeled by line 13. Note that the positive end of the capacitor (node 7) appears first in the identification, followed by the negative end (node 0), which is connected to the circuit's − terminal. The analysis of the supply will be performed twice: once under no-load conditions, when $I_L$ equals 0 amperes, and then under full-load conditions, when $I_L$ equals 1 ampere. The load resistor is specified by line 14 of the input file. To insure that the bridge rectifier diodes are always conducting, the no load condition is analyzed with a "very light load" by setting $R_L$ to 10,000 Ω. $I_L$ should be only a few milliamps under this load condition. Full-load conditions can be tested by resetting $R_L$ in the input file to 15 ohms and reanalyzing the circuit.

Circuit performance is accomplished by a transient analysis control line. As line 15 illustrates, the analysis will be performed every millisecond for a total of 34 milliseconds or about 2 complete cycles of a 60-hertz input. The output of this analysis is covered in Sec. 1.6. PSpice sometimes suffers convergence problems during a transient analysis, resulting in a failure to complete the test. This can often be overcome by using an analysis option such as the .OPTION command in line 16. The command .OPTION ITL4=50 increases the maximum number of iterations allowed at any calculation point during the transient analysis from the default value of 10 to 50. If problems with convergence still occur, increasing ITL4 to 100 (.OPTION ITL4=100) will often solve the problem, but at a significant increase in analysis time.

PROBE is a graphics postprocessor that displays the results of a simulation. The PROBE function is called by line 17 of the input file and allows the tailoring of the desired output. Both screen displays and hard copies are available when the PROBE command is used in an input file.

Line 18 terminates the input file and is a necessary command line for PSpice simulations. In Sec. 1.6, the circuit's performance is simulated by running the PSpice analysis software and using PROBE to display the results.

## 1.5 LOADING PSPICE ONTO THE COMPUTER

The authors used an IBM-compatible computer for all their PSpice development work. The first task is to load the analysis software onto the computer. To load the evaluation version of PSpice onto the hard drive (C:), a subdirectory called SPICE is first created. This subdirectory is created with the following keystrokes:

```
C:\>MD SPICE<CR>
```

Note that the underscored information is the keystrokes you must enter after the DOS prompt, and <CR> indicates either carriage return or enter.

The following is then keyed in to change to the newly created SPICE subdirectory:

```
C:\>CD SPICE<CR>
```

Next load the first 5.25-inch PSpice floppy disk into drive A:, and copy the files onto the C: drive by the following key strokes:

```
C:\SPICE>COPY A:*.* C:\SPICE<CR>
```

Repeat this last step for each floppy disk supplied with the evaluation version of PSpice.

The circuit analysis program is now loaded onto the hard drive of your computer. In the next section we use this program to analyze the unregulated supply.

## 1.6 RUNNING A CIRCUIT ANALYSIS

The authors created the PSpice input file shown in Fig. 1.4b using the DOS editor and stored it in a circuit file PWRSUP.CIR. To run an analysis on this circuit file, load the PSpice analysis program with the following keystrokes. At the SPICE subdirectory prompt type:

```
C:\SPICE>PSPICE<CR>
```

When the analysis program is loaded, you will be prompted to enter the name of the input file to be analyzed. Enter

```
Input file name[.CIR]?PWRSUP<CR>
```

Note that only the file name is entered after the prompt, followed by a carriage return <CR>. Entering the extension .CIR is unnecessary. You will next be prompted to enter the output file name. Answer yes to this question with a carriage return <CR>.

```
Output file name[PWRSUP.OUT]?<CR>
```

With the PROBE control card as part of the input file, the software will perform the analysis, and then display a blank graph. **Add_trace** will appear in reverse video below the graph. Since we want to display a trace of the output voltage, press <CR> to select this

(a) Graph of load voltage vs. time

**Figure 1.5** PSpice hard-copy output for $R_L$ = 10 KΩ: (a) $V_L$ vs. time, (b) $I_L$ vs. time.

(b) Graph of load current vs. time

option. Then key in F4 to display a list of voltages and currents that can be graphed. The voltages in this list are referenced to the 0 node and the currents are as indicated.

From Fig 1.4a we note that the load voltage is at node 7. Select V(7) by first pointing to it with the mouse, and pressing the *left* (select) key on the mouse. A hard copy of the resulting graph is shown in Fig. 1.5a. Since the analysis was done at no-load conditions ($R_L$ = 10K), we learn that the raw supply will have a maximum average dc output voltage of about 19 volts and essentially no ripple ($V_r$ = 0 volts) under no-load conditions.

Unregulated Power Supply

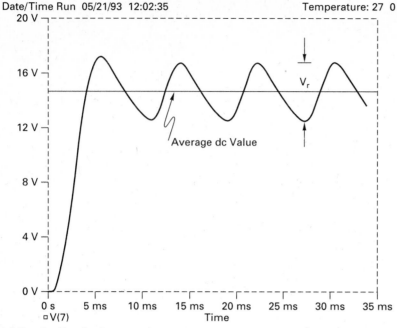

(a) Graph of load voltage vs. time

(b) Graph of load current vs. time

**Figure 1.6**  PSpice hard-copy output for $R_L$ = 15 ohms.

Figure 1.5b shows the PSpice output graph of $I_L$ versus time for the same no-load conditions. This graph was generated using the steps and keystrokes in this section. Under these no-load conditions, $I_L$ does not exceed about 2 milliamperes.

To obtain the full-load graphs, we used the DOS editor to modify $R_L$ in the PWR-SUP.CIR file to 15 ohms. Again using the procedures in this section, we generated the data shown in Fig. 1.6. Under full-load conditions, note that $V_L$, the minimum average dc load voltage, drops to about 15 volts, and the peak-to-peak ripple voltage increases to about 5 volts. Figure 1.6b further demonstrates that the supply is delivering an average dc load current of about 1 ampere. These are the data we need to design the switching regulator in Sec. 1.10.

## 1.7 THE NEED FOR A REGULATOR

The data developed using PSpice to analyze the performance of the unregulated section of the project are plotted in Fig. 1.7. This plot, referred to as a *load regulation curve*, characterizes circuit performance. The solid line reveals that the dc output voltage varies drastically from 19 to 15 volts as load current increases from approximately 0 to 1 ampere. Further the ripple voltage, shown as a pair of enveloping dashed lines, increases from 0 at no load to about 5 volts peak-to-peak under full-load conditions. The load regulation curve dramatically illustrates that load voltage is not constant, and that the ripple voltage, at full load, is excessive.

To obtain an output, constant at 5 volts dc, and reduce the ripple voltage of the raw supply to less than 2% of its dc output voltage (<100 millivolts), a regulator is needed. In the next section we look at the two available regulator types and compare and contrast the advantages and disadvantages of each.

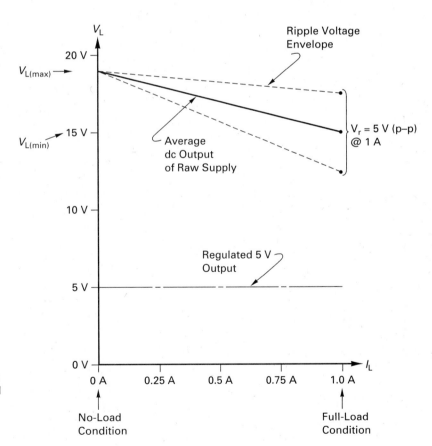

**Figure 1.7** The average dc output of an unregulated supply varies from a no-load value of 19 volts to a full-load value of 15 volts. As $I_L$ increases, $V_r$ also increases.

## 1.8 LINEAR AND SWITCHING REGULATORS COMPARED

The output of a raw supply like the circuit shown in Fig. 1.2 can be regulated by choosing either a linear regulator or a switched-mode dc-to-dc converter. Three-terminal linear regulator ICs have been used widely since the early 1970s, gaining immediate acceptance and becoming a spectacular success as a low-cost solution to the problem of regulation. They are available with both positive and negative outputs with either fixed or adjustable voltages. Since all the engineering has been completed by the device manufacturer and packaged into a simple three-terminal case, they are essentially a "drop-in" solution, often requiring only the design of a heat sink and the addition of bypassing capacitors to become completely useful. While three-terminal linear regulators do not rival discrete designs, they do have good specifications for line and load regulation, output ripple, and transient response.

Linear regulators, however, have disadvantages that must be addressed if an informed regulator selection is to be made. First among these is the problem of efficiency. As shown in Fig. 1.8a, the pass transistor of the linear regulator appears in series with the load and is between the output of the raw supply and the load resistor. The pass transistor operates continuously in its linear region, always conducting load current. Under worst-case conditions the raw supply delivers 15 volts dc and 1 ampere. Since the input power is $P_{in} = 15 \text{ V} (1 \text{ A}) = 15 \text{ W}$ and the output power is $P_{out} = 5 \text{ V} (1 \text{ A}) = 5 \text{ W}$, the efficiency of the linear regulator is

$$\% \text{ Efficiency} = (P_{out}/P_{in}) \, 100\% = (5 \text{ W}/15 \text{ W}) \, 100\% = 33.3\%$$

(a) The pass transistor of a linear regulator is always on, conducting current, and dissipating power

(b) The pass transistor of a switching regulator is switched on and off, power dissipation

**Figure 1.8** The efficiency of a switcher can approach 90%.

In other words the linear regulator package must waste

$$P_{reg} = (15 \text{ V} - 5 \text{ V}) (1 \text{ A}) = 10 \text{ W}$$

This translates into heat, and a corresponding increase in packaging size (heat sink) and cost. In effect the linear regulator is wasting 66.6% of the input power to perform the task of regulation.

Another drawback is that linear regulators are step-down devices only, requiring a larger input voltage from the raw supply that it will output. Step-up is not possible in linear regulator technology.

While not quite a drop-in solution, switching regulator designs have advanced tremendously in the last few years. This is due primarily to the demand for high-efficiency regulators in battery-powered equipment, such as laptop computers and cellular telephones, to name just two. Wasting 66.6% of a battery's power on regulation is unacceptable in portable applications. Also switched-mode dc-to-dc converters or switching regulators lend themselves to step-down (buck), step-up (boost), inverting (buck–boost), and dual-output (flyback) modes of operation.

However, the main reason for selecting a switching regulator is its inherent efficiency. As with the linear regulator, the pass transistor in a switcher is in series with the load (Fig. 1.8b). Again, under worst-case load conditions, when the raw supply is delivering 15 volts at 1 ampere, we evaluate the performance. The pass transistor in a switching regulator is never operated in its linear region, but switched on (into saturation) or off (into cutoff). Since the on resistance of a transistor is quite small (fractions of an ohm in large power applications), the power wasted in the pass element is low even at high currents ($P_{reg} = I^2R$). When the pass element is switched off, the current is low (approaching 0 A), and again the power dissipated in the regulator is low. Switching regulators can achieve efficiencies in excess of 90%, wasting less than 10% of the input power on regulation.

The major disadvantage of selecting a switching regulator is the larger values of output ripple voltage $V_r$. It is not uncommon for a switching regulator to have an output ripple component in excess of 2% of its regulated dc output voltage. Further, this ripple component is at the switching frequency of the IC and can vary from 40 kilohertz to 1 megahertz, depending on the device selected. High-frequency, high-energy spikes are difficult to remove from the output of a switcher.

What makes the selection of a switching regulator an attractive alternative today is that some manufacturers now supply their devices with a software design and analysis package that makes the task of selecting the correct IC switcher, and then designing the filter section, both simple and straightforward. Gone are the days when the design of these nonlinear circuits was best left to a small coterie of engineers that specialized in this area.

We introduce just such a design solution in the sections to follow.

## 1.9 NATIONAL SEMICONDUCTOR'S SWITCHERS MADE SIMPLE SOFTWARE

This section introduces a software package from National Semiconductor for the computer-aided design (CAD) of Simple Switcher-based dc-to-dc converter. It was developed to aid the designer working with National Semiconductor's LM2575, LM2576, and LM2577 switching regulators. The software runs on an IBM-compatible computer with MS-DOS version 2.0 or higher and requires 512K of memory.

The first task is to load the software onto the computer. At the DOS prompt, switch to the A: drive by typing the following:

```
C:\>A:<CR>
```

**Linear**

Design boost(1), flyback(2), buck(3) or buckboost(4) converter? (1/2/3/4)

    This program supports four types of power supplies.

Boost:

    Used to step up the input voltage, e.g. Vin = 5V, Vout = 12V

Flyback:

    Used for multiple output voltages, positive or negative, with
    the possibility of isolation. Both step up and step down are
    possible. High output voltages may be achieved. A
    transformer is required instead of an inductor.
    E.g.  Vin = 5V, Vout1 = 15V, Vout2 = –15V, or
          Vin = 5V, Vout1 = 15V, Vout2 = 12V, or
          Vin = 20V, Vout = 100V

Buck:

    Used for stepping down a voltage, e.g. Vin = 10V, Vout = 5V

Buck-Boost:

    Used for generating a negative voltage from a positive one
    without isolation, e.g. Vin = 5V, Vout = –5V

        To view the helpfile, PRESS &lt;F1&gt; NOW.
        To exit this program at any time, press &lt;ESC&gt;

**Figure 1.9**   The main menu of SWITCH-ERS MADE SIMPLE provides the user with the option of designing four different types of switching regulators.

Load the 5.25-inch floppy disk titled SWITCHERS MADE SIMPLE, available free from the manufacturer, into drive A: and then type

<code>A:\>SIMPLE&lt;CR&gt;</code>

To get to the main menu from the displayed company logo, press any key twice. The main menu is shown in Fig. 1.9. A review of the information presented shows that this software package supports four types of switching supplies: boost, flyback, buck, and buck–boost, which is a mode of operation that generates a negative output from a positive input voltage without the use of an isolation transformer.

    Before using the software to design a switching regulator, you can view the helpfile by pressing &lt;F1&gt;. Key in the information within the &lt; &gt; brackets to access this information file. The supplemental information in this README.DOC file is quite extensive and useful. You use the &lt;PgDn&gt; and &lt;PgUp&gt; keys to scroll through the file and press &lt;END&gt; to exit back to the main menu.

## 1.10  DESIGNING A SWITCHING REGULATOR FOR A 5.0-VOLT, 1.0-AMPERE OPERATION

Again refer to the main menu shown in Fig. 1.9. Since our unregulated supply outputs a voltage larger than the output of the switcher, press &lt;3&gt; to select a buck mode design. You are then prompted for information, some related to the unregulated supply previously analyzed with PSpice, and some information about the ambient temperature extremes and output design requirements. The following data were entered for our design using the arrow keys to move down the list and &lt;CR&gt; to enter the data:

Vinmin = 15.00 V
Vinmax = 19.00 V
Tamax = 40.00 C
Tamin = 10.00 C
Vout = 5.00 V
Ilmax = 1.00 A
Diode = Schottky

The first two entries are from the PSpice analysis of the unregulated supply, and the ambient temperature extremes that follow are typical operating environment values. The regulator's output is set with the data entered for Vout and Ilmax. Finally, select Schottky for the *catch* diode.

Press <End> to load the information and answer yes <Y> to the question *Use standard inductor?* While a custom inductor could be wound on a magnetic core, we will choose a standard off-the-shelf part. At this point the software will select the appropriate regulator IC and design the input and output filter component values. Figure 1.10 is a display of the hard-copy output information. We will return to this information, but for now note that the all component values are presented and that the design predicts a ripple voltage less than 82 millivolts, well within the 2% of Vout required for this design.

Next you are prompted to answer several questions. Each is listed below with the authors' answers.

Change any component value? N<CR>

Perform thermal check? Y<CR>

Will you use a K(TO-3), T(TO-220), M(SO24), N(16 L N)? T<CR>

Note that by selecting a TO-220 plastic package, the internal junction temperature of the switching regulator will rise to only about 72.2°C and no heat sink will be required. The software ability to do a thermal check, prior to fabrication, greatly simplifies the overall design. Next the software prompts for the following information:

Will a heat sink be used? N<CR>

Modify this design. N<CR>

Would you like to save this design? N<CR>

We answered no to the last question because we will get a hard-copy output from the software.

After all questions are answered, pressing any key will instruct the software to draw the switching regulator with accompanying components identified. Figure 1.11 is the complete circuit schematic displayed for this switcher's design. If a printer is available, a hard-copy output can be obtained. After the schematic is displayed, the software prompts you for a parts list, available by pressing any key. The printout is shown in Fig. 1.10.

## 1.11 COMPONENTS FOR THE BUCK CONVERTER

Return to the hard-copy output of the SWITCHERS MADE SIMPLE software shown in Fig. 1.10 and the circuit schematic shown in Fig. 1.11. The first subheading, *Circuit Parameters,* is a printout of both input and output specifications for the design. It is followed by the subheading *Misc calculated information,* where pertinent operating information about the design is tabulated. Of particular note is *calculated ripple voltage* (Vripple). At a specified 81.49 mV, this value is below the maximum 2% of $V_L$ required by the design. Note that $V_L$ = Vout.

------------------------------------------------

Circuit Parameters

| | | |
|---|---|---|
| Vinmin : | 15.00 | V |
| Vinmax : | 19.00 | V |
| Tamax : | 40.00 | C |
| Tamin : | 10.00 | C |
| Vout : | 5.00 | V |
| Ilmax : | 1.00 | A |
| Diode : Schottky | | |

Misc Calculated Information

| | | |
|---|---|---|
| Mode : Continuous | | |
| Peak switch current : | 1.15 | A |
| ESRmax : | 0.13 | Ohms |
| ESRmin : | 65.27 | mOhms |
| Vripple : | 81.49 | mV |
| Crossover Freq : | 8.25 | KHz |
| Phase margin: | 31.89 | Deg |
| Junction Temp : | 72.16 | C |

Component List

Cout : 330.00 uF

| | | |
|---|---|---|
| ESR : | 0.10 | Ohm |
| Vmax : | 25.00 | V |
| ECE-B1EF331 | : Panasonic | |

Cin : 47.00 uF

| | | |
|---|---|---|
| Vmax : | 27.00 | V |

L : 330.00 uH

| | | |
|---|---|---|
| Imax : | 3.45 | A |
| 430-0635 : | AIE | |
| PE-53117 : | Pulse | |
| RL2447 : | Renco | |

D1 : 3.00 A

| | | |
|---|---|---|
| Vmax : | 20.00 | V |
| 1N5820 : | Motorola | |
| MBR320P : | Motorola | |

U1 :

LM2575T-5 : National Semiconductor

**Figure 1.10** The hard-copy output of our design with all values specified.

The final subheading, *Component List,* itemizes the switching regulator components. These 5 elements, which combine to create a switched-mode dc-to-dc converter, are the (1) input filter, Cin; (2) controller, U1; (3) inductor or choke, L; (4) output filter capacitor, Cout; and (5) a "catch" diode, D1. The selection of each item was made by the software and is considered in some detail here. General information on topics such as component size, construction material, and device configurations is given. The prime objective is to acquaint the reader with the kinds of components that are suitable for use in switching regulators along with those that are not acceptable.

**Figure 1.11** Hard-copy printout of the buck-type switching regulator designed by the SWITCHERS MADE SIMPLE software.

**Input Filter (Cin)**   The input filter is not critical and a standard 47-microfarad aluminum electrolytic, with a WVDC rating of 35 volts, is chosen. During the package design phase, however, we have to insure that the lead length of this device is kept as short as possible.

**Controller (U1)**   U1 is chosen to be the LM2575T-5 switcher from National Semiconductor. Packaged in a TO-220 case, the switcher develops a fixed output of 5.0 volts dc. This low-power switching regulator employs a principle called pulse-width modulation (PWM). In this PWM IC, a constant-frequency square wave oscillator regulates output voltage by increasing switch on time when Vout drops below a preset value, and decreases switch on time when Vout goes above the same preset value. The output voltage Vout is sensed by the feedback pin of the IC. Switching frequency is normally set at 52 kilohertz.

PWM controllers do have larger output ripple voltages than other types of controllers, and the PWM controllers draw more quiescent currents than other types, but they are far less critical to circuit performance and the coil's inductance value. One necessity of all switching regulators is that they be supplied with a low-inductance, low-impedance path to ground. This precludes the use of conventional wiring techniques and requires a PCB ground plane for effective operation.

**Inductor (L)**   The inductor or coil used in a switch-mode regulator functions primarily as an energy-conversion element. It stores energy in its powdered-iron (or Mo-Permalloy) core material, in the form of a magnetic field, during the charging phase and then delivers the stored energy to the load during the discharge phase. Selecting an inductor value often includes many considerations. First among them might be maximizing output current capacity while minimizing physical size to optimize a design. This type of optimization, however, often requires nonstandard coils, which is not our intention. We will adopt a more practical approach using only off-the-shelf components and accepting the value of inductor chosen by the software.

In general, larger values of inductance provide for a larger output capacity (due to higher core storage capacity) and lower ripple currents. Lower inductor values have the effect of reducing maximum output capacity and raising ripple current but are physically smaller. We choose the PE-53117 coil from Pulse Engineering, with a 330-microhenry inductance value, for our design.

**Capacitor (Cout)**   The main purpose of the output capacitor is to reduce the output ripple voltage. Both value and component type are important when selecting a filter capacitor for switch-mode regulators. Typically, capacitor values will run the gamut from 10 to 10,000 microfarads for this type of application, with the larger valued components achieving the lower ripple voltages. To understand the reason for this, note that the model of a typical filter capacitor includes capacitance ($C$), effective series inductance (ESL), and equivalent series resistance (ESR). In switching regulator designs, the value of capacitance $C$ and ESR both contribute to total ripple, but ESR is often the most important factor. Lowest ripples are achieved with large values of $C$ and low ESRs. Since the value of ESR is the major contributor of ripple in a switching regulator, every effort should be made to reduce it.

ESR is directly proportional to capacitor volume in cubic inches, not the value of capacitance in microfarads. All else being the same, the ESR of a larger value filter capacitor is usually smaller than that for a smaller value capacitor of the same physical construction. Standard aluminum capacitors have an ESR in the range of 0.05 to 0.9 ohms for capacitor values between 100 and 1000 microfarads. One way to reduce this relatively high value of ESR is to parallel capacitors. This has the dual effect of reducing resistance and increasing effective capacitance. The down side of this approach is increased pc board space. A better method is to choose one of the high-frequency or low-ESR parts with an effective series resistance of less than 0.1 ohm as specified by the software. Several manufacturers, such as Cornell Dubilier, Mallory, Panasonic, and Sprague, make capacitors for switching regulators. The Panasonic ECE-B1EF331, 330-microfarad capacitor was selected for this project.

When selecting a capacitor, avoid Tantalum dielectrics. With ESR values below 0.05 ohms they can cause instability and unwanted oscillations. Finally, the capacitors best suited for switched-mode applications are those that have extended-foil construction (for low ESR) and radial leads (for low ESL).

Since capacitor lead length contributes to total resistance (ESR), lead length must be kept as short as possible. Layout is important if minimum ripple is to be realized. Additionally, a 0.1-microfarad disk-type capacitor across the output filter can be very useful in reducing high-energy spikes generated during power transistor switching. Ceramic types of this value have an ESR value of about 0.1 ohm or lower.

The filter capacitor working voltage rating (WVDC) for Cout should be a minimum of 25 volts.

**Diode (D1)**   In the buck converter of Fig. 1.11, the catch diode D1 provides the return path for output current when the switcher is off. Fast-recovery or Schottky-type diodes are often chosen as the catch diode in these designs. Using the ubiquitous 1N4001 as a catch diode will prove disastrous. Its very slow recovery time specification generates excessive internal heat that will quickly destroy the devise. Schottky diodes are often a better choice for the catch diode in low-output voltage switch-mode regulators. These diodes are not pn junction devices, but use a silicon-to-metal technology to create rectification employing the Schottky barrier principle. The symbol for a Schottky diode is shown in Fig. 1.11.

Lightly doped n-type silicon material (cathode) is bonded to a "barrier" metal like aluminum, which acts as the p-type (anode) region. When a forward bias is applied to a Schottky diode, electrons in the n-type material gain sufficient energy to cross the metal-to-silicon interface, where they quickly lose energy and become free electrons in the metal. Forward current flow is due only to majority carriers (electrons). Since there is no

**Figure 1.12** A complete circuit schematic of the 5-volt at 1-ampere switching regulator power supply.

minority carrier (holes) current flow, there is no stored charge and no direct reverse recovery time. When a Schottky diode is reverse biased, conduction ceases as quickly as 10 nanoseconds ($10 \times 10^{-9}$ seconds). Schottky power diodes have very low forward turn-on voltages. At low currents, $V_F$ can be as small as 0.25 volt but will rise to about 1 volt at its maximum current rating. The one drawback to these devices is their relatively low breakdown voltage, which never exceeds much more than 60 volts, prohibiting their use in high-voltage designs.

Another practical consideration worthy of note is diode lead length, which is especially true in high-power designs. The lead length between the diode, capacitor, and power switching transistor should be made as short as possible.

Finally, we selected the 1N5820 Schottky diode from Motorola for this project.

## 1.12 FINAL POWER SUPPLY CIRCUIT SCHEMATIC

A complete circuit schematic showing the unregulated power supply section combined with the switch-mode dc-to-dc converter is shown in Fig. 1.12. All parts are identified on the circuit and important packaging and wiring information is provided. In the following two chapters of this unit, we look at the packaging problem, trade-offs, and decisions that are made to develop a high-quality, reliable prototype package.

### EXERCISES

#### A. Questions

1.1   List at least three common names used to refer to the U-ground terminal of a standard household receptacle.

1.2   List the main components of an unregulated (raw) power supply.

1.3   Explain the difference between no-load and full-load conditions as they relate to power supply load current.

1.4   Modify line 15 of Fig. 1.4 to perform a transient analysis every half-cycle of a 50-hertz power frequency.

1.5   Explain the steps necessary to load PSpice onto the hard drive (C) of an IBM-compatible computer.

1.6   Discuss why linear regulators are less efficient devices than switched-mode dc-to-dc converters.

1.7   Identify and explain the main drawback in using a switching regulator over an equivalent linear regulator.

1.8   Explain pulse-width modulation (PWM) as it relates to switched-mode regulators.

1.9   Discuss the reason for selecting low-ESR-rated output filter capacitors for switching regulator applications.

1.10  Explain why a standard 1N4001 rectifying diode cannot be used as the catch diode in a switching regulator design.

#### B. True or False

Circle T if the statement is true, or F if any part of the statement is false.

| 1.1 | In the United States, household wiring is performed in accordance with the National Electric Code. | **T** | **F** |
|---|---|---|---|
| 1.2 | The narrow slot of a household receptacle is connected to the neutral wire and fused at the service entrance panel. | **T** | **F** |
| 1.3 | A step-down transformer's secondary winding voltage is less than its primary winding voltage. | **T** | **F** |
| 1.4 | During maximum load current conditions, an unregulated power supply will exhibit maximum ripple voltage. | **T** | **F** |
| 1.5 | The load regulation curve of an unregulated supply dramatically illustrates the need for an output regulator. | **T** | **F** |
| 1.6 | Linear regulators can support both step-up and step-down designs. | **T** | **F** |
| 1.7 | Linear regulators are less efficient than switched-mode dc-to-dc converters. | **T** | **F** |
| 1.8 | The switching frequency of an LM2575T-5 switcher from National Semiconductor is approximately 52 hertz. | **T** | **F** |
| 1.9 | Ripple voltage is reduced by lowering the ESR rating of the output filter capacitor. | **T** | **F** |

## C. Multiple Choice

Circle the correct answer for each statement.

1.1 In the United States, household power is delivered at 115 volts ac at a frequency of (*50 Hz, 60 Hz*).

1.2 A diode bridge rectifier (D1 to D4) converts a transformer's secondary voltage output into a (*pulsating dc, filtered dc*) output.

1.3 A typical raw supply exhibits its worst-case ripple under (*no-load, full-load*) conditions.

1.4 The typical filter capacitor for a 1-ampere raw supply is (*1000 uF, 1000 pF*).

1.5 Battery-powered portable equipment usually demands a (*linear, switching*) regulator.

1.6 The inductor or coil used in a switched-mode regulator circuit functions primarily as an (*energy storage, power dissipation*) device.

1.7 At low forward-current levels, Schottky power diodes turn on at typically (*0.25 V, 0.6 V*).

## D. Matching Columns

Match each item in column A to the most appropriate item in column B.

| COLUMN A | | COLUMN B |
|---|---|---|
| 1. Diode bridge | a. | Post processor |
| 2. Schottky diode | b. | White |
| 3. Unregulated | c. | Turns ratio |
| 4. PSpice | d. | Switching regulator |
| 5. Mo-Permalloy | e. | Filter |

|  COLUMN A | | COLUMN B |
|---|---|---|
| 6. | Neutral | f. | Catch diode |
| 7. | Probe | g. | Resistance |
| 8. | Dc-to-dc converter | h. | Raw |
| 9. | 5:1 | i. | Analysis software |
| 10. | ESR | j. | Isolation |
| 11. | Transformer | k. | Rectifier |
| 12. | Capacitor | l. | Henry |
| 13. | *L*pri | m. | Magnetic |

# Planning and Designing Electronic Equipment 2

*Upon completion of this chapter on the planning and designing of electronic equipment, the student should be able to*

- Understand the electrical criteria involved in packaging an electronic system.
- Select the appropriate type of chassis.
- Place components and hardware in their optimum positions in a package.
- Incorporate maintainability in package design.
- Incorporate human engineering in package design.

## 2.0 INTRODUCTION

A high-quality piece of electronic equipment can result only from thorough, careful planning and application of the fundamental mechanical and electronic design factors. Of these factors, *reliability* is one of the most important. A highly reliable piece of equipment will result if all the design factors that govern packaging are considered during the preliminary planning stages.

In this chapter we examine the numerous, common packaging problems involved in the preliminary planning stage of constructing electronic equipment. Consideration is given to *space* and *weight requirements, component* and *hardware selection, chassis materials* and *configurations, component* and *hardware positioning, maintenance,* and *human engineering.* Specific problems are related to the packaging of a *switching regulator power supply* that serves as an example packaging problem in the first unit of this book. Each step of the packaging phase is applied, in sequence, to this power supply, which illustrates some common packaging problems.

## 2.1 SPACE AND WEIGHT REQUIREMENTS

Once the need for a system has been determined, the mechanical and electronic design engineers will stipulate the many specifications for the completed unit to ensure that it will be compatible with its predicted environment. For example, packaging for *vehicular* installation often introduces complex dimensional and weight restrictions. If space is at a premium and size and weight must be kept to a minimum, the system must be packaged in subassemblies or adapted to a modular design that will fit into the various-shaped spaces that are characteristic of vehicles such as ships and air or space craft. *Stationary* equipment, such as that used in homes and laboratories, is normally subject to fewer severe environmental constraints and is designed around chassis, cabinets, racks, panels, and con-

**Figure 2.1** Eight-bit microcontroller system developed as a double-sided pc board.

soles that are commercially available. Modern stationary equipment makes extensive use of double-sided and multilayer printed circuit boards (Fig. 2.1).

## 2.2 COMPONENT AND HARDWARE SELECTION

Once the component values of the system have been designed, the circuits are tested in experimental or *breadboard* form. As noted in Chapter 1, breadboarding can effectively be accomplished with software analysis packages such as PSpice. It is generally the responsibility of the design engineer to determine the final selection of component values and ratings. Technicians must also be acquainted with the factors that govern these selections in order to enable them to choose the best available component for the specific applications. The final choice of components and hardware will be somewhat dependent on *cost* and *availability,* but the prime concern should be reliability. *Size, shape,* and the *finished appearance* of the unit also dictate part selection.

All electronic equipment should be as compact as possible, without reaching the point at which extreme component density prohibits assembly and service. Very often, the size of the unit is predetermined by its application, and the technician is compelled to work within established limits. The variation in size and mounting characteristics among commonly available components of the same electrical ratings makes it possible to satisfy even extreme dimensional restrictions.

The first criterion used in selecting components is, of course, that they meet the circuit's electrical demands with high reliability. Manufacturers' rated tolerance must be within design values. The closer the tolerance to the rated value of the component, the more it costs. Since cost is always a consideration, a technician should never choose components with closer tolerances than are necessary for the particular application. For example, if a resistor value is designed for 220 ohms with 10% tolerance, it certainly would be uneconomical to select a 1% resistor, since it may not appreciably improve circuit performance or reliability.

*Voltage, current, power levels, frequency,* and *continuous* or *intermittent operation* are the major circuit parameters that govern the selection of both the component value and type. In prototype construction the maximum electrical ratings (*stress levels*) of the components must be *derated* to ensure reliability. Derating provides a sizable safety factor so that components will operate well within their rated stress limits. In general, resistors are operated about 50% below their power stress levels. As an example, if a circuit resistor is required to dissipate $\frac{1}{4}$ watt, the resistor actually chosen would have a $\frac{1}{2}$-watt rating. On the other hand, a limit must be placed on overderating or compactness will be compromised. Using a 2-watt resistor where a $\frac{1}{4}$-watt resistor will operate within derating values or an electrolytic capacitor with a working voltage of 100 volts where 10 volts would be more than adequate would sacrifice size, weight, and expense without improving circuit performance.

*Environmental factors* and *human engineering* also come into play in the determination of which component or style of hardware to use. To illustrate this, hermetically sealed components would be necessary for protection against extremely humid conditions. From the standpoint of safety, external accessibility, and appearance, an extractor post fuse holder would be selected over a fuse block (Fig. 2.2).

It is important that the technician realize and appreciate that in many cases, component and hardware selection is one of compromise. This compromise comes about through the combination of availability, cost, size, and tolerance. Regardless of the choice of components, in the final analysis the system must function within the established operating specifications. Otherwise, it is of little value, even though it may be well packaged.

Hardware items are sometimes a matter of personal choice. However, the size of such items must be realistic and fit the general scheme of the system. Miniature and subminiature hardware should not be selected when high component density is not required. In many instances, the selection of hardware such as switches, indicator lamps, connectors, jacks, and plugs is determined by those styles and types that are conventionally used

**Figure 2.2** Extractor post fuse holder and fuse block.

in similar applications. If the unit being designed is to be used in conjunction with other units in an existing system, much of the hardware would be selected on the basis of mechanical and electrical compatibility and similar physical appearance. As an example, an existing system whose interconnections are terminated in *Cannon* connectors would require a special cable if a newly designed addition to the system used *Jones*-type connectors. Again, if a system makes use of *pc* (printed circuit) board connectors, additional boards used in the same system must be of a similar type (Fig. 2.3).

With the ever-widening use of integrated circuits, packaging has become more complex and a heavy reliance on printed circuitry has resulted. *Printed circuits* provide the only realistic solution to the problems associated with mounting transistors and the intricate wiring that is characteristic of integrated circuits. In the preliminary planning stage of construction, the technician is faced with the problem of using partial or total printed circuitry in a design. The solution to this problem often dictates the selection of the most suitable component and hardware types and styles. Although complex, printed circuits have reduced many packaging problems involving size, weight, and production uniformity. Since such a large percentage of components are designed for use with printed circuit boards, the technician will encounter little problem in finding components available. Certain designs mandate printed circuit boards for functionality. An example of such a design is the switching regulator power supply introduced in the previous chapter. The return (ground) path of all components in a switching design must be a large copper ground plane for effective operation. Short lead length and low inductance paths are also requirements for proper circuit functioning.

Some circuits require variable components, which are used only for initial adjustments for optimum circuit performance. Once these adjustments are made, ready access to the component is not necessary. These components are normally chassis-mounted or adapted to printed circuit design, as shown in Fig. 2.4.

**Figure 2.3**　Modern printed circuit board connectors.

**Figure 2.4**  Screw-adjustment potentiometer for printed circuit applications.

## 2.3  CHASSIS MATERIALS AND CONFIGURATIONS

Sound mechanical design is an essential element of electronic equipment packaging. Components must be provided with a sturdy mounting base that will hold them in a fixed position so that they may be electrically connected. The metal frame on which the components are normally mounted is called the *chassis.* The overall dimensions of the chassis are determined by the final component layout of the circuit. The chassis configuration is selected through an evaluation of the following factors: *circuit function, size, area restrictions, rigidity, component accessibility,* and *available manufacturing and fabricating facilities.*

The two basic types of chassis are the *box* and the *flat plate* configurations. Other chassis shapes are variations of these. Figure 2.5 illustrates several common commercially available chassis that are widely used in industrial and consumer product applications.

Factors that determine the style of chassis to use in a particular application are *weight of the heaviest components; use of controls and indicators,* such as switches, potentiometers, meters, and lamps; *circuit and operator protection; portability;* and *appearance.* If heavy items are to be mounted, a simple U-shaped chassis, as shown in Fig. 2.5a, may not provide enough strength, especially if aluminum is the chassis material. A more sturdy design, such as the box chassis in Fig. 2.5b, should be considered. Here, tabs on the long sides of the chassis are folded behind the short sides and allow for the corners to be mechanically secured by sheet metal screws, rivets, or spot welds. Small self-contained circuits are packaged in chassis-enclosure combinations, as shown in Fig. 2.5c. Some circuit applications, particularly test equipment, require extensive use of meters, knobs, and indicator lamps and in addition must be portable. A reasonable selection for this application is the chassis and enclosure configurations shown in Fig. 2.5d or e, which provide a front panel for mounting controls and indicators, and, in some cases, handles for improved portability. Figure 2.6 illustrates a typical test equipment package.

Chassis are generally fabricated from sheet metal but may also be cast. The most common types of materials used for chassis fabrication are *aluminum, aluminum alloys, low-carbon cold- or hot-rolled steel,* and *molded plastics.* Special applications may call

**Figure 2.5**  Commercial chassis configurations: (a) U-shaped; (b) standard box; (c) small utility box; (d) front panel and enclosure; (e) portable equipment case. *Courtesy of Bud Radio, Inc., Willoughby, OH.*

for chassis constructed of *copper, brass,* or *magnesium.* The choice of materials for chassis depends on *strength, weight, environmental conditions, finish, electrical and thermal requirements, cost,* and *special circuit demands* such as shielding. Aluminum chassis are the most readily available commercially in a variety of height, width, and depth dimensions with *gauges* (metal thicknesses) ranging from Nos. 14 to 20 in a wide selection of alloys and *tempers* (metal hardness). (See Appendix I for a table of metal gauges and their decimal equivalent thicknesses for aluminum and steel.) Aluminum used in chassis construction is light, strong, easy to work, and relatively inexpensive. It also requires no protective finish, since it does not normally corrode. Steel chassis need to be finished with a protective coating to prevent corrosion and to improve the appearance. Although steel chassis are more difficult to work with than aluminum and require additional care and finishing, they are more rugged than a comparable gauge in aluminum.

**Figure 2.6**  The front panel of a digital multimeter package.

If circuit design characteristics include frequencies such as *VHF* (very high frequencies—30 to 300 megahertz) and *UHF* (ultrahigh frequencies—300 to 3000 megahertz), copper or copper-plated chassis are used because their conductivity is higher than that of aluminum or steel. This characteristic is especially desirable if the chassis is used as part of the electrical circuit.

Once the chassis configuration and material have been determined, the next step is to establish the size of the finished chassis. Factors that determine the size are the *number* and the *dimensions* of the components and the *component density* and *positioning*.

## 2.4 COMPONENT AND HARDWARE PLACEMENT

The packaging of an electronic system involves the positioning of parts and components to determine their optimum density and location, finished chassis size, and a balanced layout. Both electrical and mechanical factors are taken into consideration when packaging. Logical sequence dictates that mechanical criteria are applied first.

Large, heavy components such as transformers and filter chokes should be placed near the corners or edges of the chassis, where the greatest support exists. If several heavy components are being used, they should be evenly distributed over the entire chassis. Power and filament transformers, like the one used for the switching regulator power supply, are normally positioned to the rear of the chassis to avoid the ac line cord running near or through critical circuit areas. Fuses, cable connectors, and infrequently used binding posts and jacks for external antennas or remote speaker connections also should be positioned to the rear (Fig. 2.7a). Control knobs, switches, frequently used binding posts or jacks, meters, dials, and indicator lamps are generally located to the front of the chassis (Fig. 2.7b). These items are mounted not directly to the chassis but to a front panel or cabinet. Electrical connections between the panel-mounted and chassis-mounted components are then made directly or to intermediate terminal points. Panel-mounted hardware and components should be positioned to provide balance in their appearance and to ensure ease of operation, unobstructed visibility, and a neat and orderly wired package.

Whenever possible, speakers should be front-positioned so that the sound is directed toward the operator. At times, however, when tone quality is important, a larger speaker is required and it may be necessary, because of physical size, to position it elsewhere. Sacrificing position for tone quality is not an uncommon practice.

Electrical criteria often impose restrictions on component placement. Common circuit problems that must be considered in packaging to ensure proper circuit performance are *thermal radiation* from components such as resistors, switching regulator ICs, and power transistors; *ac hum* introduced into an audio circuit; and *distributed lead capacitance and inductance,* which can drastically affect audio circuits and rf (radio-frequency) circuits. These circuit problems are discussed in the following paragraphs.

Low-power transistors and integrated circuits are heat-sensitive devices and must be positioned a reasonable distance from any component that generates considerable heat. Components that give off a large amount of heat include power transistors and switching regulator ICs. In many applications, normal air convection does not provide for adequate heat dissipation and *heat sinks* are necessary. Heat sinks are metal configurations that, when brought into physical contact with the transistor case, absorb and dissipate a large portion of the generated heat. The chassis itself can be used as a heat sink, but power transistors are manufactured with the collector electrically connected to the case. If this creates an electrical problem when mounted directly to the chassis, an insulating washer is used. This washer electrically insulates the case from the chassis without causing severe adverse effects to the thermal conductivity between the case and the chassis. This mounting arrangement is shown in Fig. 2.8. Thermally conductive grease is also used on both sides of the insulating washer to increase thermal conductivity. If chassis sinking alone would not provide adequate heat dissipation, fin-type sinks like those shown in Fig. 2.9a

**Figure 2.7** External hardware and component placement: (a) typical rear-mounted hardware; (b) typical front-mounted components and hardware.

**Figure 2.8** Correct mounting arrangement of a power transistor to a metal chassis.

(a)

(b)

**Figure 2.9** Typical heat-sink configurations: (a) chassis-mountable heat sink for medium- and large-power devices; (b) heat-sink style for low-power devices. *Courtesy of Thermalloy, International Electronic Research Corporation and Wakefield Engineering Company, Inc.*

are employed. If low-power devices require heat sinking, the type shown in Fig. 2.9b, which is not chassis-mounted, is often used. The proper size and configuration of sink required is predetermined during the circuit design stage. The problem of providing adequate space for the sink is left to the technician.

What is commonly referred to as *ac hum* or 60-hertz hum, is a condition common to most electronic circuits employing power transformers. A 60-hertz signal, usually originating at the power transformer, can be induced into the input stages of the circuit. This induced signal is then amplified along with the desired signal, resulting in a steady hum. Shielding the signal lead or, by proper placement, making the input lead very short, will minimize this problem. It is advisable, however, to make it a common practice to locate power transformers, ac leads, or any source of electromagnetic radiation away from the signal input leads.

If circuitry employing radio frequencies is to be packaged, special attention must be given to the lead lengths required by the component and hardware positions. Because of distributed lead capacitance and inductance, which become a problem at these higher fre-

**Figure 2.10** Typical rf package design.

quencies, wires used for electrical connections must be kept as short as possible. This is especially true at the input stages, as in the case of a receiver's local oscillator section, wherein the leads are extremely susceptible to noise pickup. Figure 2.10 shows an example of a compact rf package with the components positioned so as to keep the leads as short as possible.

If antenna positioning is involved in the circuit layout, it is important to locate the antenna input as close to the input circuit as possible to avoid stray pickup. In addition, if the antenna is designed to be totally contained within the unit, it should be mounted on a nonmetallic surface and kept as far away from metal elements as possible. This avoids the problem of picking up undesired frequencies. Component and hardware placement is by far the most involved and time-consuming phase of planning a package. Performance and reliability of the system will depend heavily upon this phase of design.

## 2.5 MAINTENANCE

Maintainability is a measure of the ease, accuracy, and time required to restore a circuit or system to its normal mode of operation once a failure or abnormal function is detected. Maintenance covers two general procedures. First is the repair of failures and defects. Second is *preventive maintenance,* which deals with normal or routine checks and servicing of equipment. The object of preventive maintenance is to keep equipment operating trouble-free and at optimum performance. Certain precautions should be taken in the design of any package to minimize unnecessary maintenance. Whenever leads must pass through a hole in a metal chassis or bend around the edge of a bracket, either a grommet or clamp should be used to prevent abrasion of the lead insulation, which would cause a short circuit.

High ambient temperatures within enclosures can eventually cause components to overheat and fail. To ensure cool, trouble-free operation, approximately 6 cubic inches of space per watt of power to be dissipated is necessary if normal air convection is to be relied upon for cooling a unit. Vents or holes must be provided, preferably at the top of the enclosure surrounding the circuit. If units are stacked in a rack, those units that dissipate the most heat should be located in the uppermost position to prevent the heating of components in other units. Whenever possible, liquid cooling and blowers should be avoided. Liquid cooling systems are subject to failure, and blowers could cause air circulation that will cool certain components but overheat others. Air filters are often required when blowers are used, thus adding another maintenance dimension to the unit. Liquid cooling and blowers also add bulk and weight to the unit.

Enclosures, in addition to allowing circuits to function without overheating, must also be tight enough to isolate the components and wiring from dust or other contaminants that could accumulate in the circuit and cause failures. If large holes have to be formed in an enclosure for speakers, the entry of dust can be minimized by using a *speaker cloth.* For additional protection of the speaker cone, a grid or wire screen can be used with the cloth.

Well-designed enclosures minimize maintenance problems through the various types of circuit protection they afford. They can, however, present problems in terms of accessibility. Ease in maintenance is directly related to component accessibility. By locating access plates, doors, and openings near those components that are most subject to maintenance needs, repair and service can be achieved easily, quickly, and accurately. Safety precautions should be built into the design when providing access to electronic circuits. Safety devices may have to be incorporated with the removable elements to avoid the possibility of electrical shock when working on the circuit through any access opening. Providing for maintenance in packaging design is a compromise involving three factors: *sufficient environmental protection, accessibility,* and *safety.*

## 2.6 HUMAN ENGINEERING (ERGONOMICS)

In electronic packaging design, human limitations associated with the use of electronic equipment must be considered. These human limitations involve such things as vision, arm reach, and hand manipulation, which must be considered in order to select the most suitable controls and visual displays. The ease and accuracy with which the unit is operated depends on the proper selection and placement of these components.

Meters should not be so small as to make the reading of major and minor scales difficult. The type of scale used should fit the function of the unit. For example, if only circuit conditions are to be monitored rather than the finite values of electrical parameters, the use of smaller meters having fewer scale divisions is desirable. If, on the other hand, accurate measurements are required, a larger meter with numerous minor scale divisions is preferable.

Glare must be avoided when continuous dial and scale readings are necessary. If reflected light in the unit's environment is excessive, glare shields should be employed. In addition, the panel should be finished in a color possessing low reflective characteristics (see Chapter 24).

Control knobs and switches should be mounted within easy and comfortable reach of the operator and should not be crowded. If they are, accurate individual adjustments will be difficult to make and other settings may be disturbed. Keeping with accepted conventions in mounting panel elements is very important. For example, toggle switches that are to be operated in a vertical manner should be on when the lever is in the upward position. Potentiometer adjustments in a clockwise direction should cause an increase in the magnitude of the electric quantity. In rack- or console-mounted equipment, all visual displays and indicators, such as meters and indicating lamps, should be mounted as close as possible to the eye level of the operator.

Stability is essential, especially if the unit is portable and relatively light. Some controls, especially rotary switches, require substantial torque to actuate and may cause the unit to become unsteady while in use. This problem can be reduced or overcome by either mechanically securing the unit to a mounting surface or attaching rubber pads to the bottom corners. The pads not only provide greater friction but also offer protection to the working surface, in addition to serving as electrical insulators for the unit.

The consideration of human limitations and capabilities, then, must be of concern to the technician designing a package. Many manufacturers employ human engineering consultants to render advice about the placement of control elements. A unit may function properly, but it will not be acceptable if it is difficult to control, read, or adjust.

In the next chapter, we use the basic concepts discussed here to plan the package design for the switching regulator power supply project designed in Chapter 1.

## EXERCISES

### A. Questions

2.1 List at least five design factors that go into the planning of a piece of electronic equipment.

2.2 Of those factors listed in Question 2.1, which do you consider the most important, and why?

2.3 What electrical criteria impose restrictions on component placement?

2.4 What are the two basic types of chassis from which all other variations are derived?

2.5     What is the purpose of a *heat sink?*

2.6     What is meant by *human engineering* in package design? Why is it important?

## B. True or False

Circle T if the statement is true, or F if any part of the statement is false.

| | | |
|---|---|---|
| 2.1 | Stationary equipment design imposes fewer restrictions than does that for vehicular equipment. | **T**  **F** |
| 2.2 | A U-shaped chassis provides more strength than a box chassis. | **T**  **F** |
| 2.3 | Large and heavy components should be placed near the corners or edges of a chassis. | **T**  **F** |
| 2.4 | The most common material used to make chassis is copper. | **T**  **F** |
| 2.5 | In designing high-frequency circuits, the lead lengths should be as short as possible. | **T**  **F** |

## C. Multiple Choice

Circle the correct answer for each statement.

2.1     One of the most important design factors is (*reliability, appearance*).

2.2     In circuit design involving ultrahigh frequencies, the chassis should be made of (*copper, plastic*).

2.3     Transistors and integrated circuits (*are, are not*) affected by heat.

2.4     In the design of rf packages, the antenna must be positioned (*close to, far from*) the signal input circuit.

# Packaging Design

3

## LEARNING OBJECTIVES

*Upon completion of this chapter on the packaging design of a switching regulator power supply, the student should be able to*

- Understand the function of each subcircuit of the switching regulator power supply system.
- Select the appropriate chassis configuration to house the power supply system.
- Differentiate between *chassis ground* and the *high-quality ground* of the power supply.
- Understand the selection and positioning criteria for the components and hardware assigned to the front panel.
- Understand the requirements of critical components and the materials and techniques available to support them.
- Develop a preliminary hand sketch of the chassis and enclosure elements and the approximate placement of all the major components of the switching regular power supply project.

## 3.0 INTRODUCTION

A switching regulator power supply, designed in Chapter 1 and employing many different components and devices, is used as a means of illustrating the practical applications of packaging techniques discussed in the first unit of this book. In this way, familiarity is gained with the application of the current design considerations, conventions, and practical procedures for sound electronic packaging.

Since a basic understanding of electronic circuitry involved is essential to fully interpret and apply the necessary procedures for packaging, a comprehensive discussion of the switching regulator's circuit operation was given in Chapter 1. That information is revisited here with special attention paid to general packaging considerations, chassis design, and placement of critical components.

## 3.1 BLOCK DIAGRAM

A block diagram of the switching regulator power supply designed in Chapter 1 of this unit is shown in Fig. 3.1. This type of diagram identifies the function and interrelationship of each of the system's subcircuits. Refer back to Fig. 1.12 for a detailed circuit schematic of this switching regulator power supply project. The three main subcircuits of the switching regulator power supply system are (1) the step-down transformer and associated input circuitry, (2) a bridge rectifier and filter to complete the unregulated or "raw" supply circuit, and (3) the dc-to-dc converter or switching regulator.

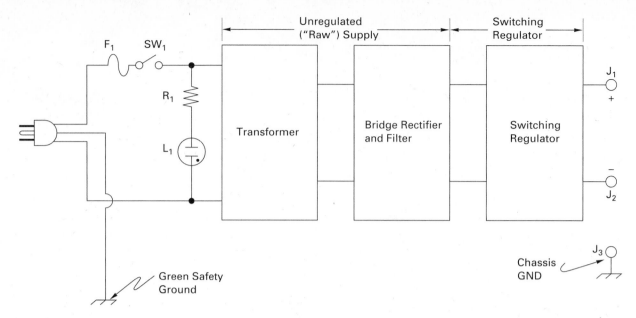

**Figure 3.1** Block diagram of the switching regulator power supply.

The three main subcircuits act together as follows: The 115-volts ac household power is first reduced in amplitude by the step-down transformer and then converted to an approximate dc level, with considerable ripple, by the bridge rectifier and capacitive filter or "raw" supply. Then, the output of the raw supply is regulated to a precise 5 volts at the output of the switching regulator circuit.

## 3.2 POWER SUPPLY PACKAGE DESIGN

Initially, a parts list will be prepared during the design phase of a project. The parts list for this switching regulator project is shown in Fig. 1.12 as part of the actual circuit schematic and in Fig. 1.10 as part of the hard copy from using the SWITCHERS MADE SIMPLE software. The parts list identifies components by part number, and usually gives tolerances and electrical ratings for each item. However, hardware, chassis material selection, and finish are decisions often left to the package designer.

Cost and availability of parts often serve as secondary factors during the circuit design phase, but should be carefully considered when designing the final package. Forced substitutes with different dimensions or tolerances can cause considerable rework of a package, resulting in lost time and effort. It often falls to the package designer to consider alternative components and devices when developing a new design, but all options should be reviewed with the circuit designer for compatibility before any changes are made final.

Normally, a switching regulator power supply has few space or weight restrictions. Therefore the first phase in the package design will be the selection of the chassis material. The type of aluminum that will be used for the chassis elements of this project will be *16-gauge No. H-1100*. This choice is made on the basis of the desirable characteristics of this material—easily worked, sturdy, lightweight, conductive, and readily available. The heaviest component, a transformer, weighs less than 22 ounces and is easily supported by this material. In addition, aluminum readily lends itself to the basic finishing techniques discussed in Unit 5, Chapter 24.

Two common styles of chassis and enclosures that lend themselves to the switching regulator are shown in Fig. 3.2. A box-type chassis to which is mounted a separate front

**Figure 3.2** Custom chassis and enclosure designs. (a) Box chassis with several enclosure, panel, and plate elements. (b) Rigid three-piece chassis and enclosure configuration chosen for the switching regulator power supply project.

panel is shown in Fig. 3.2a. In this configuration, all the parts not positioned on the front panel are placed on the box chassis element. A cover, bottom plate, and back panel are then required to complete this type of package. Although this package style provides good strength, it does have several undesirable features. First, unless a cover, bottom, and back plates are provided, there will be exposed wiring, which could result in a hazardous condition. Second, the height of the box chassis would have to be very low to accommodate the transformer and still maintain an aesthetically pleasing, low silhouette design. The three-element configuration shown in Fig. 3.2b overcomes these undesirable features and will be used to provide a stable platform onto which the switching regulator project can be built.

The three-piece chassis and enclosure shown in Fig. 3.2b has a *subchassis* element fastened vertically to the *front-and-side* element. These elements, together with the three-sided *cover* shown, result in a completely enclosed package. All items not mounted on the front panel will be placed on the sides (same piece of sheet metal as the front) or on the subchassis element. The cover fastens to both the subchassis and the front-and-side panel element to form a rigid enclosure. Unlike the box-type chassis, removal of the single cover element will expose all the components for wiring.

A preliminary series of sketches of the component layout is undertaken next. These sketches are made simultaneously with the technique of positioning the components and hardware. Each of these subcircuits and accompanying components are discussed as they relate to package and chassis design.

### 3.2.1 The Front-and-Side Panel
Two primary factors need to be considered in the design of a front panel: *visibility* and *ease of use*. The determination as to which components will be mounted to the front panel is often dictated by those components that are used most often in the convenient operation of the equipment. A review of the block diagram of the project in Fig. 3.1 (and Fig. 1.12) suggests that the following components be considered for front-panel mounting:

1. $F_1$ (fuse holder and fuse)—*optional*
2. Line cord—*optional*
3. $J_1$, $J_2$, $J_3$, (+, −, and chassis ground terminals)
4. $SW_1$ (on/off switch)
5. $L_1$ (pilot lamp—power *on* indicator)

We begin the front-panel layout by first considering the optional items to determine if it is necessary for front-panel mounting. Of the components listed above, both the line cord and fuse with holder are classified as optional. It is common practice in laboratory equipment to provide easy access to the circuit's fuse by placing it on the front panel for ease of replacement. This is especially important when frequent circuit overloads are expected. However, overloads are not a problem in this project. The use of the LM2575-5 IC, with its internal protection circuitry, makes fuse replacement rare, occurring only under extreme fault conditions. For this reason, $F_1$ will be positioned on the right side of the front-and-side panel, directly forward of the line cord (refer ahead to Fig. 3.5). Note that it is good engineering safety practice to first fuse household power, available at the three-pronged receptacle, so as to protect all components within the system.

The line cord should be provided with a strain-relief bushing for safety, and a three-lug terminal strip should be mounted inside and directly above the line cord. The *hot* and *neutral* leads can thus be wired to the two outside insulated lugs, and the center lug, directly connected to the chassis, will be used to attach the green "safety" ground lead directly to the chassis. In this way the chassis will be properly grounded when the power supply is plugged into an ac receptacle. In addition, the chassis (conductive aluminum) will be used to make the electrical connection to jack $J_3$ on the front panel.

One front-panel design that meets the requirements of *visibility* and *ease of operation* is shown in Fig. 3.3. This is but one of several arrangements that could meet the desired criteria.

The three terminal jacks, $J_1$ (+), $J_2$ (−), and $J_3$ (chassis ground, GND), are positioned central and toward the bottom of the front panel so that leads extended from these

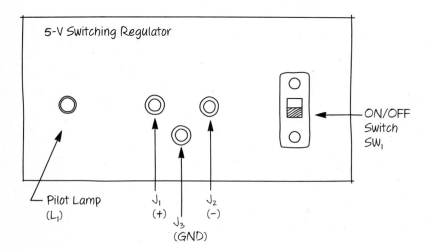

5-V Switching Regulator

ON/OFF Switch $SW_1$

Pilot Lamp ($L_1$)

$J_1$ (+)  $J_3$ (GND)  $J_2$ (−)

**Figure 3.3**  Suggested front-panel design that provides easy access to jacks and on/off switch.

terminals to any load will not interfere and obscure the remaining element of the front panel. Note that the power supply's high-quality ground or negative terminal, $J_2$, must be electrically insulated from the chassis (and chassis ground, GND) with a nylon shoulder washer.

The on/off switch ($SW_1$) is mounted directly to the right of the terminal. It is given this prominent position because it is the most frequently used hardware in the project. It is common convention in the mounting of single-pole, single-throw (SPST) switches to have power applied to the circuit when the switch lever is *up* ( or to the right in horizontal mounting) and the power off when the switch is in the *down* position (or to the left). This switch is associated with lamp $L_1$, which serves as a pilot light to indicate when power is on. Because the lamp requires no adjustment or access, it is positioned to the left on the front panel.

### 3.2.2 The Subchassis Element

Note the location of the three-pronged line cord and extractor-type fuse holder, both on the right side of the front-and-side panel. This positioning dictates that the step-down transformer be placed on the right side of the subchassis to minimize internal wire length. With the on/off switch located on the right side of the front panel, a twisted pair of leads between the on/off switch, transformer primary, and indicator lamp will be required. The twisting of leads tends to cancel the generated magnetic field about the leads, thereby reducing 60-hertz hum.

The remaining components associated with both the bridge rectifier and filter as well as the switching regulator are fragile and require the rigidity of a printed circuit board. Figure 3.4 illustrates the major components and hardware placements on the subchassis element. Note that the PCB will be mounted in a rectangular chassis cutout with machine screws and that connections to the secondary of the transformer and access to the output jacks are made via turret terminals swaged to the PCB.

**Figure 3.4** Major component and hardware placement for the subchassis element and the double-sided pc board.

A double-sided PCB with conductor paths on the top side and a large *ground plane* of copper on the bottom side will provide the low-impedance, low-inductance paths necessary for good switching regulator operation (refer to Unit 2, Chapter 8 for additional information on double-sided printed circuit board design). Also, a ground plane PCB design facilitates short component lead lengths for the Schottky diode, inductor, and output capacitor of the dc-to-dc converter, a requirement dictated by this type of switching regulator design.

### 3.2.3 The Cover Element

As discussed in Chapter 1, the LM2575-5 switching regulator IC will dissipate some heat under high-load conditions. The cumulative heat developed by this component must be adequately removed to maintain proper circuit operation. Heat removal is accomplished by providing a series of ventilation holes in both the top and bottom sections of the cover element. These holes are shown in Fig. 3.5, an exploded view of the three chassis elements. In addition, rubber feet are mounted to the bottom surface of the cover to insure that the bottom ventilation holes are not blocked. Thus, cool air will circulate up through the power supply, carrying the hot internal air out through the top ventilation holes.

**Figure 3.5** Engineering sketch of all chassis elements including major components and hardware placement for the switching regulator power supply.

## 3.3 DETAILED CONSTRUCTION DRAWINGS

Figure 3.5 shows the results of the major component positioning and mounting-space requirements. From this layout the optimum-size front-and-side panel, subchassis element, and cover can be designed and fabricated. To convert the final layout into useful detailed drawings for actual construction of the unit, accurate measurements of components and their locations must be made. Procedures for determining accurate dimensional information and the application of this information in the preparation of detailed construction drawings are discussed in the next chapter.

## EXERCISES

### A. Questions

3.1 What are the significant differences between block diagrams and circuit schematics?

3.2 List the three main subcircuits of the switching regulator project and explain the function of each.

3.3 What are the primary factors to consider in designing a front panel?

3.4 What is the difference between the chassis ground and the high-quality ground of a power supply?

3.5 Explain one method of providing adequate ventilation to an enclosed chassis.

### B. True or False

Circle T if the statement is true, or F if any part of the statement is false.

3.1 The position of individual components can quickly be determined from a block diagram of the system.     **T   F**

3.2 Block diagrams show the continuity of wiring connections between all components.     **T   F**

3.3 The purpose of a raw supply is to convert ac into an approximate dc voltage.     **T   F**

3.4 The purpose of the switching regulator is to convert the approximate dc voltage of the raw supply into a constant value.     **T   F**

3.5 The size and style of all components can always be determined from the parts list.     **T   F**

### C. Multiple Choice

Circle the correct answer for each statement.

3.1 The exact electrical connections for each component in an electric circuit is best shown by a (*block, schematic*) diagram.

3.2 The ripple of a raw supply is often (*low, high*).

3.3 The negative terminal of a power supply is referred to as (*chassis, high-quality*) ground.

3.4    Good design dictates that the fuse of a switching regulator power supply be mounted on the (*front, side or rear*) of the package.

3.5    Aluminum is selected for the chassis of our power supply system because it is a good (*conductor, insulator*).

## D. Matching Columns

Match each item in column A to the most appropriate item in column B.

| COLUMN A | | COLUMN B | |
|---|---|---|---|
| 1. | Fuse | a. | Chassis |
| 2. | Twisted pair | b. | 60-hertz |
| 3. | H-1100 | c. | Unregulated |
| 4. | Ease of use | d. | Safety |
| 5. | Switching | e. | LM2575-5 |
| 6. | Raw supply | f. | Design element |

# Preparing Detailed Drawings

4

## LEARNING OBJECTIVES

*Upon completion of this chapter on preparing detailed drawings, the student should be able to*

- Make accurate measurements using common precision instruments such as scales, calipers, and micrometers.
- Produce engineering sketches for the chassis elements.
- Know the basic line conventions.
- Know the four basic dimensioning techniques.
- Select between several methods of preparing good-quality drawings.
- Produce assembly drawings showing a variety of views.

## 4.0 INTRODUCTION

Once the preliminary plans for a package are completed and engineering sketches of the chassis configuration are prepared, *detailed drawings* must be developed to facilitate construction. These drawings entail the final positioning of all components and hardware. Sizes of these components and hardware items are first measured to obtain overall dimensions necessary to select and complete optimum orientation and position. Many of these measurements will also determine the size and location of mounting holes during the fabrication of the chassis. These detailed drawings must include all pertinent information with respect to component assembly and positioning and chassis fabrication and finish. It is important to mention here that someone other than the technician who designed the original package may be assigned the task of construction. For this reason, the detailed chassis drawings must be completely informative, accurate, and clear.

This chapter contains specific information on techniques of using measuring instruments such as scales, calipers, micrometers, and templates. Also included is relevant information on engineering sketches, various modern drafting methods, dimensioning, and procedural information for developing chassis layout drawings along with assembly and auxiliary views for detail work.

## 4.1 MEASUREMENTS WITH A SCALE

In the packaging design stage, as described in Chapter 3, approximate positioning of components and hardware was estimated by physically placing the components in the desired locations. From this preliminary method, component density can be observed. It is now necessary to take all dimensions and transfer this information first into an engineering sketch and then into detailed working drawings.

**Figure 4.1** Steel rules.

To obtain most of the required measurements, a simple *steel scale* is commonly employed. Steel scales are available in a variety of styles, including *steel tape rule, steel rule,* and *hook rule.* The 6- and 12-inch steel rules and the 6-inch hook rule are the types commonly used in electronic construction (Fig. 4.1). The major scale division on these rules is the inch, with half, quarter, eighth, sixteenth, thirty-second, and sixty-fourth subdivisions. Steel scales are also available with decimal graduations having major and minor divisions of 0.1 inch and 0.020 inch, respectively. This rule is desirable when the dimensioning format of the sheet metal drawings is in decimal form. The smaller graduation scales are used where close tolerance measurements are required.

Overall length and depth measurements for the chassis determined by the final component layout can be made by alternately positioning and reading the scale along the lines bounding the area previously established in the planning stage. The height of the unit is determined by reading the vertical measurement of the tallest component together with the desired enclosure clearance (see Fig. 4.2a).

Component measurement for mounting purposes can also be made with a scale. As an example, a method for determining on-center dimensions for mounting holes is shown in Fig. 4.2b. Mounting information, however, is usually provided by the manufacturer in the form of a paper template.

For maximum accuracy when using the 6- or 12-inch steel rule, the scale should be positioned perpendicular to the surface being measured. In this way the graduation marks on the scale will be in direct contact with the work, which will eliminate the possibility of error due to the thickness of the rule if it is placed flat on the work. In addition, neither end of a rule should be used as the reference or starting edge. For accurate measurements, the work should be aligned with the first major inch graduation mark. This assures that scale-edge deformities will not cause errors.

The hook rule, shown in Fig. 4.1 (bottom), provides an accurate reference edge along the inside face of the hook. Thus, the zero graduation mark is perfectly aligned with the edge of the work, assuring more exact measurements.

## 4.2 MEASUREMENTS WITH CALIPERS

Although rules are adequate for most measurement applications, *calipers* are necessary if direct measurements are impossible. Calipers are classified as *outside* or *inside,* depending on the shape and position of the legs as well as their specific use. The main parts of the caliper, shown in Fig. 4.3, are the *legs, spring crown,* and *adjusting knob.*

The outside caliper is readily adaptable to measuring diameters of round stock, such as the shaft of a potentiometer or rotary switch. This caliper is held by the spring crown cupped in the palm of one hand with the thumb and forefinger free to apply a rotational force to the adjusting knob (Fig. 4.3a). Rotating the adjustment knob will close the legs of the caliper on the surface to be measured. Once the inside tips of the legs contact the surface to be measured, a back-and-forth motion of the caliper along the surface will indicate

(a)

(b)

**Figure 4.2** Method of determining both (a) height and (b) on-center dimensioning.

the correct measurement. The calipers are set properly if there is a *slight friction* as the inside tips of the legs move across the widest portion of the work. Practice is necessary to acquire the correct "feel" for this optimum setting. With the leg spacing set, the measurement can now be determined by placing the caliper along the edge of a scale and reading the distance between the inside of the legs. This procedure is illustrated in Fig. 4.3b.

Drill gauges can also be used as a quick means of measuring round bodies such as diodes, transistors, and integrated circuits. Figure 4.4 shows the measurement of the outside diameter of a transistor using a drill gauge.

If greater accuracy than can be obtained from the basic caliper is required, *dial calipers* are used. The 6-inch dial caliper is common in electronic construction. This precision instrument consists of *two pairs of jaws* (one for *inside* and one for *outside* measurements), a *dial indicator,* and a *thumb adjusting knob.* This instrument is capable of accuracies from 0.001 to 0.0001 inch (Fig. 4.5). Before using the dial caliper, a zero check

(a)                                        (b)

**Figure 4.3** Outside calipers used to measure round stock: (a) correct method of setting caliper tips; (b) distance between tips is determined with a steel rule.

should be made to determine if the dial indicator is properly calibrated with the jaws. First, the contacting surfaces of the outside jaws are wiped clean with a soft cloth to remove any foreign matter or dust. The thumb adjustment screw is then rotated clockwise to close the outside jaws. The dial indicator should read zero when the jaw surfaces make contact. If not, the rack and pinion must be adjusted. The two securing screws at both ends of the rack gear are loosened and the rack is slowly moved until the dial indicator reads exactly zero. The securing screws are tightened and the instrument is set for accurate measurements. Figure 4.6 illustrates the proper method of supporting the dial caliper when making measurements. With the adjusting knob rotated to close the outside jaws on the work, the correct amount of jaw pressure is obtained through a *slipping action* whereby continued rotation will result in the adjusting knob slipping with no further forward motion imparted to the jaws. This action eliminates damage to the caliper and also measurement error.

All settings should be checked for correct jaw alignment before measurements are read. To ensure accuracy, the jaws of the calipers are slightly rocked over the work (see

**Figure 4.4** Drill gauge used to measure the case diameter of a transistor.

Figure 4.5    Dial caliper.

Fig. 4.6). The minimum deflection of the dial pointer will result in the most accurate reading.

## 4.3 MEASUREMENTS WITH A MICROMETER

The micrometer, like the dial caliper, is capable of measuring to accuracies of 0.001 inch and estimates to 0.0001 inch. The 1-inch outside micrometer is typical, although larger-capacity micrometers are available in addition to special-purpose micrometers, such as screw-thread measuring instruments.

The 1-inch micrometer is used primarily to measure diameters and thicknesses. This micrometer, shown in Fig. 4.7, consists of a *frame, anvil, spindle, barrel, thimble,* and *ratchet stop.* The proper use of this micrometer is as follows. The little finger is hooked around the frame while the thumb and forefinger of the same hand are free to rotate the thimble (Fig. 4.8). Clockwise rotation of the thimble will close the spindle on the anvil, which is rigidly supported to the frame. Counterclockwise rotation of the thimble widens the distance between the spindle and the anvil. The work to be measured is held in the other hand and placed on the anvil. The thimble is rotated to close the spindle on the

Figure 4.6    Dial caliper used to obtain accurate measurements with rocking motion to determine minimum reading.

**Figure 4.7** One-inch micrometer with ratchet stop. *Courtesy of the L.S. Starrett Company, Athol, MA.*

work. Once the spindle makes *light contact* with the work, the thumb and forefinger are moved to the ratchet stop. Clockwise rotation of the ratchet stop for two "clicks" will apply the correct amount of pressure to the work. Excessive pressure may damage the finely threaded spindle or mar the machined surfaces of the anvil and spindle, thus causing measurement errors. It is good practice to maintain continuity in measurement by duplicating the applied pressure. The two "clicks" of the ratchet stop will assure this duplication of pressure.

The spindle has a precision of 40 threads per inch. Thus, one complete rotation will move the spindle exactly $\frac{1}{40}$ or 0.025 inch. Each minor graduation mark on the barrel scale is therefore equal to 0.025 inch. Each major numbered division on the barrel scale represents 0.100 inch, requiring four complete turns to move from one major numbered division to the next. The thimble has 25 graduation lines. Each line represents $\frac{1}{25}$ of $\frac{1}{40}$ of an inch, or 0.001 inch. Therefore, as the thimble is rotated from one division to the next in a counterclockwise direction, the spindle moves 0.001 inch away from the anvil.

To obtain the spindle-to-anvil opening, three readings are necessary. First, the highest-*numbered* graduation shown on the barrel is noted. (Each numbered graduation represents 0.100 inch.) The second reading is the number of minor divisions following the major number on the barrel, each minor division representing 0.025 inch. The third reading is the lowest-numbered graduation line on the thimble that most closely aligns with

**Figure 4.8** Correct positioning of the 1-inch micrometer.

**Figure 4.9** Illustrations of micrometer scale readings: (a) readings to the one-thousandth inch using the barrel and thimble scales; (b) readings to the ten-thousandth inch using barrel, thimble, and vernier scales.

(a)

Vernier Aligns at 0.0003″

(b)

the horizontal scale line on the barrel. This value is in one-thousandths (0.001) of an inch and is added to the sum of the preceding readings to obtain the final measurement.

Figure 4.9a represents a reading of 0.337 inch obtained as follows:

*Reading 1*. The largest visible number on the barrel is 3, or 0.300 inch.

*Reading 2*. The number of smaller graduations after the number 3 is 1, or $1 \times 0.025$ = 0.025 inch.

*Reading 3*. The lowest number of the graduation lines on the thimble that most closely aligns with the horizontal scale line on the barrel is 12, or 0.012 inch.

The sum of the three readings is 0.337 inch.

Some micrometers are equipped with a *vernier scale* to extend the measurement accuracy to one ten-thousandths (0.0001) of an inch. This vernier scale is immediately above the barrel scale and its graduation marks extend along the barrel to the thimble scale. The vernier and thimble scales both run horizontally. Each of the 10 graduation lines of the vernier scale, marked 0 to 9, represent $\frac{1}{10}$ of the value of the divisions on the thimble scale—that is, $\frac{1}{10}$ of $\frac{1}{1000}$, which equals $\frac{1}{10,000}$ (0.0001) inch.

To obtain a reading with a micrometer equipped with a vernier scale, the previously discussed three readings are first taken. The fourth reading is obtained by noting which graduation line of the vernier scale aligns with any line on the thimble scale. This reading, in one ten-thousandth (0.0001) of an inch, is added to the sum of the other three readings. Figure 4.9b represents a reading of 0.4583 inch, obtained as follows:

*Reading 1*. The largest visible number on the barrel is 4, or 0.400 inch.

*Reading 2*. The number of minor graduations after the number 4 is two, or $2 \times 0.025$ = 0.050 inch.

*Reading 3*. The lowest number of the graduation lines on the thimble that most closely aligns itself with the horizontal scale line on the barrel is 8, or 0.008 inch.

*Reading 4*. The number of the division on the vernier scale that aligns with a division on the thimble is 3, or 0.0003 inch.

The sum of the readings is 0.4583 inch.

## 4.4 ENGINEERING SKETCHES

A preliminary step in the preparation of the finished detailed working drawings is the development of *engineering sketches*. Figure 4.10 shows a sheet metal sketch of the switching regulator's subchassis element. Using quadruled paper results in a relatively clear picture without the need of special equipment. Notice that the engineering sketch shown in Fig. 4.10 contains not only *dimensional information* but specifications for *finish, type of material, and hole positioning*. Also useful are manufacturer's specifications that provide hole-spacing information for mounting holes, pin sizes and configurations, case dimensions, and other pertinent information to aid in layout.

**Figure 4.10** Engineering sketch of power supply subchassis element.

## 4.5 METHODS OF CREATING DETAILED WORKING DRAWINGS

To produce good-quality drawings, the technician must select from conventional drafting equipment, like that shown in Fig. 4.11, or employ a computer-aided drafting (CAD) software package, such as AutoCAD, Release 12, described in Chapter 7 of Unit 2. While some drafting work is still being done by hand techniques, it seems apparent that the trend has shifted away from this method and toward computer-generated drawings. There are many reasons for this shift, chief among them being the proliferation of low-cost, user-friendly, general-purpose software tools and personal computers. Today's technician must become more skillful in the new computer technology, while still maintaining a passing acquaintance with the older hand-drawing techniques.

What follows is a brief overview of the typical drafting equipment necessary to generate high-quality drawings by hand methods and by CAD techniques.

### 4.5.1 Hand-Drafting Equipment
Tools common to the technician wanting to develop detailed drawings by hand-drafting techniques are (1) *drafting paper and pencils*, (2) *drafting boards*, (3) *T-squares*, (4) *triangles*, (5) *protractors*, (6) *engineering scales*, and (7) *drafting instruments and templates*.

**Drafting Paper and Pencils.** Knowledge of the size and type of paper used in drawing is helpful. Commonly available drawing paper sizes are A ($8\frac{1}{2}$ by 11 inches), B (11 by 17 inches), C (17 by 22 inches), and D (22 by 34 inches). Larger sizes are available; K size (40 by 50 inches) is normally the largest used in engineering work. The paper size is, in many cases, selected to be compatible with the filing system in use. In all cases, the paper should be of sufficient size to show detail work without crowding, yet not so

**Figure 4.11** Basic drafting equipment for electronics.

large as to have excessive unused areas or overly large borders. The preferred paper for electronic drafting is the *HP* (hot-pressed) *cream* or *buff-colored* type. This paper is sufficiently heavy and hard-surfaced to be used for high-grade, accurate detailed work. *CP* (cold-pressed) paper is less expensive and is used for general drawing or sketching. The surface of this type of paper will not hold up to erasing as well as the HP type. When blueprints are to be made from pencil drawings, a white, light-weight bond paper should be used. This type of paper allows blueprints to be produced directly without the need for tracing.

Pencil lead is available in a variety of grades from extremely hard (9H) to very soft (7B), with a medium grade of H or 2H selected for general-purpose technical drawings. Mechanical supports are preferred over wood.

**Drafting Boards.** Constructed from straight-grained wood, drafting boards for electronic drawings should be flat, with true edges, and of a convenient size, 18 by 24 inches being a common choice.

**T-square.** Although available in various lengths, the 24-inch T-square is suitable for the drafting board selected previously. Blades made of hardwoods such as maple, with clear celluloid edges, are preferred.

**Triangles.** To facilitate drawing angles, the 45° and the 30 by 60° triangles are important items for hand-drafting. Most often, they are made of clear or transparent celluloid, with a 6- to 8-in.-long leg.

**Protractors.** When angles cannot be obtained with the common triangles, the protractor is used. For most applications, the clear or shaded celluloid protractor, shown in Fig. 4.11, is satisfactory.

**Engineering Scales.** Mechanical engineer's scales are the most useful in electronic drafting. These scales provide full-size, one-half, one-fourth, and one-eighth divisions that are satisfactory for most size reductions.

**Drafting Instruments and Templates.** Drawing instrument sets, such as the one shown in Fig. 4.11, contain *dividers, compasses,* and *inking pens,* and are necessary to produce quality, accurate plates. Occasionally, however, time is more important than pre-

**Figure 4.12** Drafting and electronic layout templates.

cision. The use of templates, such as those shown in Fig. 4.12, can aid in producing good quality drawings with a considerable savings of both time and effort. Templates are used extensively as a quick means of reproducing electronic symbols and other shapes, particularly circles. They consist of various shaped cutouts in a piece of clear or shaded plastic. When these cutouts are outlined with a pencil or pen, the result is a uniform and neatly shaped electronic symbol or figure.

**4.5.2 Computer-Aided Drafting Software**   The basic building blocks for a typical computer-aided drafting system consist of a personal computer with color monitor, a keyboard and mouse for input peripherals, and a printer or plotter as the output peripheral. This hardware, together with the appropriate CAD software package, form a complete drafting/design system. Two of the more popular software packages used in industry are AutoCAD and VersaCAD. Both of these products are very powerful and are designed for general-purpose drafting work, with a large selection of features and options.

AutoCAD, Release 12, was selected to illustrate this book and is described in some detail in Chapter 7. Since it is a general-purpose drafting package, the primary operations of drawing lines, rectangles, and circles, as well as dimensioning, labeling, making modifications, and saving and printing files, are generic and apply equally to developing the chassis drawings described in this chapter as well as designing the PCB artworks of Chapter 7. Refer to Chapter 7 for a discussion of the basic methods and techniques employed in developing drawings using AutoCAD, Release 12.

## 4.6  BASIC LINE CONVENTIONS FOR DRAWINGS

Since the primary objective in engineering drafting is to transmit detailed information pictorially, a basic knowledge of the various types of lines used and their application is essential. The symbolic lines conventionally accepted are shown in Fig. 4.13. Whether by conventional drafting means or by CAD, the line width will vary depending on the type of symbol used. Thin lines are used for *dimension, center, extension, long-break,* and *section* symbols. Medium thickness is used for *hidden* and *ditto* lines. Thick lines are reserved for *object, break,* and *cutting-plane* lines.

## 4.7  DIMENSIONING FOR CHASSIS LAYOUTS

Several acceptable dimensioning techniques are used in chassis layouts, the choice of which depends on the configuration being dimensioned and the degree of accuracy required. These techniques are datum-line or base-line, center-line, continuous, and tabulation dimensioning.

| | Application |
|---|---|
| | Construction Line |
| | Dimension Line |
| | Center Line |
| | Extension or Datum Line |
| | Long–break Line |
| | Section Lining |
| | Hidden or Bend Line |
| | Ditto Line to Indicate Repeated Detail |
| | Object Line to Outline Parts |
| | Short–break Line |
| | Cutting–plane Line |

**Figure 4.13** Basic line conventions used in engineering drafting.

*Datum-line* or *base-line dimensioning* uses any two edges that are perpendicular to each other. These reference edges are called *datum lines*. Each dimension made uses one of the datum lines as a reference. This technique eliminates cumulative error, since each dimension is independent of all others. This method of dimensioning is often used for close-tolerance work, especially when mating parts are to be fabricated. Figure 4.14 illustrates base-line dimensioning.

In *pure center-line dimensioning*, the stock is first equally divided vertically and horizontally. The locations of all holes are made from these reference lines. No other lines are used as a reference. In electronic layout, datum-line dimensioning is preferable when locating holes, since edge reference facilitates this layout, as described in Chapter 20.

When holes are to be formed that match a particular pattern on a component, such as the mounting holes on a speaker, a variation of pure center-line dimensioning is preferable. Base-line or pure center-line dimensioning would create a larger tolerance in this application. For this reason, independent dimensioning of related holes or of hole patterns should be avoided. In this technique the center of the largest or major hole in the pattern is located from datum lines. All other hole centers are dimensioned using these major hole center lines as a reference. Figure 4.15 shows an example of center-line dimensioning.

As drawings become more complex, it is necessary to use both datum-line and center-line dimensioning, especially when the exclusive use of either one would adversely affect clarity. Figure 4.16 shows an example of combining both techniques on one drawing. Notice that the use of both techniques reduces the number of dimension lines crossing. Although it is acceptable practice for dimension lines to cross, dimensions may become difficult to read if this practice is used extensively.

When tolerances are not critical, *continuous dimensioning* may be used. This method is also shown in Fig. 4.16, and is commonly used in combination with center-line dimensioning. This technique of dimensioning can cause large tolerances and should be used only when critical dimensions are not required.

When dimensioning complex, high-density hole patterns, hole-center locations may be presented in *tabular* form, with no dimension lines shown on the drawing. Two datum lines are selected and identified with an appropriate designation, such as $X$ and $Y$, for the horizontal and vertical datum lines, respectively. All hole centers are then identified literally, the same designation being used for common-size holes. A table identifying each hole, with its size and distance from each datum line, is then prepared and attached to the drawing.

When a high degree of dimensional tolerance is required, measurements should be taken in decimal form. Since this degree of accuracy is not normally required in prototype construction, fractional measurements are used throughout this section. With fractional

**Figure 4.14** Base-line dimensioning of the power supply's subchassis element.

56

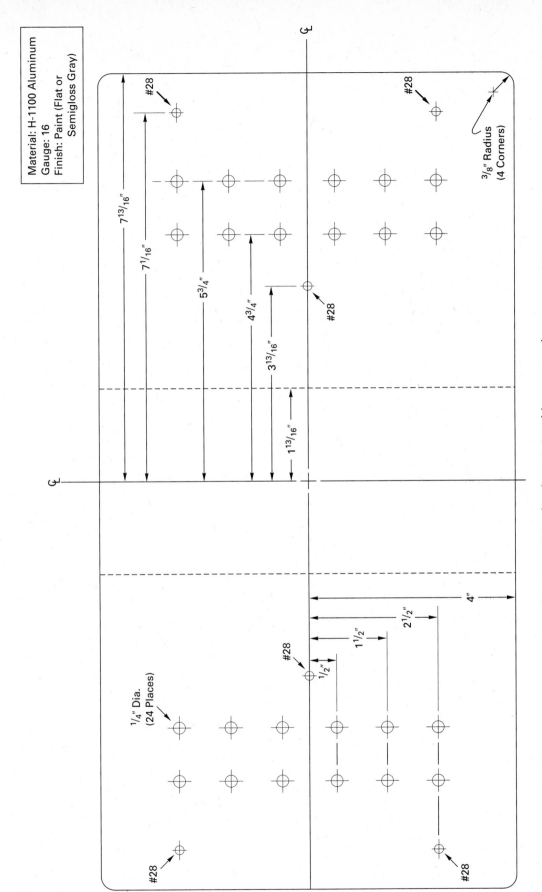

**Figure 4.15** Center-line dimensioning of the power supply cover.

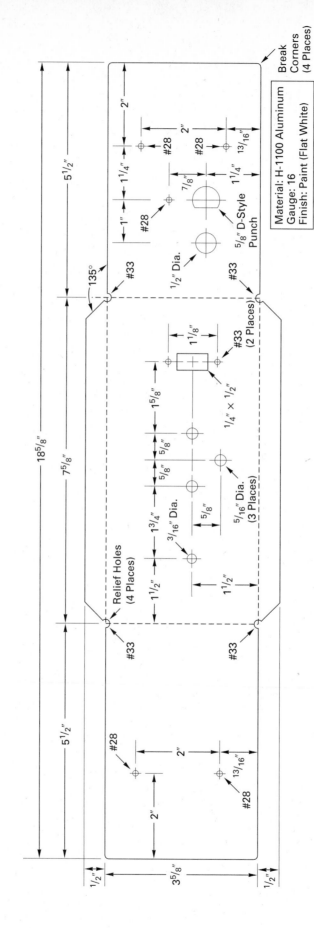

**Figure 4.16** Continuous dimensioning of the power supply front-and-side chassis element.

dimensioning, accuracy to $\frac{1}{64}$ inch is easily obtainable. Maximum allowable variation above and below the nominal value by which the given dimension may differ from the drawing to the actual work appear with the dimension where applicable.

## 4.8 ASSEMBLY DRAWINGS

After a prototype has been completed, final drawings and photographs are made of all assemblies and subassemblies. These drawings may be of several variations: *two-dimensional top, bottom* and *side views; three-dimensional views; isometric projections;* and *exploded views.* The designated views must contain all pertinent information concerning the proper location and orientation of each component and mechanical part. All parts shown in assembly drawings should be identified and labeled consistent with the designations used on the schematic diagram. A common arrangement is a numbering system relating to the parts and accompanying the drawings.

Auxiliary views are frequently necessary, especially when complex or high-density packaging obscures some of the major sections of the package. An auxiliary view is a drawing of the obscured portion of the package after eliminating obstructions of the primary views. Figure 4.17 shows a partial assembly layout for a complex amplifier project, including an auxiliary view.

Photographs are particularly advantageous because of their high quality and three-dimensional perspective characteristics. In addition, nomenclature and reference information can be superimposed on photographs. Drawings, however, are absolutely indispensable for fabrication, assembly, and wiring because of the excessive amount of explanatory information necessary.

The next phase of development, after the package planning has been finished and the detailed working drawings for the sheet metal elements are completed, is to design printed circuit boards. In Unit 2, information is given on designing both single-sided and double-sided printed circuit boards, an important element in virtually all electronic packages.

Note that each unit in this book is a separate entity in itself. The switching regulator power supply has been used as an instructional vehicle in this unit only, and is not used in the units to follow.

**Figure 4.17** Front panel assembly drawings for a complex amplifier project.

## EXERCISES

### A. Questions

4.1     How should a scale be placed when measuring the distance between mounting holes to obtain an accurate on-center dimension?

4.2     Why is it not advisable to use the end of a rule as a reference when taking measurements? What reference should be used?

4.3     Describe two methods for obtaining the diameter of round components.

4.4     How many steps are required to obtain the reading using a micrometer equipped to read to one ten-thousandth of an inch?

4.5     What information is shown on an engineering sketch?

4.6     How are the sizes of drafting paper designated?

4.7     What type of drawing paper is preferred for electronic drafting?

4.8     Describe each of the following dimensioning techniques:

      a. Datum-line

      b. Center-line

      c. Continuous

      d. Tabulation

4.9     Which dimensioning technique results in large tolerances?

4.10    When a high degree of dimensional tolerance is required, in what form should measurements be taken?

4.11    When are auxiliary views on a drawing required?

### B. True or False

Circle T if the statement is true, or F if any part of the statement is false.

4.1     The thimble on a micrometer has 25 divisions, each representing 0.010 inch.       **T   F**

4.2     Engineering sketches are easily produced using quadruled paper without the need of special drafting equipment.       **T   F**

4.3     Pencil grades H to 2H are suitable for general-purpose work in technical drawings.       **T   F**

4.4     As paper size increases (i.e., A B C D), the height and width dimensions double.       **T   F**

4.5     Datum-line dimensioning uses any two reference edges that are perpendicular to each other.       **T   F**

4.6     When tolerances are critical, continuous dimensioning is used.       **T   F**

### C. Multiple Choice

Circle the correct answer for each statement.

4.1     For maximum accuracy in taking measurements with a scale, the work should be aligned with the (*end, first major graduation*) of the scale.

4.2     Micrometers are capable of measuring accurately to (*0.0001, 0.001*) inch.

4.3     Engineering sketches should be produced on (*plain, quadruled*) paper.

4.4     Another name for base-line dimensioning is (*datum, tabulation*).

(a)          (b)

**Figure 4.18**

4.5      It is (*acceptable, not acceptable*) to combine several dimensioning techniques to achieve clarity.

4.6      When dimensional tolerance is not critical, (*center-line, continuous, base-line*) dimensioning may be used.

### D. Matching Columns

Match each item in column A to the most appropriate item in column B.

| COLUMN A | | COLUMN B |
|---|---|---|
| 1. Engineering sketch | a. | Spindle, anvil |
| 2. Micrometer | b. | Continuous |
| 3. Outside | c. | $\frac{1}{10,000}$ inch |
| 4. Subdivisions | d. | Isometric projections |
| 5. Steel scale | e. | Caliper |
| 6. Dimensioning | f. | Quadruled paper |
| 7. Assembly drawing | g. | $\frac{1}{4}, \frac{1}{8}, \frac{1}{16}, \frac{1}{32}$, inch |
| 8. Vernier | h. | Hook rule |

### E. Problems

4.1      Read the measurements shown in Figs. 4.2 and 4.3b.

4.2      Read the micrometer setting shown in Fig. 4.7 and those in Fig. 4.18.

4.3      Read the vernier caliper settings shown in Fig. 4.19.

4.4      Measure the body and lead diameters and body lengths of $\frac{1}{4}$-watt through 10-watt resistors. Also determine the body and lead diameters of several electrolytic capacitors and the body thickness and lead spacing of various-sized disk capacitors. Tabulate all measurements for future reference.

**Figure 4.19**

Unit

# TWO

# Designing Printed Circuit Boards with CAD Techniques

# 5 Printed Circuit Board Materials

## LEARNING OBJECTIVES

*Upon completion of this chapter on printed circuit board materials, the student should be able to*

- Know the classification of printed circuit boards.
- Be familiar with the characteristics of printed circuit cable.
- Be familiar with the various grades of rigid printed circuit laminates and their characteristics.
- Understand the properties of copper foil on rigid printed circuit boards.
- Select the appropriate grade of printed circuit laminate and copper foil thickness to satisfy the specifications for a variety of applications.

## 5.0 INTRODUCTION

With the advent of the transistor and the more recent heavy reliance on integrated circuits, electronic packaging and wiring problems have been compounded. No longer can the technician rely solely on hand-wiring methods; now designs must be oriented toward more extensive use of printed circuitry. Many modern electronic systems would be virtually impossible to package without incorporating printed circuits into their design.

Printed circuits are *metal foil conducting patterns,* usually copper, bonded to a *substrate* (insulating base material) for support. The metal pattern serves as the connecting medium for the electrical components that are assembled on the opposite side of the board. Component leads are fed into holes that are drilled or punched through the base material and foil. These leads are soldered to the conducting pattern to form the complete printed circuit.

The advantages of printed circuits over conventional wiring methods are that they (1) adapt easily to miniaturization and modular design, (2) provide for uniformity in production, (3) cost less, (4) virtually eliminate wiring error, and (5) minimize assembly and inspection time. Printed circuit techniques lend themselves readily to mass production, resulting in a highly reliable package. Although the format of this text is focused on prototype construction, it must be assumed that eventually the package will be released for production. Consequently, the technician should design the package with an eye toward this end by selecting materials and configurations that will reduce construction problems.

To completely explain the sequential processes involved in printed circuit board design and fabrication, the next 13 chapters are devoted exclusively to these topics. This chapter discusses the initial step in printed circuit construction—the selection of the most appropriate type of printed circuit board material. Topics discussed are classification of printed circuit boards, insulating base materials, conductive foils, bonding, and tabulated information of board characteristics to aid the technician in selection.

Printed circuits are designed as either *rigid* or *flexible* and are further classified into *single-sided, double-sided,* or *multilayer* conductors. Regardless of the type, all printed circuit boards are formed with an insulating base material on which a conducting foil is either chemically or mechanically bonded. When processed, the printed circuit board provides both an electrical wiring path and a mechanical support for the components.

Rigid printed circuit boards of the single-sided type are the most frequently used, finding extensive application in relatively uncomplicated circuitry. Figure 5.1a shows a rigid single-sided blank before processing. The completed conductor pattern ready for component mounting is shown in Fig. 5.1b.

Rigid printed circuit boards of the double-sided type consist of conducting foil bonded to *both* sides of the insulating base. This type of board is employed when circuit complexities make it difficult or impossible to develop practical wiring layouts on one side only. Double-sided boards also reduce the overall size of a comparable circuit constructed on a single-sided board, since components are mounted on both sides. Figure 5.2 shows an assembled double-sided board. Interconnections among the conductive patterns on both sides of the board are typically accomplished by plated-through holes, which is a manufacturing process discussed in Chapter 8. With the use of these interconnections, wires or component leads can be soldered between the foil patterns on opposite sides of the base materials.

Multilayer rigid printed circuit materials were developed to satisfy the needs of the more recent state of the art, which demands extremely complex wiring layouts in a relatively small space. Multilayer board material selection as well as fabrication are discussed in detail in Chapter 14.

*Flexible printed cables* consist of conductors laminated between layers of insulation and exposed for soldering only at the terminal points, as shown in Fig. 5.3. Flexible printed cables are used in a manner similar to ordinary cable but are extremely flexible. In addition, the flat configuration of these circuits make them desirable for system interconnections where a round cable may prove to be too cumbersome. *Teflon, polymide, polyester, polyvinyls, polypropylene,* and *polyethylene* are typical insulating materials used in flexible printed cables. Table 5.1 shows a comparison of the various characteristics for

Copper Foil

Insulating Base Material

(a)

(b)

**Figure 5.1** Rigid single-sided printed circuit boards: (a) copper-clad printed circuit blank; (b) processed single-sided printed circuit conductor pattern.

**Figure 5.2**   Assembled double-sided rigid printed circuit board.

**Figure 5.3**   Flexible printed circuit cables. *Courtesy of DuPont Company.*

**TABLE 5.1**

Insulation Characteristics for Flexible Printed Cables
(Courtesy of Insulfab Plastics, Inc.)

| | TFE Fluorocarbon | TFE Glass Cloth | FEP Fluorocarbon | FEP Glass Cloth | Polyimide | Polyester | Polychlorotri-fluoroethylene | Polyvinyl Fluoride | Polypropylene | Polyvinyl Chloride | Poly-ethylene |
|---|---|---|---|---|---|---|---|---|---|---|---|
| Specific Gravity | 2.15 | 2.2 | 2.15 | 2.2 | 1.42 | 1.395 | 2.10 | 1.38 | 0.905 | 1.25 | .93 |
| Square inches of 1 mil film per pound | 12,800 | 13,000 | 12,900 | 13,000 | 19,450 | 21,500 | 12,000 | 20,000 | 31,000 | 22,000 | 30,100 |
| Service Temp. Deg. C (Minimum) | −70 | −70 | −225 | −70 | −250 | −60 | −70 | −70 | −55 | −40 | −20 |
| (Maximum) | 250 | 250 | 200 | 250 | +250 | 150 | 150 | 105 | 125 | 85 | 60 |
| Flammability | Nil | Nil | Nil | Nil | Nil | Yes | Nil | Yes | Yes | Slight | Yes |
| Appearance | Translucent | Tan | Clearbluish | Tan | Amber | Clear | Clear | Clear | Clear | Translucent | Clear |
| Thermal Expansion × 10 inches/inch/deg. F. | 70 | Low | 50 | Low | 11 | 15 | 45 | 28 | 61 | — | — |
| Bondability with Adhesives | Good | Good | Good | Good | Good | Good | Good | Good | Poor | Good | Poor |
| Bondability to itself | Good | Poor | Good | Good | Poor | Poor | Good | Good | Good | Good | Good |
| Tensile strength PSI @77°F | 3,000 | 20,000 | 3,000 | 20,000 | 20,000 | 20,000 | 4,500 | 8,000 | 5,700 | 3,000 | 2,000 |
| Modulus of Elasticity PSI | 80,000 | 3 | 70,000 | 3 | 430,000 | 550,000 | 200,000 | 280,000 | 170,000 | — | 50,000 |
| Volume Resistivity ohms-cm | $2 \times 10$ | 10 | 10 | 10 | 10 | $1 \times 10$ | $1 \times 10$ | $3 \times 10$ | 10 | $1 \times 10$ | $1 \times 10$ |
| Dielectric Constant 10-10 cycles | 2.2 | 2.5/5 | 2.1 | 2.5/5 | 3.5 | 2.8–3.7 | 2.5 | 7.0 | 2.0 | 3–4 | 2.2 |
| Dissipation factor 10-10 cycles | .0002 | .0007/.001 | .0002 | .0001/.001 | .002/.014 | .002–.016 | .015 | .009–.041 | .0002/.0003 | .14 | .0006 |
| Dielectric strength (5 mils thickness) volt/mil | 800 | 650/1600 | 3,000 | 650/1600 | 3,500 | 3,500 | 2,000 | 2,000 | .125 in thk 750v/mil | 800 | 1,500 |
| Chemical Resistance | Excellent | Excellent | Excellent | Excellent | Excellent | Excellent | Excellent | Good | Excellent | Good | Excellent |
| Water Absorption, % | 0 | .10/68 | 0 | .18/30 | 3 | 0.5 | 0 | 15 | .01 | .10 | 0 |
| Sunlight Resistance | Excellent | Excellent | Excellent | Excellent | Excellent | Fair | Excellent | Excellent | Low | Fair | Low |

**TABLE 5.2**

Conductor Temperature Rise and Current Derating for Flat Cables:
(1) Rise in Temperature Above Ambient as a Function of Current;
(b) Approximate Derating Curve for Flat Cable

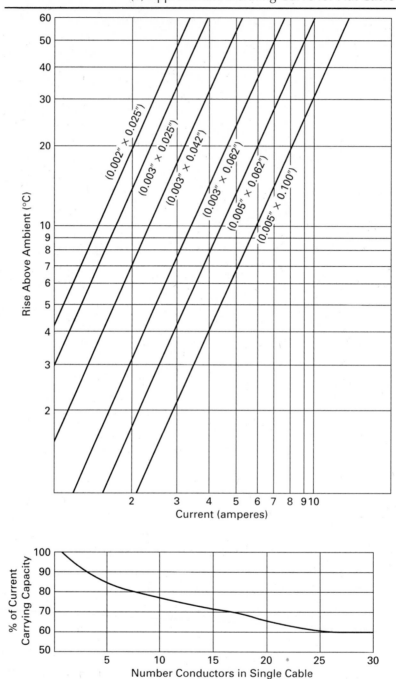

common insulations. Since various methods and adhesives are used for the internal bond of these laminates, no one method of insulation removal will suffice for all types of these cables. The insulation is usually stripped by *friction wheel strippers, knife-edge strippers, chemical dipping,* or *melting* by a concentration of hot air. Because flexible printed cables have a large flat conducting area, they can handle larger currents than conventional wire cable. Table 5.2 shows the approximate current capacities and derating values when a multiple number of conductors are laminated within a common cable.

## 5.2 INSULATING BASE MATERIALS FOR PRINTED CIRCUIT BOARDS

Printed circuit boards of the rigid variety will be discussed exclusively in the remaining sections of this text since they are by far the most common. The insulating or resin material for these boards is either a *thermosetting plastic* or a high-temperature-resistant *thermoplastic polymer* that provides the support for the conductor pattern and components. Our discussion will be limited to the thermosetting plastics since they are used almost exclusively in rigid boards.

Base materials for printed circuit boards are laminates containing a *reinforcement* such as *paper, glass,* or *fabrics,* including *cotton cloth, asbestos, nylon cloth,* and *glass cloth.* These laminates are formed under pressure and heat. The thermosetting plastic resins common to the manufacture of printed circuit boards are *phenolic, epoxy, melamine, silicone,* and *Teflon.* Silicon-glass is mechanically inferior and more expensive than either phenolic, epoxy, or melamine resin boards. Teflon is used primarily for microwave applications and is approximately fifteen times more expensive than phenolic and at least five times as costly as the epoxy and melamine resin types. The phenolic types are less expensive than the epoxy resins and for many applications are both electrically and mechanically suitable. In general, the epoxy types possess superior electrical and mechanical characteristics than phenolic or melamine. In most cases, the reliability of a particular type of board will by far outweigh the cost factor.

To aid the technician in selecting the most suitable base material for a specific application, there follows a description of some of the electrical, mechanical, and chemical qualities of the more common base materials. The classifications are in accordance with NEMA (National Electrical Manufacturers Association) standards. These standards designate phenolic resin printed circuit boards as either *X, XX,* or *XXX,* which represent the amount of resin present, the amount increasing with the number of *X's.* In general, the more resin content, the more improved are the electrical and mechanical characteristics of the board. A *P* suffix together with the *X* designation denotes that the phenolic is capable of being punched, but only at elevated temperatures, where a *PC* suffix indicates cold-punch qualities.

**Grade XX.** This grade of paper-base phenolic possesses fair electrical characteristics and good mechanical properties. It machines well but is not recommended for punching. It offers little abrasive action to drills and, as a result, does not dull these bits easily. This grade is used by many manufacturers as a reference for determining the costs of other laminate types. For example, if the price index of nylon cloth phenolic is 3.5, this indicates that its cost is that much more than grade *XX,* which has a cost index of 1 (see Table 5.4).

**Grade XXP.** This type is similar in electrical and mechanical properties to grade *XX* and it can be punched at increased temperatures of between 200 and 250°F (93 to 121°C).

**Grade XXX.** This material possesses some improved electrical and mechanical properties over the *XX* type and is recommended for use at radio frequencies. It is available in a *self-extinguishing* form, meaning that it will not support combustion if ignited. This grade possesses fair *dimensional stability* (i.e., it maintains its size within close tolerances under environmental changes such as temperature and moisture).

| Description Characteristics—Applications | NEMA Grade | Conforms to Military Spec. MIL-P | Color* | Standard Thicknesses (others available) | Mechanical — Times 1000 | | | | |
|---|---|---|---|---|---|---|---|---|---|
| | | | | | Compr. Str. (Lb./Sq. In.) $\frac{1}{2}$" Face | Flexural Str. (Lb./Sq. In.) $\frac{1}{16}$" Face—Lengthwise | Flexural Str. (Lb./Sq. In.) $\frac{1}{16}$" Face—Crosswise | Tensile Str. (Lb./Sq. In.) $\frac{1}{16}$" Lengthwise | Tensile Str. (Lb./Sq. In.) $\frac{1}{16}$" Crosswise |
| Paper/Phenolic—General Purpose. For panels, contactors, terminal blocks—high strength, low cost | X | | N.B.C. | .010–2.0 | 38.0 | 28.0 | 20.0 | 20.6 | 16.0 |
| Paper/Phenolic—For applications requiring good mechanical strength and dim. stability with elec. properties secondary | XP | | N.B.C. | .020–2.0 | 35.0 | 30.0 | 25.0 | 20.0 | 16.0 |
| Paper/Phenolic—Excellent cold punching, high strength | XPC | | N.B.C. | .015–.250 | 42.0 | 23.0 | 20.0 | 17.0 | 12.7 |
| Paper/Phenolic—An economy grade offering excellent shear and punch characteristics at room temp. | XPC | | N.B.C. | .031–.125 | 32.0 | 24.0 | 20.0 | 12.0 | 10.0 |
| Paper/Phenolic—Good mechanical, electrical and machining properties | XX | 3115, PBG | N.B. | .010–2.0 | 34.0 | 22.0 | 19.0 | 16.0 | 13.0 |
| Paper/Phenolic—Excellent punching, high mechanical and high electrical. Best all purpose | XXP | | N.B.C. | .015–.250 | 41.0 | 23.0 | 18.0 | 16.3 | 13.8 |
| Paper/Phenolic—Excellent electrical grade used where dim. stability or min. cold flow is essential | XXX | 3115, PBE | N.B. | 0.10–2.0 | 35.0 | 20.0 | 17.0 | 13.1 | 10.5 |
| Paper/Phenolic—Best quality paper base. High insulation resistance, low loss material | XXXP | 3115, PBEP | N.B. | .031–.250 | 28.0 | 19.0 | 17.5 | 12.0 | 9.5 |
| Paper/Phenolic—High quality, cold punching material. High insulation resistance | XXXPC | 3115, PBE-P | N.B. | .031–.50 | 30.0 | 19.0 | 15.0 | 14.0 | 11.0 |
| Cotton Cloth/Phenolic—General purpose. Excellent mach. and punching. For panels, wedges, gears, wear strips | C | 18324 15035, FBM | N.B. | .031–10.0 | 41.0 | 22.0 | 18.0 | 12.1 | 9.80 |
| Cotton Cloth/Phenolic—Best canvas base electrical grade. Excellent acid resistance. For panels, steam valve discs | CE | 15035, FBG | N.B. | .031–3.0 | 39.0 | 21.0 | 19.0 | 13.5 | 9.80 |
| Cotton Cloth/Phenolic. A finer weave than Grade C. Better for fine punching, close machining. Mechanical usage | L | 15035, FBI | N.B. | .020–4.0 | 39.0 | 21.5 | 18.0 | 11.9 | 7.9 |
| Cotton Cloth/Phenolic—Best insulation of cloth grades. Excellent resistance to moisture, mild acids, alkalies | LE | 15035, FBE | N.B. | .010–3.0 | 38.0 | 22.0 | 17.0 | 15.0 | 10.6 |
| Fine Weave Cotton Cloth/Phenolic, with graphite added. Non-insulator, used for thrust bearings, pump bearings | Linen Graphite | 5431 (AER.) | Green-Black | .008–6.0 | 37.0 | 20.0 | 16.0 | 15.0 | 11.0 |
| Asbestos Paper/Phenolic—Suitable for continuous use at 150°C. Good mechanical, poor electrical strength | A | 8059 (USAF) | N.B. | .015–1.0 | 45.0 | 28.0 | 21.0 | 18.0 | 13.0 |
| Asbestos Cloth/Phenolic—High heat resist., dimensional stab., low water absorp., low thermal expan., good impact resist. | AA | 8059 (USAF) | N. | .047–4.0 | 40.0 | 30.0 | 19.0 | 9.2 | 8.1 |
| Nylon Cloth/Phenolic—Excellent electrical and good mechanical properties under humid conditions | N1 | 15047-NPG | N. | .015–1.125 | 40.0 | 13.5 | 10.5 | 10.0 | 6.0 |
| Glass Cloth/Phenolic—Used in applications where stability of elec. and mech. properties is essential | G3 | | N. | .031–2.0 | 55.0 | 20.0 | 18.0 | 23.0 | 18.0 |
| Glass Cloth/Melamine—Excellent mechanical and electrical properties. Good arc resistance, flame retardance | G5 | 15037, GMG | Gray-Brown | .008–3.5 | 70.0 | 50.0 | 43.0 | 39.0 | 36.0 |
| Glass Cloth/Silicone—Suitable for high temp. (Class H) applications. Low loss electrical properties | G7 | 997, GSG | White | .010–2.0 | 62.0 | 38.0 | 32.0 | 33.0 | 31.0 |
| Glass Cloth/Melamine—An arc and flame resistance laminate with outstanding elec. prop. to meet MIL-P15037C, GME | G9 | 15037C, GME | Gray-Brown | .008–3.5 | 62.0 | 83.0 | 61.0 | 48.0 | 35.0 |
| Glass Cloth/Epoxy—High in. resist., very low absorp. Highest bond strength of glass laminates. High stability in humidity | G10 | 18177, GEE | Green | .010–2.0 | 50.0 | 70.0 | 55.0 | 49.0 | 38.0 |
| Glass Cloth/Epoxy—High flexural strength retention at elevated temperatures. Extreme resistance to solvents | G11 | 18177, GEB | Green | .010–2.0 | 58.0 | 70.0 | 65.0 | 48.0 | 37.0 |
| Paper/Epoxy—Designed for optimum balance of mechanical and electrical properties, with excellent machinability | FR-3 | 22324, PEE | Ivory | .031–.50 | 32.0 | 26.5 | 22.0 | 12.0 | 9.0 |
| Glass Cloth/Epoxy—A flame resistant epoxy laminate with excellent machining and electrical properties | FR-4 | 18177, GEE | Green | .010–.125 | 55.0 | 82.0 | 75.0 | 48.0 | 37.0 |
| Glass Cloth/Epoxy—High strength retention at elevated temperatures. Self-extinguishing | FR-5 | 18177, GEB | Green | .010–1.5 | 50.0 | 58.0 | 52.0 | 49.0 | 36.0 |

*These are AVERAGE and not GUARANTEED properties. Guaranteed values are NEMA or applicable MIL-P specifications.*

Sheet Sizes: Range from 36″ to 48″ wide × 36″ to 120″ long, depending on Grade and Thickness. Consult nearest Insulfab office for additional information.

*Colors: Symbols N, B, or C refer to Natural, Black, or Chocolate. "N" ranges from light tan to dark brown. Other referenced colors are "Natural for grade." MIL-P specifications approve Natural colors only.

**TABLE 5.3a**

Insulating Material: Grades, Uses, and Properties (*Courtesy of Insulfab Plastics, Inc.*)

| Properties | | | | | | Electrical Properties | | | | | | | Physical and Thermal Properties | | | | | | |
|---|---|---|---|---|---|---|---|---|---|---|---|---|---|---|---|---|---|---|---|
| Mod. of Elas. (Lb./Sq. In.) 1/16" Lengthwise (Times 10⁶) | Mod. of Elas. (Lb./Sq. In.) 1/16" Crosswise (Times 10⁶) | Izod Impact (Ft.Lb./In. Notch) 1/16" Build-Up Edge—Lengthwise | Izod Impact (Ft.Lb./In. Notch) 1/16" Build-Up Edge—Crosswise | Bonding Strength 1/2" (lbs.) | Rockwell Hardness M Scale | Diel. Str. (V/Mil) Perp. To Lam. 1/16" S/S @ 23°C | Diel. Breakdown KV Par. To Lam. 1/16" S/S @ 23°C | Dissipation Factor @ 1 Mc. 1/16" As Received | Dissipation Factor @ 1 Mc. After 24 Hr. In Water 1/16" @ 23°C | Diel. Constant @ 1 Mc. 1/16" As Received | Diel. Constant @ 1 Mc. After 24 Hr. In Water 1/16" @ 23°C | Continuous Operation Temp. Limit Deg. C | Coef. Of Expansion In./In./C° Lengthwise × 10⁻⁵ 1/16 | Coef. Of Expansion In./In./C° Crosswise × 10⁻⁵ 1/16 | Deformation Under Load 1/16" (Percentage) @ 70°C | Thermal Conductivity BTU/Hr./Sq. Ft./In. Thick/F° | Water Absorp. % 24 Hr. @ 23°C 1/16" Thick | Specific Gravity | NEMA Grade |
| 1.70 | 1.40 | 1.00 | 0.80 | 1030 | 102 | 580 | 45 | 0.042 | 0.047 | 5.4 | 5.6 | 120 | 2.0 | 2.5 | 0.98 | 1.6 | 1.70 | 1.38 | X |
| 1.50 | 1.10 | 0.70 | 0.60 | 1100 | 110 | 550 | 62 | 0.054 | 0.080 | 6.5 | 7.5 | 120 | 2.0 | 2.5 | 0.60 | 1.6 | 3.00 | 1.44 | XP |
| 1.00 | 0.80 | 0.65 | 0.60 | — | 93 | 470 | 65 | 0.046 | 0.078 | 5.1 | 6.5 | 120 | 1.6 | 3.0 | 1.69 | 1.7 | 5.00 | 1.36 | XPC |
| 1.40 | 1.00 | 0.90 | 0.75 | — | 85 | 550 | 73 | 0.051 | 0.073 | 5.1 | 5.8 | 120 | 1.6 | 3.0 | 2.00 | 1.7 | 2.60 | 1.37 | XPC |
| 1.26 | 0.88 | 0.60 | 0.50 | 900 | 102 | 625 | 60 | 0.040 | 0.046 | 5.0 | 5.4 | 120 | 1.1 | 2.6 | 0.93 | 1.7 | 1.53 | 1.35 | XX |
| 1.10 | 0.80 | 0.80 | 0.70 | — | 105 | 555 | 65 | 0.036 | 0.044 | 5.0 | 5.3 | 120 | 1.5 | 1.7 | 1.59 | 1.7 | 1.25 | 1.36 | XXP |
| 1.36 | 0.90 | 0.60 | 0.50 | 960 | 110 | 745 | 68 | 0.034 | 0.036 | 5.0 | 5.4 | 120 | 1.7 | 3.5 | 0.56 | 1.8 | 0.78 | 1.35 | XXX |
| 0.62 | 0.47 | 0.5 | 0.5 | — | 98 | 740 | 90 | 0.028 | 0.030 | 3.8 | 3.9 | 120 | 1.06 | 1.21 | 1.64 | 1.7 | 0.40 | 1.27 | XXXP |
| 0.70 | 0.60 | 0.60 | 0.50 | 1050 | 104 | 740 | 70 | 0.028 | 0.032 | 4.1 | 4.3 | 120 | 1.1 | 1.2 | 1.67 | 1.7 | 0.40 | 1.28 | XXXPC |
| 1.05 | 0.79 | 2.80 | 2.40 | 2100 | 103 | 310 | 40 | 0.053 | 0.089 | 5.2 | 5.8 | 120 | 1.8 | 2.2 | 1.00 | 2.3 | 2.00 | 1.33 | C |
| 1.14 | 1.00 | 2.50 | 2.20 | 1800 | 104 | 400 | 60 | 0.045 | 0.050 | 5.0 | 5.6 | 120 | 1.8* | 2.2 | 1.30 | 2.3 | 1.52 | 1.35 | CE |
| 1.00 | 0.80 | 1.70 | 1.40 | 1820 | 105 | 300 | 50 | 0.055 | 0.070 | 5.8 | 6.5 | 120 | 2.2 | 3.5 | 1.20 | 2.2 | 1.60 | 1.33 | L |
| 0.91 | 0.75 | 2.00 | 1.60 | 1820 | 108 | 430 | 60 | 0.045 | 0.050 | 5.6 | 6.0 | 120 | 2.2 | 3.5 | 0.94 | 2.2 | 1.20 | 1.33 | LE |
| 0.84 | 0.70 | 1.87 | 1.15 | 1700 | 105 | — | — | — | — | — | — | 120 | 2.2 | 3.5 | 0.96 | 2.9 | 2.00 | 1.40 | Linen Graphite |
| 1.30 | 1.10 | 3.00 | 2.80 | 1250 | 100 | 120 | 5 | — | — | — | — | 150 | 1.0 | 1.3 | 1.06 | 2.3 | 1.00 | 1.56 | A |
| 1.42 | 1.06 | 5.00 | 3.50 | 2100 | 99 | 90 | 10 | — | — | — | — | 170 | 0.9 | 1.4 | 1.30 | 3.5 | 1.20 | 1.64 | AA |
| 0.40 | 0.35 | 4.00 | 2.50 | 1500 | 85 | 330 | 70 | 0.030 | 0.040 | 3.7 | 3.8 | 120 | 3.0 | 5.0 | 3.40 | 2.2 | 0.50 | 1.17 | N1 |
| 1.70 | 1.50 | 8.00 | 6.00 | 1010 | 103 | 500 | 60 | 0.018 | 0.090 | 4.9 | 8.0 | 150 | 1.0 | 1.5 | 0.30 | 1.8 | 2.10 | 1.73 | G3 |
| 2.10 | 1.90 | 13.00 | 10.00 | 1800 | 123 | 395 | 34 | 0.011 | 0.065 | 6.1 | 7.2 | 150 | 1.0 | 1.1 | 0.60 | 2.4 | 2.20 | 1.90 | G5 |
| 1.80 | 1.60 | 16.00 | 16.00 | 850 | 102 | 400 | 70 | 0.002 | 0.003 | 4.1 | 4.2 | 250 | 1.0 | 1.0 | 0.25 | 1.4 | 1.77 | 0.10 | G7 |
| 3.20 | 2.20 | 15.00 | 13.00 | 2100 | 115 | 400 | 65 | 0.012 | 0.015 | 7.0 | 7.2 | 150 | 1.5 | 1.8 | 0.50 | 2.4 | 0.68 | 1.92 | G9 |
| 1.90 | 1.80 | 10.00 | 8.00 | 2300 | 105 | 510 | 65 | 0.021 | 0.022 | 4.5 | 4.6 | 130 | 1.0 | 1.5 | 0.26 | 1.8 | 0.20 | 1.80 | G10 |
| 2.50 | 2.40 | 11.00 | 9.00 | 2100 | 114 | 600 | 60 | 0.018 | 0.022 | 4.8 | 5.0 | 150 | 1.0 | 1.4 | 0.10 | 1.8 | 0.20 | 1.77 | G11 |
| 1.00 | 0.90 | 0.70 | 0.65 | — | 95 | 550 | 65 | 0.031 | 0.032 | 4.3 | 4.4 | 120 | 1.3 | 2.5 | 1.50 | 1.6 | 0.40 | 1.42 | FR-3 |
| 2.80 | 2.50 | 11.00 | 9.00 | 2100 | 107 | 500 | 65 | 0.016 | 0.018 | 4.2 | 4.4 | 130 | 1.2 | 1.5 | 0.25 | 1.8 | 0.20 | 1.93 | FR-4 |
| 2.50 | 2.40 | 13.00 | 9.00 | 2000 | 109 | 490 | 60 | 0.013 | 0.019 | 4.5 | 4.6 | 150 | 1.0 | 1.5 | 0.10 | 1.8 | 0.13 | 1.95 | FR-5 |

| *Physical Properties* | MIL-P-18177<br>MIL-P-13949 | *G-10* | *FR-4* | *1226F* |
|---|---|---|---|---|
| Water Absorption %<br>029/23 . . . . . . . . . . . . . . . . . . . . . . . . . . . . . . . | *0.35 (max.)* | *0.14* | *0.13* | *0.13* |
| Specific Gravity. . . . . . . . . . . . . . . . . . . . . . . . . . . . . . . | — | — | — | — |
| Rockwell Hardness<br>M Scale . . . . . . . . . . . . . . . . . . . . . . . . . . . . . . . | — | — | — | — |
| Flexural Strength<br>Lengthwise condition A-lbs./sq. in. . . . . . . . . . . . . . . . . . . . .<br>Crosswise, condition A . . . . . . . . . . . . .<br>Tested   150°C after 1 hour   150°C<br>Lengthwise . . . . . . . . . . . . . . . . . . . . . . . . . . .<br>Crosswise . . . . . . . . . . . . . . . . . . . . . . . . . . . | *50,000 (min.)*<br>*40,000 (min.)*<br><br><br>*N/A* | *79,000*<br>*60,000*<br><br><br>*N/A* | *73,100*<br>*65,100*<br><br><br>*N/A* | *71,000*<br>*63,500*<br><br><br>*N/A* |
| Tensile Strength, psi<br>Lengthwise, condition A<br>Crosswise, condition A | —<br>— | *54,370*<br>*45,565* | *54,470*<br>*45,065* | *59,170*<br>*44,765* |
| Compressive Strength, flatwise, psi | — | *33,033* | *33,033* | *33,033* |
| Izod Impact, edgewise, ft. lbs./in., $^1/_8$ in. thick<br>Lengthwise, condition A . . . . . . . . . . . . . . . . . . . . . . .<br>Crosswise, condition A . . . . . . . . . . . . . . . . . . . . . . .<br>Bond, Strength, condition A lbs., $^1/_2$ in. thick. . . . . . . . . . . . . . . . . . . .<br>condition D 48/50 . . . . . . . . . . . . . . . . . . . . . . . | *7.0 (min.)*<br>*5.5 (min.)*<br>*2000*<br>*1600* | *13.3*<br>*8.5*<br>*2560*<br>*2370* | *13.3*<br>*8.5*<br>*2560*<br>*2370* | *13.3*<br>*8.5*<br>*2560*<br>*2370* |
| *Electrical Properties* | | | | |
| Dielectric Strength, parallel, KV (360 sec.)<br>Condition A . . . . . . . . . . . . . . . . . . . . . . . . . . .<br>Condition D48/50 . . . . . . . . . . . . . . . . . . . . . . . | *35 (min.)*<br>*30 (min.)* | *55*<br>*50* | *55*<br>*50* | *55*<br>*50* |
| Dissipation factor, 1 MHz<br>Condition D24/23 . . . . . . . . . . . . . . . . . . . . . . . | *0.030 (max.)* | *0.020* | *0.021* | *0.021* |
| Dielectric Constant, 1 MHz<br>Condition 24/23 . . . . . . . . . . . . . . . . . . . . . . . | *5.4 (max.)* | *4.6* | *4.8* | *4.8* |
| Insulation Resistance, megohms, condition C96/35/90 . . . . . . . . . . . . . . . . . | | $.76 \times 10^7$ | $2.38 \times 10^7$ | $2.2 \times 10^7$ |
| Arc Resistance, seconds, condition D48/50 . . . . . . . . . . . . . . . . . . . . | *60 (min.)* | *75* | *82* | *78* |
| Volume Resistivity, megohms cms, condition C96/35/90 . . . . . . . . . . . . . . . . . | $10^6$ *(min.)* | $8.2 \times 10^7$ | $5.5 \times 10^8$ | $2.3 \times 10^8$ |
| Surface Resistivity, megohms, condition C96/35/90. . . . . . . . . . . . . . . . . | $10^4$ *(min.)* | $4.9 \times 10^5$ | $2.10 \times 10^6$ | $2.2 \times 10^6$ |
| Flame Resistance, seconds . . . . . . . . . . . . . . . . . . . . . . . . . . . | *15 (max.)* | *NA* | *3* | *4* |
| UL 94 Flame Classification UL file—E45456 A & B. . . . . . . . . . . . . . . . . | | | *UL94-VO* | |
| Maximum Operating Temperature, °C continuous. . . . . . . . . . . . . . . . . | | *130* | *130* | *130* |
| Blister Resistance, 1/16″ × 1″ × 1″ NEMA LI-1971 . . . . . . . . . . . . . . . . . | *Excellent* | *Excellent* | *Excellent* | *Excellent* |
| Resistance to Plating Solution: trichlorethylene vapors and etching solutions | | *Excellent* | *Excellent* | *Excellent* |

## TABLE 5.4

### Insulating Material Comparison Chart (Courtesy of Insulfab Plastics, Inc.)

| Properties | X | XP | XPC | XX | XXP | XXX | XXXP | XXXPC | C | CE | L | LE | A | AA | N1 | G3 | G5 | G7 | G9 | G10 | G11 | GPO1 | GPO2 | FR2 | FR3 | FR4 | FR5 |
|---|---|---|---|---|---|---|---|---|---|---|---|---|---|---|---|---|---|---|---|---|---|---|---|---|---|---|---|
| **ELECTRICAL** | | | | | | | | | | | | | | | | | | | | | | | | | | | |
| Insulation Resistance | 5 | 2 | 3 | 7 | 5 | 7 | 7 | 10 | 1 | 2 | 1 | 3 | 1 | * | 8 | 8 | 2 | 2 | 6 | 10 | 10 | 1 | 1 | 10 | 10 | 10 | 10 |
| Dielectric Str. Perp. to Laminations | 9 | 6 | 8 | 10 | 8 | 9 | 8 | 10 | 1 | 6 | 1 | 6 | 2 | * | 8 | 8 | 8 | 5 | 8 | 10 | 10 | 6 | 5 | 10 | 10 | 10 | 10 |
| Dielectric Str. Par. to Laminations | 1 | 2 | 2 | 6 | 8 | 8 | 8 | 10 | 1 | 4 | 1 | 6 | 1 | * | 10 | 2 | 4 | 6 | 9 | 6 | 6 | 5 | 5 | 9 | 9 | 6 | 6 |
| Dielectric Losses Radio Frequency | * | 1 | 1 | 2 | 2 | 3 | 3 | 5 | * | 1 | 1 | 1 | * | * | 6 | 2 | 5 | 10 | 4 | 6 | 6 | 5 | 5 | 5 | 5 | 6 | 6 |
| Dielectric Losses Power Frequency | 1 | 1 | 1 | 2 | 2 | 3 | 3 | 4 | 1 | 2 | 1 | 3 | * | * | 9 | 2 | 6 | 10 | 3 | 10 | 9 | 6 | 6 | 4 | 4 | 10 | 10 |
| ARC Resistance | 1 | 1 | 1 | 1 | 1 | 1 | 1 | 1 | * | 1 | 1 | 1 | * | * | 1 | 1 | 8 | 10 | 9 | 5 | 5 | 3 | 2 | 5 | 5 | 5 | 5 |
| Dielectric Properties Stability | 1 | 3 | 2 | 5 | 5 | 7 | 6 | 8 | 1 | 3 | 1 | 4 | * | * | 10 | 3 | 5 | 6 | 9 | 8 | 8 | 3 | 3 | 8 | 8 | 8 | 8 |
| **MECHANICAL** | | | | | | | | | | | | | | | | | | | | | | | | | | | |
| Mechanical Strength | 5 | 2 | 2 | 3 | 3 | 2 | 4 | 2 | 4 | 3 | 4 | 3 | 2 | 4 | 2 | 6 | 8 | 4 | 9 | 10 | 10 | 8 | 8 | 2 | 3 | 10 | 10 |
| Impact Strength | 3 | 3 | 2 | 1 | 2 | 1 | 2 | 1 | 4 | 3 | 3 | 2 | 2 | 5 | 4 | 8 | 8 | 7 | 9 | 10 | 10 | 8 | 8 | 1 | 1 | 10 | 10 |
| Bond Strength | 1 | 3 | * | 1 | 1 | 2 | 2 | * | 9 | 9 | 7 | 7 | 1 | 9 | 9 | 1 | 9 | 1 | 7 | 10 | 8 | 6 | 6 | * | * | 10 | 10 |
| Water Absorption | 1 | 2 | 2 | 3 | 5 | 6 | 5 | 8 | 3 | 5 | 3 | 5 | 4 | 4 | 9 | 4 | 4 | 9 | 7 | 10 | 10 | 6 | 6 | 8 | 9 | 10 | 10 |
| Dimensional Stability Due to Moisture | 1 | 2 | 2 | 3 | 4 | 5 | 3 | 6 | 1 | 3 | 1 | 3 | 5 | 7 | 5 | 7 | 8 | 10 | 8 | 10 | 10 | 6 | 6 | 6 | 6 | 10 | 10 |
| Heat Resistance | 3 | 3 | 2 | 2 | 2 | 2 | 2 | 2 | 3 | 3 | 3 | 4 | 7 | 8 | 3 | 8 | 9 | 10 | 9 | 6 | 9 | 5 | 6 | 2 | 3 | 6 | 9 |
| Dimensional Stability Due to Temp. | 5 | 5 | 5 | 3 | 3 | 2 | 2 | 2 | 4 | 1 | 4 | 2 | 8 | 9 | 1 | 10 | 9 | 9 | 10 | 7 | 8 | 5 | 5 | 2 | 3 | 7 | 8 |
| **PHYSICAL** | | | | | | | | | | | | | | | | | | | | | | | | | | | |
| Specific Gravity | 1.36 | 1.36 | 1.36 | 1.36 | 1.36 | 1.32 | 1.33 | 1.34 | 1.36 | 1.35 | 1.36 | 1.35 | 1.55 | 1.58 | 1.15 | 1.67 | 1.90 | 1.73 | 1.95 | 1.80 | 1.80 | 1.74 | 1.75 | 1.38 | 1.42 | 1.90 | 1.93 |
| Machinability | 4 | 4 | 6 | 5 | 6 | 5 | 4 | 5 | 7 | 7 | 8 | 7 | 4 | 1 | 8 | 2 | 2 | 1 | 2 | 4 | 3 | 3 | 3 | 5 | 7 | 4 | 4 |
| Punching | 4 | 10 | 9 | 5 | 9 | 3 | 7 | 8 | 8 | 7 | 10 | 8 | 3 | 2 | 8 | 3 | 3 | 3 | 2 | 3 | 2 | 3 | 3 | 8 | 9 | 5 | 5 |
| Flame Retardency | 1 | 1 | 1 | 1 | 1 | 1 | 1 | 1 | 1 | 1 | 1 | 1 | 5 | 5 | 1 | 1 | 10 | 10 | 10 | 1 | 2 | 1 | 7 | 7 | 8 | 10 | 10 |
| **CHEM** | | | | | | | | | | | | | | | | | | | | | | | | | | | |
| Acid Resistance | * | * | 2 | 2 | 2 | 2 | 2 | 1 | 4 | 6 | 4 | 6 | * | 2 | * | 4 | 2 | 8 | 2 | 4 | 6 | 4 | 4 | 1 | 1 | 4 | 4 |
| Alkalai Resistance | * | * | * | 2 | 2 | 1 | 2 | 2 | 1 | 2 | 1 | 1 | 1 | 3 | * | 3 | 4 | 2 | 4 | 5 | 7 | 3 | 3 | 1 | 1 | 5 | 5 |
| Res. to Oxidizing Agents | * | * | * | 2 | 2 | 1 | 2 | 2 | 1 | 1 | 1 | 1 | 1 | 1 | * | 2 | 4 | 8 | 4 | 4 | 7 | 3 | 3 | 1 | 1 | 4 | 4 |
| **GENERAL** | | | | | | | | | | | | | | | | | | | | | | | | | | | |
| NEMA Grade | X | XP | XPC | XX | XXP | XXX | XXXP | XXXPC | C | CE | L | LE | A | AA | N1 | G3 | G5 | G7 | G9 | G10 | G11 | GPO-1 | GPO2 | FR2 | FR3 | FR4 | FR5 |
| Military Type | | | | PBG | | PBE | PBEP | PBEB | FBM | FBG | FBI | FBE | | | NPG | | GMG | GSG | GME | GEE | GEB | | | | PEE | GEE | GEB |
| Base | Paper | Paper | Paper | Paper | Paper | Paper | Paper | Paper | CRS Fab. | CRS Fab. | Fine Fab. | Fine Fab. | ASB Paper | ASB Paper | Nylon Cloth | Cont Glass Cloth | Cont Glass Cloth | Cont Glass Cloth | Cont Glass Cloth | Cont Glass Cloth | Cont Glass Cloth | Mat Glass | Mat Glass | Paper | Paper | Cont Glass Cloth | Cont Glass Cloth |
| Type Resin | PHE | PHE | PHE | PHE | PHE | PHE | PHE | PHE | PHE | PHE | PHE | PHE | PHE | PHE | PHE | PHE | MEL | SIL | MEL | EPOX | EPOX | POLY | POLY | PHE | EPOX | EPOX | EPOX |
| A.I.E.E. Insulation Class | A | A | A | A | A | A | A | A | A | A | A | A | B | B | A | B | B | H | B | B | F | B | B | A | A | B | F |
| Comparative Price | .8 | .9 | .9 | 1.0 | 1.0 | 1.2 | 1.3 | 1.4 | 1.3 | 1.3 | 1.7 | 1.7 | .9 | 2.6 | 3.5 | 2.5 | 2.2 | 5.8 | 2.4 | 3.1 | 3.1 | .7 | .8 | 1.1 | 2.9 | 3.1 | 3.6 |

Use only for general selection. Consult complete data sheets for final decision. Price comparisons are only a guide. Properties improve in quality from 1 to 10.

PHE: Phenolic—MEL: Melamine—SIL: Silicone—POLY: Polyester.

*Indicates non-significant values.

**Grade XXXP.** This grade is recommended for most general-purpose applications and is similar to grade XXX, but possesses an additional advantage in that it can be punched at elevated temperatures. Frequencies up to 10 MHz are permitted with this grade.

**Grade XXXPC.** This is similar in electrical and mechanical properties to grade *XXXP* board and has higher insulation resistance and lower water absorption. These qualities make this type of board particularly useful for high-humidity applications. In addition, this material can be punched at temperatures of 70 to 120°F (21 to 49°C). For these reasons, *XXXPC* printed circuit boards are widely preferred over grades *XX* and *XXP*. However, when punching and moisture are not considerations, grade *XXXP* is generally selected.

**Grade G-5.** This is a melamine-glass base material with excellent electrical and mechanical properties. It is the least expensive glass cloth base but it is one of the most abrasive grades and, as a result, is very seldom used for general-purpose printed circuit applications.

**Grade G-10.** Manufactured with a glass-cloth reinforcement and epoxy resin, this grade possesses superior electrical and mechanical properties when compared to phenolic boards. Because of its high flexibility, low dielectric losses and moisture absorption, and high bond strength between the base and the conducting foil, this laminate is widely used and is recommended for frequencies up to 40 MHz.

**Grade G-11.** Also employing a glass-cloth reinforcement with epoxy resin, this grade is similar in characteristics to grade *G-10* but is more flexible at higher temperatures. In addition, it is more resistant to acids, alkali, and heat than *G-10,* but is more difficult to machine. The cost of both of these grades is the same.

**Grade FR-3.** This grade is designed for optimum balance of electrical and mechanical properties and is manufactured with paper reinforcement impregnated with an epoxy resin. It is considered superior to grade *XXXPC* because of its higher arc resistance, lower moisture absorption, and improved dimensional stability qualities. It is used almost exclusively as a self-extinguishing type in the higher frequency ranges, up to 10 MHz.

**Grade FR-4.** This grade is identical to grade *G-10* with the addition of flame retardants. For this reason, it is currently the type most often specified for applications in minicomputers, CB radios, and military and aerospace instrumentation equipment.

The information presented for these various types and grades of laminates is included with additional information in Tables 5.3a, 5.3b, 5.4, and 5.5. These tables provide for a complete comparison of a wider range of electrical, mechanical, and chemical properties of the described grades, in addition to many other base materials and their properties. All of these properties should be evaluated when selecting the most suitable base material for a particular application.

Single- and double-sided copper-clad laminates are available in sheets that range from 26 to 36 inches wide by 36 to 48 inches long. Smaller, precut dimensions are also available at extra cost. The thicknesses of the laminates, including the copper foil, range from $\frac{1}{32}$ to $\frac{1}{4}$ inch in increments of $\frac{1}{32}$ inch.

A characteristic of all printed circuit boards is their tendency to *warp* and *twist*. These conditions are the result of the two dissimilar materials (base and copper foil) bonded together to form the laminate. After etching, however, the boards will significantly straighten out because of the large amount of copper loss. Warp and twist describe the deviation of the board from a straight line fixed by two points at the extremities of the

**TABLE 5.5**

Copper-Clad Properties (*Courtesy of Insulfab Plastics, Inc.*)

| | | Bond Strength Lbs./Inch Width | | | | |
|---|---|---|---|---|---|---|
| | | 1 oz. | | 2 oz. | | |
| NEMA Grade | Mil-P Type | Ave. | Min. | Ave. | Min. | Hot Solder Resist. @ 500°F. Sec-Min |
| XXXP | 13949B PP | 10 | 9 | 12 | 10 | 10 |
| XXXP | | 10 | 9 | 12 | 10 | 10 |
| FR-2 | | 10 | 9 | 12 | 10 | 10 |
| XXXP | | 10 | 9 | 12 | 10 | 15 |
| FR-3 | 22324 PEE | 10 | 9 | 12 | 10 | 25 |
| G-10 | 13949B GE | 11 | 10 | 13 | 12 | 40 |
| FR-4 | 13949B GF | 11 | 10 | 13 | 12 | 20 |
| G-11 | 13949B GB | 9 | 8 | 10 | 9 | 40 |
| FR-5 | | 9 | 8 | 10 | 9 | 30 |

board. To determine the degree of warp, this line is taken along the longest edge. Twist is evaluated along a diagonal line between opposite corners of the board. These values of warp and twist are expressed in percentage of board or diagonal length, with the thinner boards tending to warp more than the thicker ones. Deviations range from approximately 12% for a thickness of $\frac{1}{32}$ to 5% for a $\frac{1}{4}$-inch thickness. Double-sided boards reduce these values by at least a factor of 2. These deviations of the boards become important considerations when designing for plug-in-type circuits, wherein only extremely small amounts of warp or twist can be tolerated.

## 5.3 CONDUCTING FOIL

Copper is the principal type of foil used in the manufacture of printed circuits. For special applications, nonstandard materials such as *aluminum, steel, silver,* and *tin* foils are used. Copper foil has the advantages of high conductivity, excellent soldering characteristics, low cost, and ready availability in a variety of widths. The copper used for printed circuits is at least 99.5% pure in order to maintain its high-conductivity properties.

The manufacturers of copper foil for printed circuit applications employ two basic production methods. These methods are termed *rolled foil* and *electrodeposited foil.* Rolled foil is formed from refined blocks of pure copper. The foil thus produced is largely free of pinholes and imperfections and has a smooth surface on both sides. Made in widths not exceeding 24 inches and thicknesses less than 0.001 inch, this rolled film has excellent tensile-strength properties. Because of its smooth surfaces, however, rolled foil requires special bonding treatments to enhance the *wetting* between the adhesive and the foil and to remove contaminants.

Electrodeposited foil, used almost exclusively in the manufacture of printed circuit boards, is produced by the *plating* of the film from solutions of *copper sulfate* or *copper cyanide* on a revolving stainless steel drum from which the foil is continuously stripped. The inner surface of the resulting film exhibits a smooth finish, whereas the outer surface is coarse, thereby promoting improved bonding with the increase of surface area. The thickness of the foil is controlled by the solution concentration and the electrical and me-

**TABLE 5.6**

Copper Foil Thickness Tolerances

| Weight (oz/ft²) | Normal Thickness (in.) | Tolerance (in.) |
|---|---|---|
| $\frac{1}{2}$ | 0.0007 | ± 0.0002 |
| 1 | 0.0014 | + 0.0004 |
|  |  | − 0.0002 |
| 2 | 0.0028 | + 0.0007 |
|  |  | − 0.0003 |
| 3 | 0.0042 | ± 0.0006 |
| 4 | 0.0056 | ± 0.0006 |
| 5 | 0.0070 | ± 0.0007 |

chanical parameters of the plating process. Thicknesses of less than 0.001 inch and widths in excess of 5 feet are obtainable by this method.

Foil thicknesses, regardless of the manufacturing process, are specified for printed circuit boards in *ounces* of foil per *square foot*. Foils of 1 ounce per square foot have an approximate thickness of 0.0014 inch, whereas 2- and 3-ounce foils have thicknesses of 0.0028 and 0.0042 inch, respectively. Other available thicknesses are shown in Table 5.6, which gives film thickness with the thickness tolerance.

The selection of a thickness of copper for a particular circuit application is determined principally by the amount of current that it must be able to handle. This current-carrying capability depends primarily on the thickness and width of the conductor path, in addition to the temperature. Owing to the larger radiating conductor surface area, the current capacities of printed circuit boards can easily approach 50 amperes. Table 5.7 shows the maximum current capacities for several conductor width and foil thicknesses to aid in the proper weight selection.

## 5.4 BONDING

The manufacturers of printed circuit boards choose the proper adhesives to bond the conducting foil to a particular base material in order to produce copper-clad laminates that will suit the needs of a variety of applications. The characteristics of *bond strength, hot solder resistance,* and *chemical resistance,* as listed in the tables in Sec. 5.3, largely depend on the particular adhesive used by the manufacturer for bonding and therefore will

**TABLE 5.7**

Maximum Recommended Current-Carrying Capacity for Various Conductor Widths and Thicknesses

| Conductor Width (in.) | Current (A) | | | |
|---|---|---|---|---|
|  | $\frac{1}{2}$-oz Foil | 1-oz Foil | 2-oz Foil | 3-oz Foil |
| 0.005 | 0.13 | 0.50 | 0.70 | 1.00 |
| 0.010 | 0.50 | 0.80 | 1.40 | 1.90 |
| 0.020 | 0.70 | 1.40 | 2.20 | 3.00 |
| 0.030 | 1.00 | 1.90 | 3.00 | 4.00 |
| 0.050 | 1.50 | 2.50 | 4.00 | 5.50 |
| 0.070 | 2.00 | 3.50 | 5.00 | 7.00 |
| 0.100 | 2.50 | 4.00 | 7.00 | 9.00 |
| 0.150 | 3.50 | 5.50 | 9.00 | 13.00 |
| 0.200 | 4.00 | 6.00 | 11.00 | 14.00 |

influence the technicians's choice of board. Of these characteristics, bond strength is one of the most important. This affects the reliability of a printed circuit board. If the foil should lift from the base during manufacture or in service, repair is virtually impossible. The bond must withstand the stresses involved in processing, fabricating, and servicing the printed circuit. These stresses include chemical attack from etching solutions; physical forces introduced through shearing, bending, twisting, shock, and vibration; and thermal shock during the soldering process. When shearing large sheets of single-sided copper-clad laminate into smaller blanks, the foil side should be facing *upward* with the insulating base material against the shear table. By cutting through the foil side first, there is less tendency of tearing the foil away from the edge of the blank. In addition, care must be exercised that the shear blade be allowed to return *slowly* to its normal position as it passes the newly cut edge of the board. If the blade were allowed to spring back quickly, it might snap the edge of the board just enough to crack the base material and upset the bond along that edge.

The two most common types of adhesives used in printed circuit bonding are *vinyl-modified phenolics* and *modified epoxys*. Although both adhesives display excellent bonding qualities, the epoxy resins are superior in this respect. The manufacturing process of bonding involves the application of a uniform thickness of adhesive on the foil, after which it is forced-air-dried and laminated to the base material in a press. Once the adhesive has cured, the excess foil is trimmed around the edges of the board. Finally, the foil is cleaned to remove all films and oxides resulting from the manufacturing process.

## 5.5 BOARD SELECTION

Selecting the most appropriate printed circuit board grade for a particular application depends primarily on the mechanical and electrical requirements of the circuit. The information contained in Tables 5.3, 5.4, and 5.5 will dictate which grade of board is most suitable for a specific application. An evaluation of the various board characteristics will be made with reference to selecting the most appropriate grade of base material for an amplifier's printed circuits. This example will serve as a guide to illustrate the important factors that need to be considered for proper board selection.

*Example.* A printed circuit board grade is to be selected to meet the requirements of an audio power amplifier with circuit specifications as follows:

*Type of construction:* prototype

*Maximum continuous current:* 1.6 amperes

*Maximum voltage:* 36 volts

*Frequency:* audio range (maximum 20 kHz)

*Board mounting:* corner securing by machine screws to metal chassis

*Hardware mounting:* swaged terminals and machine screws

*Machine operations:* drilling and shearing only

*Soldering:* hand

Grades of printed circuit boards containing silicone or Teflon resins are not considered because of their higher cost and high-frequency characteristics that are far beyond the requirements of the amplifier. Melamine resin boards are also not considered because of their extremely abrasive characteristics, which make them difficult to drill.

One important consideration for comparison and selection is *hot solder resistance*. This characteristic expresses thermal bond strength as a function of time and temperature. Since the boards will be hand-soldered, it is difficult to accurately control soldering time and temperature. Therefore, a board grade with a moderate to high value of hot solder resistance should be selected. Table 5.5 compares some common boards. The *FR* and the *G* grades are found to be superior to the phenolic grades listed.

*Bond strength* also needs to be considered in terms of the mechanical stresses that result from the fabrication of the board. Tables 5.3, 5.4, and 5.5 show that the epoxy-glass resins have a higher bond strength than all the other grades with *FR-4* and *G-10* having high comparable values of bond strength.

Warp and twist are of no great concern, since all the boards in the amplifier circuit are to be secured to the chassis element with machine screws and nuts.

The *machinability* of the board is another important mechanical consideration. As shown in Table 5.4, the phenolics can be drilled easier than the epoxy-glass types. Finally, the base material's *mechanical strength* must be taken into account. Mechanical strength and machinability are inversely related. The epoxy glass is stronger than the phenolics. Since mounting holes will be formed in the corners of each board and terminals will be swaged to serve as circuit interconnections, fracturing of the base material is a necessary consideration. Although the machining of the glass-base epoxy materials is more difficult than that of the phenolics, this characteristic is not as significant as mechanical strength of the laminate in the selection process. Machinability refers to the ease or difficulty of drilling or punching and the abrasive action of the material, which affect the selection of tools and methods of machining. Because no punching or machining other than drilling will be performed on any of the boards used in the amplifier, the mechanical superiority of glass-base epoxy outweighs the small difference in machinability characteristics.

In considering the electrical characteristics, as shown in Tables 5.3, 5.4, and 5.5, *all* the board materials listed for the amplifier circuits have parameters that are in excess of those required. The parameters of *dielectric constant* and *dissipation factor* are the most important for high-frequency applications. Therefore, since the amplifier will operate in the low-frequency (audio) range, these parameters need not be considered here. For dc to low-frequency applications, the parameters that must be considered are *arc* and *tracking resistance, insulation resistance,* and *dielectric breakdown voltage.*

Arc resistance is the ability of a base material to prevent undesired conducting paths created by an arc among portions of the existing conductor pattern. Tracking resistance describes the formation of undesired conductor paths *through* the insulating material. These characteristics need not be considered at low-power and low-voltage applications, such as the 36 volts maximum of the amplifier.

Insulation resistance, expressed in megohms, is the measure of the total *leakage* resistance among adjacent conductors. Because of the extremely high ohmic values, as listed in Table 5.5, this parameter does not become a consideration for this application.

Dielectric breakdown voltage is a function of the spacing among conductors and, as such, need not be considered here for purposes of board material selection. This property is discussed under conductor spacing in Chapter 6.

Based on the mechanical and electrical requirements of the amplifier circuit and comparing the properties of the various laminates in the tables provided, grades *XXXP, FR-4, G-10,* and *G-11* appear to be the best choices to obtain a realistic compromise. In our selection process, we first eliminate grades *XXXP* and *G-11* because their bond strength and machinability values are lower. Also, grade *XXXP* is not as strong mechanically as *FR-4* and *G-10.*

For our amplifier circuit, either grade *FR-4* or *G-10* is suitable since both have comparable properties for this application. However, even though it is of no concern to our amplifier circuit, grade *FR-4* does have one additional advantage over grade *G-10* in that it has a much higher *flame retardancy* value, as seen in Table 5.4. Since the cost of these two grades is about the same, we will select grade *FR-4* for our amplifier circuit.

One final consideration that also influences the selection of the *FR* or *G* series over the phenolic types is the ease of tracing circuits on the completed printed circuit board. As shown in Table 5.3a, the phenolics are available in opaque *natural, brown,* or *black* colors. The brown color is used almost exclusively with printed circuit boards. The type *FR* or *G* grades are generally *translucent* with a *green tint*. This translucent characteristic is extremely helpful in inspecting circuits. It is difficult to compare component and conductor patterns with a schematic diagram when the base material is opaque. By holding the translucent board up to a strong light, the relationship among components and conductor pattern is easily seen.

With the base material selected, the final step is the choice of the most suitable base thickness and copper foil weight. As mentioned previously, warp and twist present no problem in this amplifier package. Since all boards will be firmly secured to the chassis element, no other mechanical parameters need to be considered. For these reasons, $\frac{1}{16}$-inch base material will be used. For the smaller boards, $\frac{1}{32}$-inch material may be used; but for the sake of consistency and to avoid board fracture, especially around swaged terminals and mounting holes, all boards in this package will be $\frac{1}{16}$ inch.

In selecting the foil thickness, the major consideration is the amount of current that the circuit is expected to handle. The maximum circuit current is 1.6 amperes. From Table 5.7, for a typical conductor width of 0.030 inch, the maximum current capacity for 1-ounce copper foil is approximately 1.9 amperes. This standard 1-ounce foil, which will easily meet the circuit current demands, is therefore selected. In summary, the solution to the problem of board selection for the amplifier package is $\frac{1}{16}$-*inch 1-ounce FR-4 single-sided copper-laminated printed circuit board.*

The printed circuit board having been selected, the technician must next prepare the necessary artwork for transforming the schematic diagrams of the various circuits into conductor patterns that will be later processed. In the following chapter we develop the techniques for producing printed circuit conductor pattern artwork.

## EXERCISES

### A. Questions

5.1    List the advantages of printed circuit boards over conventional wiring methods.

5.2    What does the abbreviation NEMA represent?

5.3    Describe the meaning of NEMA suffixes *X, P,* and *C*.

5.4    What is the temperature rise of a 0.005- by 0.062-inch conductor if the current through it is 4 amperes?

5.5    Using Table 5.3a, select the laminate that provides the optimum balance of mechanical and electrical properties.

5.6    Referring to Table 5.4, match the grades of insulation independently in terms of the least and most desirable characteristics of cost, machinability, punching quality, water absorption, bond strength, and arc resistance.

5.7    Using Table 5.7, select the minimum conductor width for $\frac{1}{2}$-, 1-, 2-, and 3-ounce copper foil if the maximum dc current rating is to be 2.5 amperes.

5.8    What is the advantage of using flexible printed cables as compared to round cables?

5.9    What causes pc boards to warp and twist?

5.10   Referring to the tables provided in this chapter, determine the optimum pc insulating material and minimum conductor thickness and width to meet the following requirements: 1-ampere dc current, a frequency of 20 kHz, drilling and shearing operations only, wire-wrap terminals, and finger-type connectors for lead and conductor path connections.

## B. True or False

Circle T if the statement is true, or F if any part of the statement is false.

5.1    Grade *FR-4* pc boards have a greater flame retardency than grades *G-10* or *G-11*.       **T**   **F**

5.2    As the copper foil weight in ounces per square foot doubles, the foil thickness will also double.       **T**   **F**

5.3    The more *X*s in NEMA standards for designating phenolic resin in pc boards, the higher the resin content and the better the punching quality.       **T**   **F**

5.4    Foil thicknesses are designated in ounces per square inch.       **T**   **F**

5.5    When shearing blanks of single-sided pc boards, the copper foil side should be facing upward.       **T**   **F**

5.6    The current-carrying capacity of a printed circuit conductor path is dependent on temperature, foil thickness, and path width.       **T**   **F**

5.7    For any given foil thickness, as path width increases, current capacity decreases.       **T**   **F**

5.8    Mechanical strength and machinability of pc boards are inversely related.       **T**   **F**

5.9    A *PC* suffix on the NEMA standards indicates that the pc board may be punched at elevated temperatures.       **T**   **F**

## C. Multiple Choice

Circle the correct answer for each statement.

5.1    Printed circuit board material manufactured from (*epoxy, phenolics*) offers little abrasive action during drilling.

5.2    Printed circuit board material used for microwave applications uses a base material made from (*Teflon, epoxy*).

5.3    The principal foil material used in the manufacture of pc boards is (*aluminum, copper*).

5.4    A copper foil having a thickness designation 2 oz/ft$^2$ has a thickness of (*0.0014, 0.0028*) inch.

5.5    The maximum recommended current-carrying capacity of a pc board conductor made from $\frac{1}{2}$-oz/ft$^2$ copper foil with a width of 0.050 inch is (*1.5, 2.5*) amperes.

5.6    The copper used to manufacture the foil on pc boards must be at least (*80%, 99.5%*) pure.

5.7    The most commonly used pc board is the (*flexible, rigid*) type.

## D. Matching Columns

Match each item in column A to the most appropriate item in column B.

| COLUMN A | COLUMN B |
|---|---|
| 1.  *XXXPC* | a.  Copper purity |
| 2.  Double-sided | b.  Friction wheel stripper |
| 3.  99.5% | c.  Epoxy resin |

COLUMN A

4. *G-10*
5. Multilayer PCB
6. 1 oz/ft$^2$
7. Flexible PCB
8. *FR-4*

COLUMN B

d. 0.0014 inch
e. Two foil layers
f. Cold-punching quality
g. Flame retardent
h. Four foil layers

# 6 Single-Sided Printed Circuit Board Design: The Preliminary Sketch

## LEARNING OBJECTIVES

*Upon completion of this chapter on the preliminary sketch of a single-sided printed circuit board design, the student should be able to*

- Assemble the basic materials required to prepare a preliminary sketch of a single-sided printed circuit board.
- Assemble the required information necessary to prepare a preliminary sketch.
- Establish a two-sided viewing format for the layout sketch.
- Be familiar with sketching common component and device outlines.
- Select the appropriate sizes of pad diameters.
- Determine on-center distances between pads.
- Properly sketch pad outlines on grid intercepts.
- Determine the optimum positioning of components.
- Become familiar with component layout for automatic insertion applications.
- Properly route conductor paths for electrical interconnections.

## 6.0 INTRODUCTION

After the engineer has completed the circuit schematic diagram, the printed circuit designer converts this into drawings and artworks that will result in a functional packaging arrangement that will be fabricated into a PCB. The *component layout sketch*, which includes conductor path routing, is first completed.

Component layout sketches are scaled two-dimensional representations showing the size, shape, and position of all components, devices, and hardware that will appear on the finished board as well as the overall size and shape of the board. Also included on the layout sketch is the routing of all conductor paths that form the electrical connections specified in the schematic. In addition, *terminal pads* are shown at each lead access hole. These pads will be used to electrically connect (by soldering) all component and device leads to the circuit pattern. The component layout sketch is a composite view showing *both* sides of the PCB. It is therefore essential that these drawings be carefully constructed so that each side is clearly distinguishable from the other.

The component layout sketch is typically developed in two stages: the preliminary sketch and the finished drawing. Depending on the complexity of the circuit, the experienced designer may not need to prepare the sketch but may draw the finished drawing directly using *computer-aided drafting*, which is discussed in Chapter 7. In general, however, most electronic systems are of such complexity that they do require a preliminary working sketch to serve as a guide in preparing the final drawing. It is rare, and unnecessary, for the beginning designer to attempt a finished drawing without the aid of a sketch. The sketches not only help in establishing the overall layout plan but also result in a more technically correct drawing with fewer oversights and errors.

This chapter contains the basic information required to produce a preliminary sketch. The development of a finished component layout drawing is discussed in Chapter 7. The information provided is in three major categories: (1) how to draw the body outlines for components and devices, (2) how to use the schematic diagram to position the components and devices for a single-sided board design, and (3) how to route the electrical interconnections properly so that the circuit will function in accordance with the circuit schematic.

## 6.1 MATERIALS REQUIRED FOR DEVELOPING A COMPONENT LAYOUT "WORKING SKETCH"

The term *sketch* often implies a freehand drawing, made to no particular scale. We will not be using the term in this common form for the generation of the "working sketch" for a PCB component layout drawing. This type of sketch is drawn freehand and to a 2:1 scale. It needs to be emphasized that the working sketch is a guide to be followed closely when producing the finished component layout drawing. It is therefore essential that scale and thoroughness be maintained in all phases of generating this preliminary working sketch.

The basic implements required to develop the preliminary sketch are (1) pencils, (2) erasers, (3) decimal scales, (4) decimal and fractional circle templates, and (5) ruled grid paper.

The pencils used may be mechanical pencils with thin leads, 0.5 or 0.7 millimeter, or standard wooden drafting pencils with a lead hardness of at least 2H. For standard drafting pencils, appropriate sharpeners are necessary to maintain their points. For single-sided board design, two lead colors are required, the standard black and a contrasting color such as blue or red. The eraser used should be compatible to the leads and the type of paper used.

For making measurements, a rule with major graduations of 0.1 ($\frac{1}{10}$) inch is required. Fractional scales do not lend themselves to component layout drawings and should be avoided. The length of the rule will depend on the overall size of the design. Generally, a length of 12 or 18 inches is suitable.

Decimal and fractional plastic circle templates are used to draw terminal pads and round-bodied components. The templates selected should have hole sizes ranging from 0.1 to 1.00 inch and from $\frac{1}{32}$ to about 2 inches.

The drawing media on which the sketches are to be produced should be a good-quality grid paper in a size large enough to easily accommodate the overall size of the scaled drawing. The color of the grid lines is preferably blue, although black is acceptable. The grid is a series of equally spaced vertical and horizontal lines and serves as an invaluable aid in the positioning of components and the placement of pads and conductor paths. The grid spacings should be 0.10 inch for both the vertical and horizontal lines. Grids with spacings of more than 0.1 inch should not be used in printed circuit layout. Figure 6.1 shows a piece of 10-division-per-inch graph paper with highlighted vertical and horizontal grids on every one-inch increment.

## 6.2 BASIC INFORMATION FOR BEGINNING THE LAYOUT SKETCH

To begin producing the layout sketch, the PCB designer must have information available relative to the circuit and the parts. Primarily, a detailed circuit schematic with all parts identified either on the circuit drawing or on an accompanying parts list is initially required. An example of a detailed schematic is shown in Fig. 6.2, which is a basic

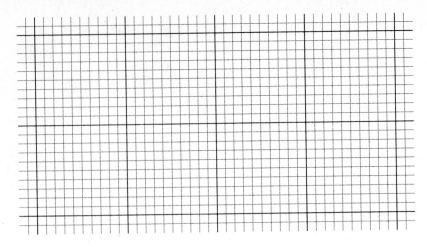

**Figure 6.1** 10-Division-per-inch grid system.

common-emitter transistor amplifier. The parts list provided with the schematic identifies each of the components.

In addition to the circuit schematic, each of the parts or data sheets, which provide the physical size and shape of all components, devices, and hardware items, needs to be available. Also, electrical data sheets for all devices showing the pin arrangements are necessary.

Typically, the design engineer will provide the draftsperson with additional information regarding electrical and mechanical restrictions, such as the size and shape of the board, positioning of critical components, forbidden areas of the board for certain parts placement, required slots or cutouts, and placement of external interconnections.

**Parts list**
R1–47 kΩ, ¹/₂W, 10%
R2–2.7 kΩ, ¹/₂W, 10%
R3–2.2 kΩ, ¹/₂W, 10%
R4–180 Ω, ¹/₂W, 10%
C1–1.0 μF, 15 WVDC Electrolytic Capacitor
C2–0.33 μF, Disk Capacitor
C3–100 μF, 3 WVDC Electrolytic Capacitor
Q1–2N3053, NPN Transistor,
    TO–5 Case Style

**Figure 6.2** Standard common-emitter transistor amplifier circuit schematic with parts list.

## 6.3 FORMAT OF THE LAYOUT SKETCH

Printed circuit boards are typically flat, rigid surfaces on which components are mounted and are electrically connected. For single-sided boards, components, devices, and hardware are normally mounted on the *insulated* side, termed the *component* side or *side 1*. The terminal pads and conductor paths are processed into the copper foil on the opposite side, referred to as the *circuit* or *foil* side or *side 2*. Because the layout sketch represents a composite view of both sides, a direction of viewing the board must be adopted before layout can begin. It is common practice for PCB designers to view the layout from the component side of the board. This is because the components are considered to be the dominant elements in a circuit. In addition, the intended layout must not only result in a functional system but also should have the components well arranged so that the board is appealing to the eye. To achieve this, the components should be uniformly spaced using all of the available board area. By careful parts placement, the resulting board will have a balanced distribution of components with no crowding and/or open areas that are void of parts. It can be seen from this discussion that the logical viewpoint for laying out a PCB is the component side. Thus initial attention will be focused on the placement of the components with the routing of the conductor paths to be considered after a balanced parts layout has been achieved.

With the direction of viewing established, it must be noted that, even though the components are positioned first, conductor routing must be considered at each step; that is, the conductor paths throughout the component arrangement must be routable. Thus the components and devices must be placed in such a position as to allow them to be conveniently connected electrically by a conductor pattern on the foil side of the board. To avoid confusion in distinguishing each of the board sides on the layout, a color-coded system needs to be established. For example, all detail appearing on the component side of the board, such as component and device bodies, hardware, and outside board edges, will be drawn using *black* lead. All detail appearing on the circuit side of the board, such as terminal pads and conductor paths, will be drawn using *blue* lead. Because this book does not have multicolor illustrations, all conductor paths in this chapter are shown as single, solid black lines connecting the circles that represent the terminal pads. It should, therefore, not be difficult to differentiate the linework on one side from that on the other.

Finally, the scale to which the drawing will be made must be considered. Industrial layouts are typically drawn to either a 2:1 or 4:1 enlarged scale. A 2:1 scale is used in this chapter to show techniques used to make enlarged scale drawings. Scale drawings are necessary to reduce small unavoidable layout errors.

## 6.4 DRAWING COMPONENT AND DEVICE BODY OUTLINES

With all of the necessary information available and the format established, we can now begin developing the working sketch. Before the components can be positioned onto the sketch, we must first learn how they are shown as two-dimensional body outlines that replace the graphic symbols in the circuit schematic. These outlines will be drawn to a 2:1 scale and will be shown so as to represent the components and devices as they will appear on the layout drawing, viewed from the component side.

The circuit schematic shown in Fig. 6.2 will be used to illustrate the drawing of the component body outlines. In the parts list of the amplifier, there are four $\frac{1}{2}$-watt resistors (R1, R2, R3, and R4), one tubular electrolytic capacitor (C1), one ceramic disk capacitor (C2), one vertically mounted electrolytic capacitor (C3), one npn transistor (Q1), and five external connections [IN, GND(in), +12 Vdc, OUT, and GND(out)].

The exact dimensions of each component must first be determined. The axial-lead-style $\frac{1}{2}$-watt resistors are found to have a diameter of 0.160 inch and a length of 0.416 inch with lead diameters of 0.032 inch. This style of component is mounted to the board surface by first bending the two leads at right angles at a specified distance from the ends of the body. The leads are then passed through drilled access holes with sufficient length to make connection to the terminal pads on the circuit side of the board. It is common practice when drawing the body outline of axial lead components not to show the lead access holes or the lengths of leads between the body ends and the hole. Rather, since this is a composite view, only properly positioned terminal pads that will accept each lead are drawn beyond the ends of the body outline. The diameter of the pads and the distance between them must first be determined.

Terminal pad diameters for *unsupported* holes (not plated-through) on single-sided boards are dependent on the diameter of the lead to be inserted, the diameter of the drill hole, and the amount of copper foil required to remain about the pad after the hole is drilled. To provide the necessary clearance for our resistor leads, which have a 0.032-inch diameter, the drill hole should be from 0.010 to 0.020 inch larger in diameter. This would require a hole size of diameter between 0.042 and 0.052 inch. It is general practice for unsupported holes that the terminal pad diameter be at least 0.040 inch larger than the drill hole. For our example, a pad diameter of between 0.082 and 0.092 inch would be acceptable on a 1:1 scale. Commonly available pad diameters in this range are 0.075, 0.080, 0.093, and 0.100 inch in diameter. It should be noted that after a board has been processed, the bonding of the conductor paths and terminal pads is related to the amount of foil remaining after the hole has been drilled. For this reason, it is recommended that the size of the terminal pad be made as large as possible. Making them too large, however, reduces the available space for conductor path routing. The final size selection thus becomes a trade-off between conductor spacing and maximum diameter of terminal pads.

Additional factors that must be considered at this time are (1) reducing the number of different hole sizes, which results in fewer drill bit changes during fabrication, and (2) selecting only a small number of terminal pad diameters to satisfy all lead diameters in the design, thus reducing the required inventory of pad sizes. Every effort should be made to select the fewest drill and pad sizes that will satisfy the total range of hole and pad size requirements.

In selecting pad diameters for single-sided boards, the following guidelines are recommended. For leads with diameters of less than 0.020 inch a 1:1 scale pad diameter of 0.075 inch may be used. Lead diameters of between 0.020 and 0.040 may effectively use a pad diameter of 0.100 inch.

The on-center distances between pads is determined by both the component body length and the lead diameter. Since an individual pad must be provided for each lead, one is shown at each end of the body outline. The preferred distance that the leads should extend outward from the component body before the 90-degree bend is made is 0.060 inch. To determine the overall bend distance $X$, a bend radius should be added. The minimum bend radius for each lead should be approximately 0.030 inch for lead diameters up to 0.030 inch, and 0.060 inch for lead diameters of 0.031 to 0.050 inch. Lead diameters greater than 0.050 inch should have a bend radius of two times the lead diameter. Refer now to Fig. 6.3. It can be seen that the formula for calculating the on-center distance ($S$) for terminal pads is

$$S = 2X + B \qquad (6.1)$$

where $X$ is the bend distance (extension distance + bend radius) and $B$ is the body length. From the discussion above, and with reference to Fig. 6.3, the *minimum* value of terminal pad spacing for axial-lead-style components using maximum lead diameter is 2(0.060 + 0.060) plus the body length ($B$).

Referring again to the $\frac{1}{2}$-watt resistor shown in Fig. 6.3. and using the maximum values given for $X$ and $B$ in the accompanying table, we calculate the value of $S$ as follows:

| Dimension | Inches | | Millimeters | |
|---|---|---|---|---|
| | MIN | MAX | MIN | MAX |
| B | 0.344 | 0.416 | 8.74 | 10.6 |
| D (DIA) | 0.115 | 0.160 | 2.92 | 4.06 |
| X * | 0.090 | 0.120 | 2.29 | 3.05 |
| d (Dia.) | 0.026 | 0.036 | 0.66 | 0.91 |
| S | 0.524 | 0.656 | 13.3 | 16.7 |

* Depends on application. Minimum values for dense layouts.

**Figure 6.3**  Tabulating dimensional information on detail drawings.

$$S = 2X + B$$
$$= 2(0.060 + 0.060) + 0.416$$
$$= 0.240 + 0.416 = 0.656 \text{ inch}$$

This figure would normally be increased to 0.70 inch so that the on-center pad dimensions will fall on grid.

The terminal pads shown in Fig. 6.4a have been drawn on grid with an on-center dimension of 1.4 inch and diameters of 0.2 inch at 2:1 scale. Although the pads are shown in black, they should be drawn with blue pencil so as to more readily distinguish their true position as being on the foil side of the board.

Note in Fig. 6.4 that no center lines are used in drawing the terminal pads. Simply drawing the outline of the pad with a circle template is sufficient. The underlying grid should be used as a guide to locate and center the pad. It is important to center the pads on

(a) Pad spacing for ½ watt resistor

(b) Body and pad sizes for ½ watt resistor

(c) Two–dimensional view of electrolytic capacitor

**Figure 6.4**  Drawing the two-dimensional component view of axial lead resistors and capacitors.

*6.4 Drawing Component and Device Body Outlines*

grid intercepts and construct the component body outlines off-grid as required by their dimensions. For precise artwork generation and board fabrication, it is more important for the pads to be positioned accurately on grid than it is for the component body outlines.

When mounted to the component side of a PCB, axial lead components normally rest flush against the insulated board surface. As such, their two-dimensional body outlines, as viewed from the component side (side 1), would appear as rectangles. Using the adopted color-code system, all of the linework for side 1 would be drawn with black lead. Since our layout will be drawn to a 2:1 scale, each full-size dimension must be multiplied by a factor of 2 before drawing the body outlines. For a $\frac{1}{2}$-watt resistor, the maximum dimensions for the body length and diameter given in the table of Fig. 6.3 would be doubled, resulting in a rectangle drawn as close to 0.32 inch × 0.83 inch as the grid will permit. This is shown in Fig. 6.4b. Note that the grid lines provide a useful guide for measuring and drawing the body outlines by simply counting squares. When using a 0.1 grid to show the outline of this resistor, the body is to be laid out to a rectangle of just over 3 squares (0.32 inch) by a little more than 8 squares (0.83 inch). Note also that the body outline is neatly drawn and reasonably centered between the two terminal pads using a straightedge even though the layout is termed a *sketch*.

Capacitor C1 in Fig. 6.2. is the electrolytic type with axial leads and with a body diameter of 0.25 inch. Its body outline, shown in Fig. 6.4c, is drawn similar to that of a resistor, the only difference being the dimensions of the lead diameter, body diameter, and length. Because its maximum lead diameter is slightly less than that of a $\frac{1}{2}$-watt resistor (0.031 versus 0.032 inch), the diameter of its pads may be the same as those used with the resistor (i.e., 0.2 inch at 2X scale). Also, since the lead diameter is 0.031 inch, the bend radius will be 0.060 inch, the same as that for the resistor. The total body length of the electrolytic capacitor is 0.625 inch. Therefore, the on-center pad spacing is $S = 2(0.060 + 0.060) + 0.625 = 0.865$ inch. To be on grid, this becomes 0.90 inch between pad centers at a 1:1 scale. As shown in Fig. 6.4c, the body outline is centered between 1.8-inch spaced pads and is shown as a rectangle of approximately 0.5 × 1.35 inch at a 2:1 scale. Note the plus sign (+) inside the body outline near one end. Because electrolytic capacitors are polarized, they need to be correctly assembled onto the board. The plus sign serves as a key for proper installation.

Another component commonly used in electronic circuits is the radial lead ceramic type *disk* capacitor, which is shown in Fig. 6.5a properly mounted to a PCB. Capacitor C2 in Fig. 6.2 is a disk type. It is seen from its side view that the body is approximately circular and that it has two unpolarized leads which extend directly through the board without bends. Above the side view shown in Fig. 6.5a is the two-dimensional shape of the disk capacitor as viewed from the top or as it would appear on the component layout drawing and viewed from side 1. Accurately drawing this complex shape, which tapers almost to points at its outside edges with maximum body width at its center is time-consuming and unnecessary for PCB layout. Rather, the less difficult shape of a rectangle has been adopted even though it does not exactly represent the body shape.

To construct the component view of disk capacitor C2, the on-center lead spacing is first measured and found to be 0.375 inch. At 2X scale, this dimension becomes 0.750 inch. This is then increased to 0.80 inch for the on-grid locations of pad centers. The lead diameters of the component are 0.025 inch. With this information, the pads, at 2X scale, will be drawn with 0.20-inch diameters on grid. This is shown in Fig. 6.5b. Remember that these pads would normally be drawn with blue pencil. The diameter $w$ and thickness $t$ of the capacitor are next measured and found to be 0.590 and 0.156 inch, respectively. At 2X scale, these dimensions become 1.18 and 0.312 inch. The two-dimensional view of the component is thus represented by a rectangle that is slightly over 3 squares by approximately 12 squares and centered between the terminal pads. Note in Fig. 6.5c that the terminal pads are on grid and are *inside* the body outline. This is not always the case. Other types of orientations of body outlines to terminal pads are shown for different sizes of disk capacitors in Fig. 6.5d.

**Figure 6.5** Drawing the two-dimensional component view of a disk capacitor: (a) viewed from the top and from the side; (b) terminal pads drawn on grid; (c) pads shown inside the body outline; (d) orientation for different sizes of disk capacitors.

Capacitor C3 in Fig. 6.2 is an electrolytic type with a tubular shape having two radial leads protruding from the *same* end of the body. This style is designed to be mounted vertically onto the PCB with its leads extending straight into the access holes. This orientation is similar to that of the disk capacitor except that the base of the electrolytic type is seated flush with the surface of the board. This is shown in Fig. 6.6a. As with other PCB-mounted components, the pad positions and size will first be determined. The 1:1 on-center lead spacing and lead diameters are found to be 0.2 and 0.025 inch, respectively.

**Figure 6.6** Drawing the two-dimensional component view of an electrolytic capacitor with radial leads: (a) top and side views; (b) terminal pads drawn on grid; (c) body outline drawn around the terminal pads.

Post

Shoulder

Shank

Side 1

PCB

C⃗L

Mounting Holes

(a) Pictorial view of
turret terminal

(b) Top view of turret terminal assembly

Side 1

Side 2

PCB

Copper

Swage and Solder (3 Places)

(c) Side view of terminal assembly

**Figure 6.7**   Turret terminal detail.

Thus the pads will be drawn on grid at 2X scale with a 0.4-inch on-center dimension and diameters of 0.2 inch. This is shown in Fig. 6.6b. The case diameter is measured to be 0.335 inch, and at 2X scale it becomes 0.67 inch. Using a 0.7-inch hole in a circle template, the body outline is centered and drawn around the two terminal pads (see Fig. 6.6c). As seen from the component view, this capacitor appears as a circle since it is vertically mounted. Because it is polarized, a plus sign is placed beside the appropriate pad outside of the body outline to ensure correct orientation onto the board.

The five external connections to the points labeled IN, GND(in), +12 Vdc, OUT, and GND(out) in Fig. 6.2 will each be made with turret terminals for our example design problem. Pictorial, top, and side views of a turret assembly are shown in Fig. 6.7.

Turret terminals are a common method of providing points for soldering wires to make electrical connections between the PCB circuit and other external points. Each of the five turret terminals will require a pad in order to make electrical connections to a path on the circuit side of the board. The two-dimensional representation of a turret terminal is quite simple, as shown by Fig. 6.8a. The inner circle represents the diameter of the turret section, the next larger circle the shoulder, and the outer circle represents the terminal pad. Actually, all that is required is to show the terminal pad diameter for each of the terminals. This is shown in Fig. 6.8b. Appropriate labeling on the layout sketch will make it clear which of the pads will be used for making external connections requiring turret terminals. In addition, their larger diameters and locations near the board edges make them easy to identify.

To serve as an example, we will use a turret terminal with a shank diameter of 0.1 inch. If an access hole of 0.01 inch greater than the shank diameter is selected for ease of

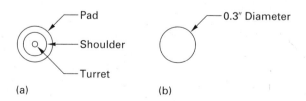

Pad

Shoulder

Turret

0.3″ Diameter

(a)

(b)

**Figure 6.8**   Drawing the two-dimensional representation of a turret terminal pad: (a) top view of turret terminal; (b) terminal pad only.

installation, a drill diameter of 0.11 inch would be required. Refer to Appendix II for a table of common drill bit sizes. It can be seen from the table that a No. 35 drill bit would be used for this hole. To satisfy the requirement of selecting a pad diameter that will be 0.040 inch larger than the drilled hole, a pad with a 1:1 scale diameter of 0.150 inch, having a 2X scale diameter of 0.300 inch, would be required. This is shown in Fig. 6.8b. This size pad would be shown for all external connections requiring the installation of turret terminals having a shank diameter of 0.1 inch.

The final component of Fig. 6.2 to be drawn is the transistor Q1. The body outline and pad arrangement for transistors are more complex in their drawing than the components discussed thus far. Although all of the critical dimensions can be measured with a scale, this is a type of component where dimensional information is best obtained from the manufacturer's specification sheets. Refer to Fig. 6.9a for the manufacturer's outline drawing of the TO-5 transistor case specified for Q1. Note that the maximum lead diameter is 0.019 inch. Therefore, a 1:1 scale pad diameter of 0.075 is required. When drawing the pads to a 2X scale, the diameter used will be 0.150 inch.

The transistor leads are numbered 1 (*emitter*), 2 (*base*), and 3 (*collector*). These leads are positioned in a triangular arrangement on a 0.2-inch-diameter dotted-lined circle with the base lead centered but elevated a distance of 0.1 inch above the in-line leads of the emitter and the collector. The emitter lead is closest to the case tab and the collector lead is farthest away from the tab with a separation of 0.2 inch. The three terminal pads

**Figure 6.9**   Drawing the two-dimensional component view of a TO-5 style transistor using manufacturer's specifications.

that will accept the transistor leads for straight entry into the board are drawn with 0.15-inch diameters at a 2X scale and laid out at the dimensions shown in Fig. 6.9b. Note that the arrangement of the pads is inverted from that of the bottom view shown for this case style. This is necessary to represent the true orientation of the device as viewed from the top of the case as it is positioned on the component side of the board. Care must always be exercised in arranging pad positions for devices from the component view since the manufacturer's specifications commonly provide lead designations from the lead side or bottom view of the device.

Referring again to the specifications for the TO-5 case, the largest body diameter is specified as 0.37 inch at a 1:1 scale or approximately 0.75 inch at a 2X scale. This circular body is drawn concentric to the dotted-line lead circle. For ease of drawing, the device tab is shown as a square approximately 0.1 by 0.1 inch and is positioned closest to the pad numbered 1 (emitter). This square is drawn outward from the body outline at a 45-degree angle to the center of the horizontal grid line which passes through the centers of the emitter and collector pad circles.

The complete outline drawing of the TO-5 case with leads coded and with the identification tab is shown in Fig. 6.9c. Note from this view that the emitter, base, and collector are read in a counterclockwise direction from the tab.

It is not possible to demonstrate here the drawing of body outlines for all styles of components and devices and it is hardly necessary. As emphasized in this section, the basic objective is to draw the top view of the part as seen from the component side of the board. As much detail as necessary is shown to illustrate the important features of the part. Drawing excessive detail not only may be a waste of time but also may cause confusion.

In summary, terminal pads are first positioned and drawn on grid after their scaled dimension (in our case, 2X) and spacing have been determined. The component body outline is then drawn in its relative position to its pads. The following sequence should be followed: (1) calculate the pad diameter from the lead diameter and the hole size; (2) determine the 2X pad spacing for the component or device; (3) position and draw the 2X diameter terminal pad on grid for each lead; and (4) center and construct the body outline in its proper position. Finally, because the pads will be processed from the copper foil on the circuit side of the board, they should be drawn in a contrasting color, normally *blue*. The body outlines, which appear on the component side of the board, should be drawn with *black* lead.

## 6.5 POSITIONING COMPONENTS ON THE LAYOUT SKETCH

Now that you have become familiar with the drawing of common components and devices typically mounted on a PCB, we will begin the discussion of the most creative and interesting aspect of printed circuit design. This is the positioning of the parts on the board and the routing of the electrical connections (conductor paths) between terminal pads in accordance with the requirements of the circuit schematic. The typical concern faced by beginning designers is that after several parts have been tentatively positioned on the board, connecting paths between some of their pads will be unroutable. That is, there may be no possible routes available to run a conductor path between pads that must be connected without crossing another drawn path that is not associated with it. This problem, however, will be eliminated as proper layout skills are developed. This also emphasizes the need for drawing the preliminary sketch prior to preparing the final component layout. The sketch is used to resolve any problems of components being positioned so as to create unroutable paths.

It needs to be understood initially that printed circuit design is a creative task and, as such, does not lend itself to any single "correct solution." There are many acceptable layout practices, some with obvious reasons for specific component positioning and some where the reasons are somewhat obscure but are based on past practices that have proven successful. Your unique designs may be just as acceptable and correct as long as the basic guidelines initially established are not violated.

For our first design, we will construct a preliminary sketch of the amplifier circuit shown in Fig. 6.2 using the skills for drawing component body outlines developed in the preceding section. A statement of the design problem follows.

> *Design Problem.* Using a 2:1 scale and a 0.1-inch grid system, design a component layout sketch of the amplifier circuit of Fig. 6.2. There are no restrictions on final board size or shape except that the finished layout maximizes the total board area. There are also no restrictions on the position or orientation of any external connections. All terminal pads are to be drawn on grid. The component view should show all necessary labeling using black lead to show the component side and blue lead to represent all pads and conductor paths on the foil side of the printed circuit board.

It is seen from the design problem that the designer has few restrictions, especially in component positioning and external connections. This is not typical, but it is felt that your first design should provide layout practice without rigid requirements.

We begin the design with the circuit schematic shown in Fig. 6.10 which is essentially the same as that shown in Fig. 6.2 with the addition of specific points identified with letters and numbers to serve as guides for locating portions of the circuit as they are discussed.

Next, a sketch, such as that shown in Fig. 6.11, is made of the overall board size with usable component area. The overall board outline is indicated by penciled corner brackets. A solid continuous line, called the *component border,* is drawn inside the overall board outline. This line shows the usable area on the board for placing components. The margin around the edges of the board is called the *forbidden area.* This space is reserved for drilling holes to accommodate hardware for mounting the board without interfering with components or conductor paths. For this design, an initial board size of $5 \times 6$ inches is selected. This size is at a 2:1 scale.

There is no easier way to begin the component layout sketch than to position components and hardware inside the component border as they appear on the schematic. This is termed the *schematic viewpoint* and will provide ease of conductor path routing. Rearrangement of initial component placement is usually necessary to obtain a well-balanced layout having a uniform distribution of parts in the usable area with no crowding

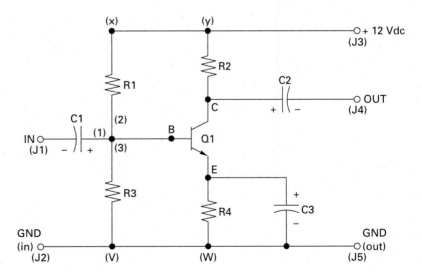

**Figure 6.10** Schematic labeled for use as a guide for parts placement and conductor path routing.

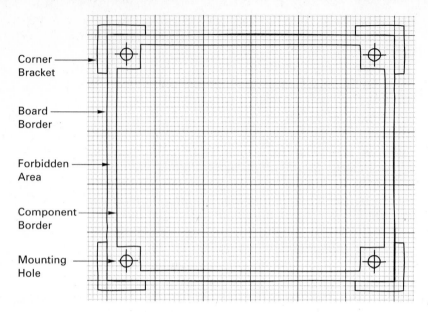

Corner Bracket

Board Border

Forbidden Area

Component Border

Mounting Hole

**Figure 6.11**   Initial border layout sketch.

or large wasted spaces. Precut component outlines are commercially available to aid in obtaining optimum layouts rapidly for final sketching. These precut outlines, called *dolls* or *puppets,* can be conveniently positioned and repositioned until the desired layout is achieved. It is possible, by using the computer-aided drafting techniques discussed in Chapter 7, to produce component outlines that can be cut into dolls.

Using the schematic viewpoint, the dolls are placed onto the 0.1-in. grid inside the component border. The result may appear as shown in Fig. 6.12. All component body outlines and terminal pads are to a 2:1 scale. All of the parts are labeled with the same designations as on the schematic. Note that the parts are positioned approximately as they appear in the schematic. For example, R1 is directly above R3 and to the left of R2, which is above Q1. Some components, like capacitor C1, require special orientation. This capacitor must be positioned with its negative side, which is not labeled, closest to the external terminal marked IN and its positive terminal, labeled +, facing toward Q1. The input, out-

**Figure 6.12**   Initial component layout using dolls and the schematic as a guide.

put, ground, and power terminals are positioned toward the ends of the board, which follows their position on the schematic. This method of solving the problem of arranging components to make path routing for input and output connections with the least amount of difficulty is termed the *peripheral viewpoint*.

Another method of ensuring optimum component location is called the *central viewpoint* and is similar to the schematic viewpoint. This viewpoint, however, focuses attention on the placement of passive components that support the operation of an active device such as Q1. The initial location of Q1 should be central to the resistors and capacitors that support its operation as shown in the schematic.

Observe that there are no crossover leads in the circuit schematic. Since our parts placement follows that of the schematic, it is possible to make all of the electrical connections as shown in the schematic. Even though the design is routable as shown, it does not meet the specified requirement of maximizing the total board area. The bold outline drawn around the positioned dolls in Fig. 6.12 represents the component border. Notice the large unused areas of the board. These areas are shown as a, b, and c. This gives the appearance that the board is much larger than required for this design. To overcome this problem, we will slightly reposition some of the dolls until a more reasonable balance is achieved. Compare Fig. 6.12 with Fig. 6.13. To begin, C1 will be rotated 90 degrees with the GND (in) and IN terminal pads positioned directly above and below this capacitor while still remaining centrally positioned. Also, note that for better component balance, moving Q1 to the right will allow R2 and R4 to be brought inward so that they will be adjacent to R1 and R3. This will reduce the overall width of the board. C3 can then be placed to the right of R4 bringing it directly under Q1. Finally, C2 is rotated 90 degrees and positioned to the right of R2 and above Q1. Wherever possible, centers of component bodies should be aligned as was done with R1, R2, R3, and R4. There are trade-offs, however, in order to reduce the total required space and to improve the overall appearance of the board. For example, C1 is positioned midway between the top and bottom edges of the board. C2, Q1, and C3 are spaced vertically to avoid needless crowding, but this does not allow C2 to center on R2 nor C3 to center on R4. These types of trade-offs in rearranging components help to optimize the overall layout in regard to uniform distribution. Note also that no body outlines are positioned on a diagonal to the horizontal and vertical grid lines. This is to prevent the generation of wasted board space. In addition, any component placed at a diagonal would look out of place compared to the other vertically and horizon-

**Figure 6.13** Finalized component and hardware positioning.

tally placed parts. The final component layout incorporating all of the techniques discussed is shown in Fig. 6.13.

Even though parts have been rotated and repositioned, two important points need to be emphasized. First, no part has been moved any great distance from its initial position in the schematic. The resulting path routing distances are thus as short, or in some cases shorter, than in our initial parts placement. Keeping interconnecting paths as short as possible is normally a prime requirement of printed circuit design. Second, each component move or rotation has resulted in reducing the amount of required board space, generating a more compact layout which is also required in a good design.

Once the components have been repositioned to result in an improved layout, the external connections on both ends of the board can then be moved inward to reduce the required length of the board. After the final component and hardware positions are obtained, each doll is carefully traced with a pencil and removed from the layout. Care must be taken to sketch the pads associated with each doll in the correct location. Remember that all pad centers are to be placed on grid.

We now summarize how optimum component placement is accomplished. We first establish an initial point of reference by positioning the parts as they appear in the circuit schematic. The schematic is a concise representation of all electrical connections between parts. It contains minimum crossovers and, as such, part placement based on this scheme also results in minimum crossovers. The parts may then be rearranged to (1) reduce path routing problems, (2) result in a more compact design to reduce wasted board space, and (3) improve the appearance of the board by arriving at a uniform distribution of parts. In general, conductor path routing is simplified when the component layout parallels that of the circuit schematic.

## 6.6 LAYOUT CONSIDERATIONS FOR AUTOMATIC ASSEMBLY

When designing a component and conductor pattern layout of a pc board that is to be assembled by automatic insertion, component placement is critical and is governed by specific design criteria. To make automatic assembly cost effective, the designer must standardize the components selected for the circuit. By reducing the number of different size and shape axial and radial lead components, the variety of lead access hole spacings will be kept to a minimum. It is also important to arrange component positions on as few axes as possible, with rows and columns having uniformity of lead spacing and orientation. The optimum layout is that in which all components have the same lead span between terminal pads and are placed in columns and rows having all of their axes parallel. When more than one span is part of the layout, grouping of components of similar span where possible increases assembly efficiency by reducing worktable travel.

In most systems, the insertion head does not rotate. Because of this, component axes placed perpendicular to others reduce assembly efficiency since the board must be rotated 90 degrees to install these parts. In addition, no component axis should be positioned at a diagonal to any board edge.

As much as possible, component layouts should orient keyed components uniformly. For example, all notches on ICs that key pin 1 should face the same direction as well as placing the ICs in precise columns and rows.

For automatic insertion designs, boards must be provided with two tooling holes from which all lead access holes are placed on a fine grid in increments of 0.025, 0.050, and 1.00 inch. These tooling holes are later used to position the board on the automatic insertion machines. As in other designs, pad centers must be precisely located on grid.

Lead access hole sizes for automatic insertion are, in general, made 0.015 inch larger than those made for manual component assembly. This means that the diameters of the terminal pads should be increased by 0.030 inch to keep the same amount of copper surrounding the lead access holes. This surrounding copper is called the *annular ring* and is generally given a minimum-width specification, such as 0.012 inch.

In Chapter 16, Fig. 16.20 shows an automatic insertion head together with finger-to-lead and finger-to-body clearance for parallel-positioned axial components. The typical length and width clearances required for each guide finger are 0.100 and 1.30 inch, respectively. When axial components are placed side by side, as in Fig. 16.20a, essentially no clearance is needed. However, to locate insertion holes on grid, the body center-line spacing of adjacent components is rounded off to the nearest 0.100 inch. When positioned end to end, a minimum of 0.100 inch between insertion holes is required.

The pad centers for adjacent ICs should have a minimum spacing of 0.150 inch. When they are positioned end to end, the spacing can be extremely small—as small as two times the manufacturer's body tolerance.

## 6.7 CONDUCTOR PATH ROUTING

The final process in the completion of the component layout sketch is to interconnect correctly the terminal pads associated with each of the components and devices in accordance with the circuit schematic. To continue in our layout of the amplifier circuit shown in Fig. 6.10, conductor paths between the pads shown in Fig. 6.13 will be drawn.

The width of the conductor paths is primarily related to (1) the maximum current they are expected to handle, (2) circuit density, and (3) fabrication limitations. Because it is the purpose of this chapter to introduce the basic techniques for routing conductor paths, we will focus only on this topic. (Refer to Chapter 5 for determining conductor path width and to Chapter 7 for the criteria for the spacing of these paths.)

Conductor paths are drawn on the sketch with a straightedge and are shown as single lines, typically 0.5 or 0.7 mm thick. Blue lead should be used to indicate that these paths are on the foil side of the board together with the pads. Since the conductor paths form the electrical connections, they must originate at a pad or at the intersection of other paths and terminate at a pad or intersect with another path. The conductor paths should be drawn on grid lines whenever possible since the pads are centered on grid. They should extend toward the pads as if to bisect them. For clarity, the conductor line drawn should cut across the pad circumference and extend approximately halfway to its center point. See pad a in Fig. 6.14. Changes in path direction are drawn in 45-degree angle steps, as shown in the paths between pads a and b and between pads b and d in Fig. 6.14. When routing a conductor path that must intercept another path for electrical connection, it is drawn only at right angles, as shown in the path labeled e in Fig 6.14. Finally, no more than four paths should originate or terminate at one terminal pad. Space permitting, paths entering a pad should be separated by an angle equal to or greater than 90 degrees. See pad c in Fig. 6.14.

The task of conductor path routing begins with drawing as many of the short interconnect points as possible, leaving the longer paths until later. The circuit schematic of Fig. 6.10 shows an electrical connection between points *x* and *y*. A single solid blue line in drawn horizontally between the pads labeled *x* and *y* to form this connection. This is shown in Fig. 6.15. As each connection is made on the layout sketch, it should be checked off or lined out on the circuit schematic. This is done to ensure that all connections have been correctly made in accordance with the schematic. The procedure is repeated for the connecting points labeled *v* and *w*.

A grouping of short connections is labeled as 1, 2, 3, and B on the schematic. This indicates that a single electrical connection is to be made from the + side of C1 (point 1) to the bottom pad of R1 (point 2) and the top pad of R3 (point 3) as well as to the base of Q1 (point B). Points 2 and 3 are first connected with a single vertical line. The connection between point 1 can then be made to either point 2 or 3 since they are now shown as elec-

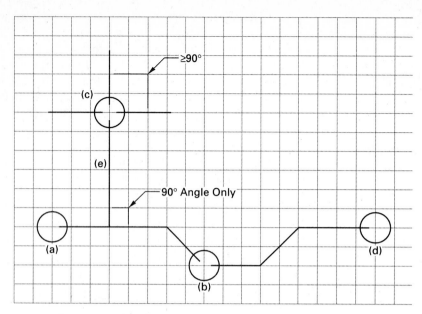

**Figure 6.14** Techniques for routing conductor paths.

trically connected. However, it is recommended that this connection be made to the pad of point 2 since it is closer to the + side of C1 and thus will result in a shorter path (see Fig. 6.15). Note that this path is partially routed through the body outline of C1. This is normal practice since the conductor path (in blue) is on the foil side of the board and the component body is on the component side, which prevents mechanical or electrical interference problems. Even though conductor paths are generally made to follow grid lines, long diagonal paths are acceptable as long as they contribute toward high circuit density and leave no large unused board areas. To complete this connection, a line is drawn to the base pad of Q1, which is perpendicular to the path between points 2 and 3. It is important to note that this path was positioned approximately midway between the bottom pads of R1 and R2 and the top pads of R3 and R4. Following is an explanation of this design cri-

**Figure 6.15** Routing conductor paths using the schematic as a guide.

terion. The conductor paths and terminal pads represent copper conductors which are separated by an insulating path or gap. As such, conductor path positioning is always planned so as to maintain as uniform an insulating gap as possible between adjacent paths and terminal pads as allowed by proper grid positioning. The insulating space between conductor paths and pads and between adjacent paths is of major concern in PCB layout. Needless crowding between conductors is to be avoided. A conductor path may have unequal insulation spaces on either side as it is routed between other conductors, but this should not result in a severe difference in their proportions. Conductor paths must not be routed along the very edge of the board. A space must be allowed between the outside edge of the path and the board edge to prevent damage to the path when the board is processed into its finished size.

The finished layout sketch is shown in Fig. 6.16 properly labeled and coded. Following is a summary of the basic steps used in conductor path routing:

1. Route conductor paths along vertical and horizontal grid lines.
2. Change path direction by using short 45-degree line segments.
3. Draw conductor path lines so that they extend slightly inside the pad.
4. Avoid crowding.
5. Begin path routing by drawing the shortest and most direct routes.
6. Keep track of each path drawn by checking it off on the circuit schematic.
7. Draw path to path intersections as perpendicular lines.
8. Keep angles of paths entering common pads equal to or larger than 90 degrees.

Because of the complexities and difficulties inherent in the design of printed circuits, a flow chart is provided in Fig. 6.17 to aid the technician by showing the logical sequence of steps necessary to generate a component and conductor layout sketch.

The next chapter is devoted to demonstrating computer-aided drafting using a computer system and a computer-aided design (CAD) software package to convert the com-

**Figure 6.16** Final component and conductor pattern layout sketch.

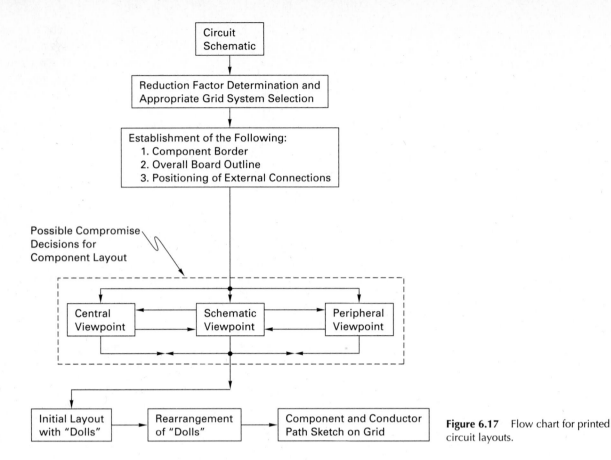

**Figure 6.17** Flow chart for printed circuit layouts.

ponent and conductor pattern layout sketch into highly accurate component layout and conductor pattern scaled drawings.

## EXERCISES

### A. Questions

6.1   What features are shown in a layout sketch?

6.2   Why is the layout sketch described as a composite view?

6.3   What are the recommended grid system and scale to be used for layout sketches?

6.4   Why are body outlines referred to as two-dimensional views?

6.5   Briefly describe the focus of each of the three viewpoints used to obtain optimum component positioning for routing conductor paths.

6.6   Why are terminal pads drawn on grid?

6.7   To what does the term *forbidden area* refer?

6.8   Give two examples of employing specific design criteria to make automatic assembly cost effective.

6.9   How much larger are the lead access hole diameters made for component assembly by automatic insertion than by manual insertion?

6.10   How are path entries to terminal pads and changes in path directions made?

## B. True or False

Circle T if the statement is true, or F if any part of the statement is false.

6.1      Layout sketches are drawn at a 2:1 scale on a 0.01 grid system.      **T   F**

6.2      Dolls are precut component outlines at a 2:1 scale that show the shape and size of the part with terminal pad requirements of location and size.      **T   F**

6.3      Resistor outlines include a keying symbol for proper orientation.      **T   F**

6.4      Conductor paths must not be drawn through component body outlines.      **T   F**

6.5      The component and conductor pattern layout allows both sides of the pc board to be viewed simultaneously.      **T   F**

6.6      A separate terminal pad must be provided for each component lead to be soldered to the conductor pattern.      **T   F**

6.7      Conductor paths are drawn only horizontally or vertically.      **T   F**

6.8      The width of the conductor path is determined by the amount of current it is to carry.      **T   F**

6.9      The angle between paths entering any terminal pad should not be less than 45 degrees.      **T   F**

6.10      Designing a pc board for automatic assembly uses the same criteria as those for manual assembly.      **T   F**

## C. Multiple Choice

Circle the correct answer for each statement.

6.1      The overall board outline is delineated by (*corner brackets, forbidden area*) on the layout sketch.

6.2      The recommended format for a single-sided layout sketch is a composite view of both sides of the board where the conductor side is considered to be the (*bottom, top*) view.

6.3      When positioning dolls, a (*balanced, symmetrical*) layout is achieved if there is a uniform distribution of components with no large areas of wasted space.

6.4      The recommended pad diameter for lead diameters between 0.020 and 0.040 inch is (*0.080, 0.10*) inch.

6.5      The on-center distance between terminal pads for a $\frac{1}{2}$-watt resistor when $X = 0.090$ inch and $B = 0.344$ inch is (*0.60, 1.10*) inch.

6.6      The outer circle shown in the top view of an installed turret terminal represents the terminal (*shoulder, pad*).

6.7      The metal tab that is used to key a TO5-style transistor is closest to the (*collector, emitter*) lead.

6.8      The initial location of an active component is best accomplished by using (*peripheral, central*) viewpoint.

6.9      As conductor paths are routed to terminal pads, direction changes should be made at (*45°, 90°*) angles.

6.10      The surrounding copper remaining after a terminal pad is drilled is called the (*annular ring, shoulder*).

## D. Matching Columns

Match each item in column A to the most appropriate item in column B.

| COLUMN A | | COLUMN B | |
|---|---|---|---|
| 1. | Grid system | a. | Axes |
| 2. | Precut component outlines | b. | Peripheral |
| 3. | Corner brackets | c. | Turret terminal |
| 4. | Keying | d. | Conductor path |
| 5. | Automatic insertion | e. | Doll |
| 6. | Shoulder | f. | +, −, metal tab |
| 7. | Routing | g. | Schematic |
| 8. | Balance | h. | 0.1 inch and 1.0 inch |
| 9. | Parts list | i. | Overall board outline |
| 10. | Viewpoint | j. | Component placement |

(a)

(b)

**Figure 6.18**  Seven-fixed-interval UJT timer.

**Figure 6.19** Series feedback regulator.

### E. Problems

6.1   Using the appropriate templates and poster paper, trace and cut 2:1 scale dolls for all resistors, capacitors, and the TO5-style transistor described in this chapter. Make a new layout sketch of Fig. 6.2 with all input and output terminals on the right side of the pc board. The order from top to bottom is to be IN, OUT, +12 Vdc and GND.

6.2   Draw the component layout and conductor pattern to a 2:1 scale for the circuit diagram shown in Fig. 6.18a. This circuit is a seven-fixed-interval UJT timer. The pin arrangement of the timer selector switch used in this circuit is shown in Fig. 6.18b. $L_1$, $S_8$, and the battery are to be chassis-mounted. External connections may be made to any board edge. The component layout design must be confined within a $2\frac{1}{2}$- by $4\frac{1}{2}$-inch border. Choose a suitable chassis, indicating lamp, and battery and battery clip styles for the design. An optional design may be considered that uses individual chassis-mounted time-interval selector switches. The seven time intervals for the circuit range from approximately 5 to 35 seconds. The pc board is fabricated in Problem 10.3 and assembled and soldered in Problems 16.3 and 17.3, respectively. Installation of the pc board into the chassis fabricated in Problem 23.4 along with the mounting of the on–off switch, indicating lamp, and battery clip is accomplished in Problem 25.4. All in-

**Figure 6.20** Tachometer.

terconnections among the chassis-mounted components and hardware and the pc board are soldered in Problem 26.5.

6.3 Construct a component layout and conductor pattern for the series feedback regulator circuit shown in Fig. 6.19. All input and output connections are to be located on one edge of the pc board and designed for finger-type connection. The maximum size of the pc board is to be 3 by $3\frac{1}{2}$ inches. The printed circuit is fabricated in Problem 10.4 with component assembly and soldering performed in Problems 16.4 and 17.4, respectively.

6.4 Design 2:1 scale component and conductor pattern layouts for the tachometer circuit given in Fig. 6.20. Maximum pc board size is to be $1\frac{1}{2}$ by 2 inches with provision for a suitable pc connector on one of the 2-inch sides of the board. Construct 2:1 artwork masters and marking and solder mask layouts for this design. All components and external connections are to be identified and keyed where necessary. The silk screens are constructed in Problem 10.2 and employed in Problem 10.5. Printed circuit board component assembly and soldering are undertaken in Problems 16.5 and 17.5, respectively. The pc board and meter are mounted in Problem 25.5. The harness for interconnections is constructed in Problem 26.6. An optional double-sided circuit design may be considered for this problem.

# Drawing PCB Artworks with Computer-Aided Drafting Techniques

## LEARNING OBJECTIVES

*Upon completion of this chapter on drawing printed circuit board artworks using computer-aided drafting techniques, the student should be able to*

- Access the main menu and establish the name for a new PCB component layout drawing and conductor pattern artwork.
- Select the SETTINGS submenu from the ROOT menu and establish the units, grid, snap, and limits for a new drawing.
- Enter the DRAW menu and use the LINE function to define the component border and corner brackets for a new PCB design.
- Use the DRAW menu functions to draw 1:1 scale two-dimensional component outlines for resistors, capacitors, devices, and integrated circuits.
- Use the TEXT and DTEXT editor to create and position component labels such as $R_1$, $C_2$, and $IC_1$.
- Draw a 1:1 scale component and conductor pattern artwork using layers to display, individually or as composite layer combinations, corner brackets with component layout, donut pattern for solder mask, and component marking mask artwork.
- Save and print the 1:1 scale artwork layers for PCB processing.

## 7.0 INTRODUCTION

The demand continues to grow for engineering drawings and PCB layouts to be done using computer-aided drafting (CAD) techniques. This industrial trend became prevalent in the early 1980s and the technology is now invading education. Two of the more popular software packages used in industry, and currently being adopted by some colleges and universities for their computer laboratories, are AutoCAD and VersaCAD. These software packages are very powerful commercially available programs with an astounding array of options, but they are relatively expensive.

To be compatible with this trend, we have selected AutoCAD Release 12 as the software package to illustrate this technology. It is used in this chapter to create the component layout drawing and conductor pattern artwork for the amplifier circuit designed in Chapter 6 and hand sketched in Figure 6.16. Using AutoCAD, we describe the primary operations of drawing lines, rectangles, circles, donuts, conductor patterns as well as scaling, labeling, making modifications, saving, and printing files.

## 7.1 INEXPENSIVE CAD SOFTWARE

The list of software packages for drafting is large and expanding. The selection process can be difficult. While a commercial software package like AutoCAD might be available at the university computer center, low-cost drafting packages are also needed for student home use. Writing in the Spring 1993 issue of the ASEE *Journal of Engineering Technology,* Professor Martin Pike of Purdue University Programs at Kokomo details the process he followed for selecting a low-cost drafting package in his article "Inexpensive CAD Software for Student Home Use." According to this article, studies have shown that many engineering students routinely purchase computers for their own unrestricted personal use and to do homework assignments. Most students select an IBM-compatible computer because of its flexibility and compatibility with university equipment. Furthermore, students often request information on low-cost, reasonable substitutes for AutoCAD and Versa-CAD that can be used to complete drafting assignments at home.

The conclusions from this article indicate that a shareware program called Draft Choice (at a cost of about $50 at this writing) and the student edition of a commercial program called AutoSketch, Version 3.0 (approximately $150), were good choices for student home use. Cited by Professor Pike as reasons for these selections are (1) the ease of installation, (2) standard dimensioning techniques for linear systems, and (3) an interactive menu system similar to AutoCAD and VersaCAD. However, if transferring files is important, AutoSketch is the choice, since its files can be saved in a file format that makes them transportable to a university-installed AutoCAD program.

Selecting the most appropriate CAD software package is becoming more difficult as the number of available packages increase. However, the selection and use of one of the available CAD software packages has today become a virtual necessity.

## 7.2 AUTOCAD SETUP FOR PRINTED CIRCUIT BOARD DESIGNS

To use AutoCAD to create a PCB component layout drawing, we must first access the main menu to begin a new drawing. A name is given to the new drawing at the appropriate main menu prompt. Then we proceed with a routine setup procedure for the drawing. The setup involves the selection of units, grid, snap, and limits. Each of these important first steps is considered in detail in this section on AutoCAD setup.

The authors understand that many institutions are using AutoCAD Release 10 and 11 as well as 12. For this reason, we have included Sec. 7.2.1 for users of Release 10 and 11. Those using Release 12 should begin with Sec. 7.2.3.

### 7.2.1 AutoCAD Main Menu
If you are using AutoCAD Release 10 or 11, access the AutoCAD subdirectory on the hard drive (C:) of your computer by keying in the following underlined keystrokes:

```
C:\CD\ACAD<Enter>
```

Then at the ACAD subdirectory prompt type:

```
C:\ACAD>ACAD<Enter>
```

You are presented first with the main menu for AutoCAD Release 10 and 11. This menu is shown in Fig. 7.1. Note that the main menu has a series of selections, most of which are self-explanatory. Selecting 0 allows you to exit AutoCAD. 1 to 4 are related to creating/editing/plotting new drawings, and items 5 to 9 are utility options. At the bottom of the main menu you are prompted to enter your selection.

```
                    ┌─────────────────────────────────────┐
                    │                                     │
                    │   Main Menu                         │
                    │                                     │
                    │        0.   Exit AutoCAD            │
                    │        1.   Begin a NEW drawing     │
                    │        2.   Edit an EXISTING drawing│
                    │        3.   Plot a drawing          │
                    │        4.   Printer Plot a drawing  │
                    │                                     │
                    │        5.   Configure AutoCAD       │
                    │        6.   File Utilities          │
                    │        7.   Compile shape/font description file │
                    │        8.   Convert old drawing file│
                    │        9.   Recover damaged drawing │
                    │                                     │
                    │   Enter selection:                  │
                    │                                     │
                    │                                     │
                    └─────────────────────────────────────┘
```

**Figure 7.1**   The main AutoCAD menu has 9 options. Selecting option 1 allows you to create, edit, and plot a new drawing. Options 5 to 9 are utilities options.

Since we wish to create a new drawing, we select item 1:

```
Enter selection:1<Enter>
```

The name of the new drawing is entered after the next prompt:

```
Enter NAME of Drawing:PCB-1<Enter>
```

Note that a valid name consists of no more than eight letters and/or numbers. AutoCAD next loads the drawing editor and presents a graphics screen and the screen menu.

**7.2.2 AutoCAD Graphics Screen**   The AutoCAD graphic screen and screen menu are shown in Fig. 7.2. Using a mouse, you can move up or down the screen menu (right side of the graphics screen) to highlight an item. Then pressing the mouse's left (pick) button will select the action you want AutoCAD to perform. This first screen menu is called the ROOT menu. Selecting any item from the ROOT menu will call a submenu of new items. We use the mouse to select SETTINGS from the ROOT menu in the next section to establish our new drawings (PCB-1) units, grid, snap, and limits.

The center of Fig. 7.2 is the graphic area and will display the PCB layout as the drawing is developed. At the top is a status line. This line identifies the layer currently being displayed, whether the ORTHO and SNAP functions (discussed in later sections) are active, and the exact X,Y coordinate location of the cursor.

At the bottom left edge of the graphics screen, a command line is shown. This command line is used to display and enter information.

**7.2.3 Setting Units, Grid, Snap, and Limits Using AutoCAD 12**   To begin a new drawing, type **NEW** at the command prompt:

```
Command: new <Enter>
AutoCAD Release 12 menu utilities loaded
```

followed by <Enter>. Type a file name, no more than 8 letters or numbers. When the dialogue box appears, notice that the cursor crossed lines change to an arrow pointer, which is moved with the mouse to make selections in the dialogue box. The filename selected in

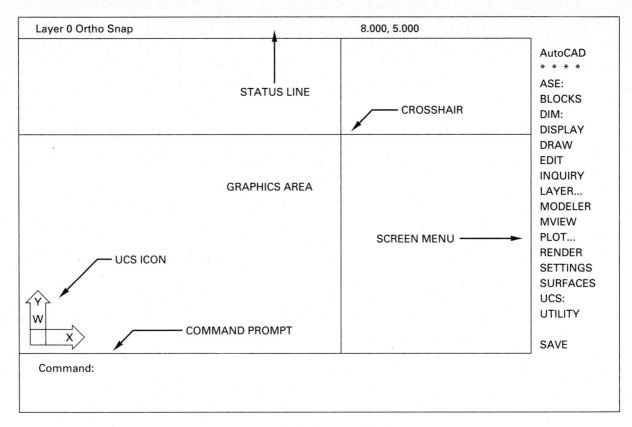

**Figure 7.2** AutoCAD's graphic screen includes a screen menu (right), status line (above), command prompt (below), user coordinate system (UCS, lower left), and graphics area.

Sec. 7.2.1, PCB-1, is used and appears beside the blinking bar in the New Drawing Name dialogue box displayed on the screen:

Command: PCB-1 <Enter>

Move the arrow pointer and click on the OK box with the mouse, or simply use an <Enter> (see Fig. 7.3).

The PCB-1 drawing will be printed on a standard sheet of $8\frac{1}{2}$ by 11-inch white paper. This paper will be oriented with its long axis along the horizontal. To establish the setting for PCB-1, highlight the SETTINGS submenu in the ROOT menu with the mouse

**Figure 7.3** Create New Drawing dialogue box in AutoCAD 12.

pointer and select it. In this text we identify a highlighted menu item with bold caps followed by <Enter> to indicate the mouse selection or pick button:

**SETTINGS**<Enter>

You are presented with the first page of the SETTINGS submenu. Since UNITS is on the next page of this submenu, highlight next:

**NEXT**<Enter>

**Units.** To format the units for PCB-1, highlight UNITS and pick it with your mouse:

**UNITS:**<Enter>

A series of options will be displayed, allowing us to tailor how AutoCAD interprets units. The first is the UNITS Report formats shown in Fig. 7.4a. Note that there are five options and that at the bottom command line, the default value is given in < > brackets. The <1>, indicating scientific units, was the default value. Since we want to express every dimension in decimals, we select 2 as follows:

```
Enter choice, 1 to 5 <1>: 2<Enter>
```

Next you are asked to select the degree of displayed precision. We enter 3 at the prompt

```
Number of digits to right of decimal point (0 to 8) <4>: 3<Enter>
```

(see Fig. 7.4a).

AutoCAD is capable of displaying angular measurements in several formats, as shown in Fig. 7.4c. The default is option 1, decimal degrees. Choose this method of measurement by keying in an <Enter> at the command prompt, since the displayed default value is also 1:

```
Enter choice, 1 to 5 <1>: <Enter>
```

Again we are asked to select the degree of displayed precision. We choose the default value 0, at the prompt, indicating that fractional degrees are not necessary (Fig. 7.4d).

Angles are measured with respect to a zero reference, with either a clockwise (CW) or counterclockwise (CCW) direction for positive angles. Normal convention dictates 0 degrees at 3 o'clock with positive angles measured CCW from that reference. We choose the normal convention in Figs. 7.4e and f, <0>: <Enter> and <N>: <Enter>, respectively.

**Grid and Snap.** At the command line prompt press <F1>. Then return to the first page of the SETTINGS submenu by first highlighting and then selecting LAST:

**LAST**<Enter>

Next highlight GRID and select it:

**GRID**<Enter>

From our discussion of grid selection (see Sec. 6.1) a 0.05-inch grid is suitable for the PCB-1 drawing to be designed in this chapter. Therefore, at the GRID command line

```
Command:  _UNITS Report formats:     (Examples)
        1.  Scientific          1.55E+01
        2.  Decimal             15.50
        3.  Engineering         1'–3.50"
        4.  Architectural       1'–3 1/2"
        5.  Fractional          15 1/2

With the exception of Engineering and Architectural formats,
these formats can be used with any basic unit of measurement.
For example, Decimal mode is perfect for metric units as well
as decimal English units.

Enter choice, 1 to 5 ⟨2⟩ :  ⟨Enter⟩
```

(a)  Decimal dimensioning is used for PCB designs.

```
Number of digits to right of decimal point (0 to 8) ⟨4⟩ : 3 ⟨Enter⟩
```

(b)  A displayed precision of 0.001 inches is suitable.

```
Systems of angle measure:              (Examples)
        1.  Decimal degrees             45.0000
        2.  Degrees/minutes/seconds     45d0'0"
        3.  Grads                       50.0000g
        4.  Radians                     0.7854r
        5.  Surveyor's units            N45d0'0"  E

Enter choice, 1 to 5 ⟨1⟩ :  ⟨Enter⟩
```

(c)  All angles will be in units of degrees.

```
Number of fractional places for display of angles (0 to 8) ⟨0⟩ : ⟨Enter⟩
```

(d)  Fractional degrees are not necessary for most basic PCB designs.

```
Direction for angle 0:
        East       3 o'clock  =  0
        North     12 o'clock  =  90
        West       9 o'clock  =  180
        South      6 o'clock  =  270
Enter direction for angle 0 ⟨0⟩ :  ⟨Enter⟩
```

(e)  All angles are measured from 3 o'clock reference.

```
Do you want angles measured clockwise? ⟨N⟩ :  ⟨Enter⟩
```

(f)  All angles are measured counterclockwise.

**Figure 7.4**  Units is one of the first necessary setups to begin a PCB drawing.

shown in Fig. 7.5a enter 0.05 followed by <Enter>. The grid is turned on or off using the F7 key.

SNAP is an AutoCAD function. It is the smallest dimension the cursor can resolve. If set to the grid setting above, the cursor can only be positioned on the grid, in both the X and Y directions. The snap is set to the grid spacing by typing 0.05 after the SNAP command line shown in Fig. 7.5b. To set SNAP equal to GRID, highlight the grd=snap function on the GRID menu:

**grd=snap**<Enter>

```
Command: '_GRID
Grid spacing(X) or ON/OFF/Snap/Aspect ⟨0.000⟩ : 0.05 ⟨Enter⟩
```

(a) Grid spacing is set at this command prompt.

```
Command: _SNAP
Snap spacing or ON/OFF/Aspect/Rotate/Style ⟨0.100⟩ : 0.05 ⟨Enter⟩
```

(b) Snap spacing is set at this command prompt.

**Figure 7.5** The Grid and Snap functions can be set equal using the grd=snap function.

Then at the command line enter 0.05 followed by <Enter>. The SNAP function can be switched on or off using function key <F9>. The command line on the graphics screen indicates the on or off state of the **SNAP** function, <Snap on> or <Snap off>.

**Limits.** Sizing the plotting paper completes the AutoCAD setup of a new drawing. Drawing size is set in the LIMITS submenu. Return to the first page of the SETTINGS submenu by highlighting LAST:

**LAST**<Enter>

Next highlight LIMITS and select it:

**LIMITS**<Enter>

Figure 7.6 illustrates that coordinates 0.000,0.000 identify the lower left corner of the paper. Press <Enter> and the upper right hand corner defaults to 12 by 9 inches, where the longer length is along the X axis. To resize to 4 by 4 inches, type 4,4 after the <12.000, 9.000> prompt. Function key F6 controls the X,Y coordinate information on the status line. This is done by entering Zoom, followed by entering All to activate the coordinates for the limits selected.

## 7.3 DRAWING THE COMPONENT BORDER AND CORNER BRACKETS

As detailed in Chapter 6, the size of a printed circuit's component border is determined after considering several factors. For the amplifier component and conductor pattern layout shown in Fig. 6.16, we need a 2.1- by 2.5-inch PCB to support the components and conductor pattern. The overall board edge and component border are defined by corner brackets. We begin the component layout drawing (PCB-1) for the amplifier by drawing the 2.1- by 2.5-inch rectangular component border and defining it with corner brackets.

```
Command: '_LIMITS
Reset Model space limits:
ON/OFF ⟨Lower left corner⟩ ⟨0.000,0.000⟩ : ⟨Enter⟩

Upper right corner ⟨12.000, 9.000⟩ : 4,4 ⟨Enter⟩
```

**Figure 7.6** Setting the limits in AutoCAD.

**7.3.1 Defining the Component Border**  Enter the DRAW submenu by first highlighting DRAW in the SETTINGS menu and then choose this selection with the mouse pick button:

```
DRAW:<Enter>
```

Then select LINE:

```
LINE:<Enter>
```

Using F6 and the COORDS displayed on the status line, we center our 2.5 by 2.1 component border in the 4- by 4-inch grid defined in the previous section. Using <Ortho on> (F8) and <Snap on> (F9), move the cursor until it is positioned at location 0.750, 1.000, as indicated on the status line. Set this point by using the pick button on the mouse. Next move to location 0.750, 3.100 and select this point with the mouse. You will note a straight line between these two points. Continue the component border construction by moving to location 3.250, 3.100 and then 3.250, 1.000, setting the points each time with the mouse pick button. Complete the box by moving back to the starting location 0.750, 1.000. Set this point with the pick button of the mouse and terminate the line construction with a space bar <SP>.

Figure 7.7a illustrates the component border just drawn with the coordinates of the corners shown. In the next section we position the corner brackets at the four corners of the component border.

**7.3.2 Defining the Corner Brackets**  To locate a 0.1-in.-wide, half- by half-inch corner bracket at each corner of the PCB we use the Zoom and Window features of Auto-CAD accessible from the pull-down VIEW menu. Moving the cursor up to the status line will display 9 menus. These pull-down menus are (1) FILE, (2) ASSIST, (3) DRAW, (4) CONSTRUCT, (5) MODIFY, (6) VIEW (7) SETTINGS, (8) RENDER, and (9) MODEL. Move the cursor to highlight the VIEW menu and select it with the mouse pick button:

```
VIEW <Enter>
```

Figure 7.7b shows the VIEW menu options for Zoom and from these pull-down menus select Zoom and Window with the mouse. The command line will prompt for the first corner. Locate the cursor at the lower left corner of a viewing window that will expose one corner of the component border. Press the mouse pick button. Then move to the upper right corner of the viewing window and select with the pick button (see Fig. 7.7c). The command lines for Zoom and Window are given in Fig. 7.7d.

To draw the corner brackets, enter the DRAW menu and again select LINE. Construct a half- by half-inch bracket, 0.1 inch wide, using the steps outlined in the last section on drawing a component border. To exit the Zoom and Window, again highlight the VIEW pull-down menu and select and pick Zoom and All to return to the full view.

Repeat the steps outlined in this section to construct corner brackets at the edges of the component border. Figure 7.7e shows the PCB component border with half-inch corner brackets.

## 7.4 DRAWING PCB COMPONENTS

The component and wiring sketch of the amplifier developed in the last chapter is now converted into a detailed 1:1 scaled component layout drawing. This two-dimensional drawing is viewed from the component side of the PCB and includes component bodies,

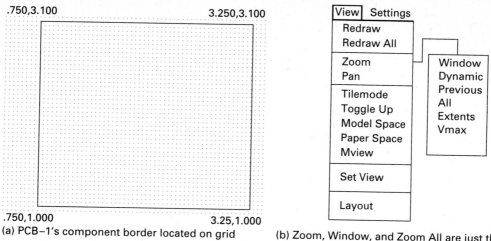

.750,3.100                3.250,3.100

.750,1.000                3.25,1.000

(a) PCB–1's component border located on grid

(b) Zoom, Window, and Zoom All are just three selections in the View pull-down menus

Other Corner
Zoom Window

First Corner

Component
Border

(c) Zoom and Window in the View menus

Command:  '_zoom
All/Center/Dynamic/Extents/Left/Previous/Vmax/Window/〈Scale(X/XP)〉 : _window
First corner:  Other corner:

(d) Command line for the Zoom and Window functions

(e) Component border (2.5 × 2.1 in.) with corner
    brackets (½ × ½ in. long × 0.1 in. wide)

**Figure 7.7**    The Line function in the DRAW menu is used to construct both the component border and corner brackets. Corner brackets are drawn using the DISPLAY menu's Zoom and Window functions.

lead access holes (circles with a diameter equal to the donut diameter of the conductor pattern artwork), and component and device labeling. Circuit wiring is also part of the component layout drawing. Using the component border developed in the last section and the sketch shown in Fig. 6.16, we begin by drawing the components in their approximate locations and to exact (1:1 scale) size.

### 7.4.1 Axial Lead Components

As an illustration, we draw a two-dimensional representation of a half-watt resistor (for this exercise, the snap will have to be changed to 0.025 inch). This component measures 0.15 inch in diameter, is 0.8 inch long, and the center of the lead access hole is 0.10 inches from the end of the body. Begin by entering the DRAW menu and calling the LINE function. Then using the procedure outlined in Section 7.3.1 draw a rectangle using the grid as a guide and the dimensions given above (see Fig. 7.8a). Next highlight and select the CIRCLE function in the DRAW menu:

**CIRCLE:**<Enter>

Select CEN, DIA: which indicates center of the circle and its diameter:

**CEN,DIA:**<Enter>

(a) Rectangular components are drawn with the LINE function

Command:__CIRCLE 3P/2P/TTR/〈Center point〉: Diameter/〈Radius〉:__D
Diameter: 0.1 〈Enter〉

(b) The CIRCLE command line prompts for the diameter of the circle

¹/₂–Watt Resistor Body

(c) Two–dimensional view of a ¹/₂–watt resistor

**Figure 7.8**  Rectangular components are generated using the Line and Circle functions in the DRAW menu.

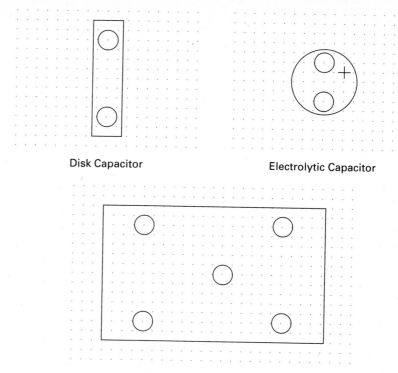

Disk Capacitor                                    Electrolytic Capacitor

Relay

**Figure 7.9** Several radial lead devices.

At the command line prompt, <center point>, locate the cursor in the center of the component body, 0.075 inch from either edge of the component body, and 0.10 inch from the end of the body. Press the mouse pick button. Then at the command line prompt for circle diameter, enter 0.1 followed by <Enter> (see Fig. 7.8b for the CIRCLE command line).

Repeat the drawing instructions for circles to add a second lead access hole circle. Figure 7.8c shows the final component configuration for a half-watt resistor.

**7.4.2 Radial Lead Components**    Figure 7.9 shows a two-dimensional view of several radial lead components. Note that they can all be adequately represented by drawing either a rectangle or circle for the body outline and circles to represent the terminal pad areas or donuts. Therefore, the techniques outlined in Sec. 7.4.1 can also be used to construct these components.

**7.4.3 Devices and Integrated Circuits**    Figure 7.10 shows the two-dimensional view of a few typical devices and integrated circuits (ICs). Note again that the bodies of these

**Figure 7.10** Some standard devices and IC packages.

8–Pin mini DIP                          TO–5 Transistor Outline

*7.4 Drawing PCB Components*

devices are represented by line segments and again circles are used to show the donuts. The student should gain some experience by duplicating the component and device outlines illustrated in this section.

### 7.4.4 Final Component Layout and Labeling

The AutoCAD drawing shown in Fig. 7.11a completes the component layout for Fig. 6.16. Note that all components and devices are positioned on grid. We next add descriptive labels to the components by using the text editor. At the command line prompt, type

`TEXT <Enter>`

You will be prompted to locate the starting point of the text:

`Justify/Style/<Start point>:`

Move the cursor to locate the lower left edge of the letter/number and select this point with the mouse pick button. When prompted for the letter/number height, it is recommended that you pick a minimum of 0.1 inch. Therefore, at the command line prompt, type

**Height<0.200>**`0.1 <Enter>`

Normally labeling is done at a zero rotation angle:

**Rotation angle <0>** `<Enter>`

The component layout drawing with all parts identified is shown in Fig. 7.11b. In the next section we add the conductor pattern or wiring traces.

### 7.4.5 Adding the Wiring Traces

Before adding the wiring traces on the component layout drawing check the status line for the condition of the ORTHO (orthogonal) function. This function can be switched on and off using the F8 function key. With the ORTHO function on, lines can be drawn only at right angles to one another and parallel to the edges of the working drawing. Switching the ORTHO function off, however, allows line segments to take any angle. Therefore, use F8 to switch ORTHO off before adding the wiring traces to the component layout drawing.

Figure 7.12 shows the finished component layout drawing for the amplifier circuit with all wire traces added. The illustrated wiring trace segment from point a to b and then to point c is completed by first entering the DRAW menu and selecting the LINE function. Then at the command line prompt

`Command: LINE From point:`

use the mouse to locate the donut center at point a and select it with the pick button. At the next command line prompt

`To point:`

move the cursor to point b, selecting it with the pick button. Finally, at the next prompt

`To point:`

again move the cursor, this time to point c, select it by pressing the pick button, and press the pick button one more time to terminate this trace segment. All wiring traces are added to the drawing in a similar fashion.

(a)

**Figure 7.11** Final component layout using AutoCAD: (a) components positioned; (b) components with labels.

(b)

**Figure 7.12** Component layout with conductor path routing complete.

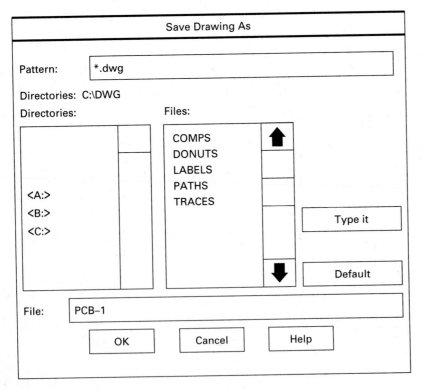

Save Drawing As

Pattern: *.dwg

Directories: C:\DWG
Directories:

Files:

COMPS
DONUTS
LABELS
PATHS
TRACES

<A:>
<B:>
<C:>

Type it

Default

File: PCB–1

OK    Cancel    Help

**Figure 7.13** Dialogue box used to save drawings.

## 7.5 SAVING THE COMPONENT LAYOUT DRAWING

During the length of time it takes to make a complete set of PCB drawings using Auto-CAD, the work should be periodically saved in the hard drive of your computer. This is to avoid losing all work completed due to either the computer hanging up or a power outage. The drawing is saved with the original filename that was specified when the new drawing was originated. It is possible to periodically save your work and remain in AutoCAD. At the command prompt, type SAVE followed by <Enter>:

```
Command: SAVE <Enter>
```

Depress the CTRL and C keys simultaneously to return to the command prompt if necessary. When SAVE is used, a dialogue box, as shown in Fig. 7.13 appears on the screen. The original filename, PCB, is typed at the command prompt and appears in the File: box. The arrow pointer is then moved to the OK box to return to the drawing.

## 7.6 CONDUCTOR PATTERN ARTWORK

Complete sets of printed circuit board drawings are made using AutoCAD layering. Layers can be thought of as a stack of transparent pieces of paper that are used to form a system of drawings. Each layer is drawn separately from any other layer. The layers can be viewed individually or in composite combinations. This gives the ability to maintain precise registration between all layers used. Layers can also be printed individually or in composite combinations.

To access Layers, pick Settings at the top of the screen. The pull-down menu shown in Fig. 7.14 will appear on the screen. Now pick Layer Control . . . in the menu and the Layer Control dialogue box shown in Fig. 7.15 will be displayed. Each layer to be drawn is given a name. The name is typed in the box with the blinking cursor, followed by an <Enter>. Once entered, pick the New box with the arrow pointer. The layer name will then appear in the Layer Name listing section of the Layer Control dialogue box. The lay-

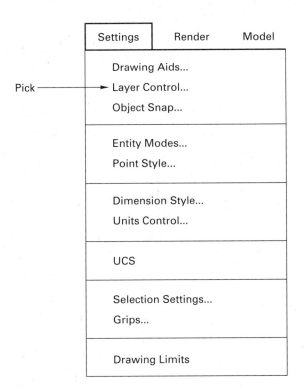

**Figure 7.14**  Settings pull-down menu.

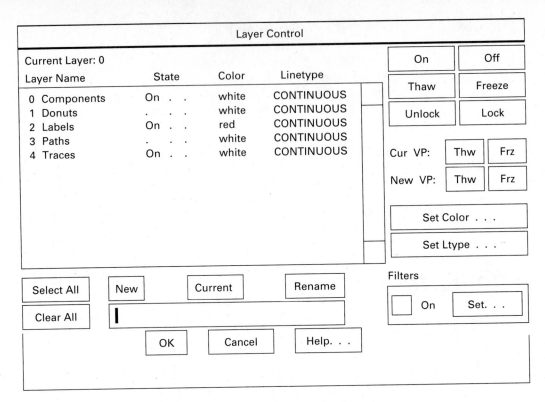

**Figure 7.15** Layer Control dialogue box.

ers to be viewed are turned on by picking the On box. Drawing entities will be produced only on the Current layer even with other layers on in a composite view. The Current layer must also be on for drawing entities to be seen on the screen as they are made. To make the layer to be drawn Current, select the layer name with the pick arrow. The name will become highlighted. Be sure that this layer is also turned on. Now pick the Current box. This returns you to the drawing screen with the Current layer name displayed at the left side of the status line. The component layout shown in Fig. 7.11a was drawn on layer 0. The labeling was produced on layer 2 with layer 0 on as a guide for positioning the literal detail. Therefore, Fig. 7.11b is the resulting display with layers 0 and 2 turned on. A composite print of these layers is often desired when a marking mask is required (see Sec. 7.8). Wiring traces are drawn on layer 4. The results in Fig. 7.12 are with layers 0, 2, and 4 turned on.

The conductor pattern artwork is the 1:1 scaled drawing that will be used to fabricate the pc board. It consists of 0.1-inch-wide solid (filled) corner brackets to define the finished size of the board, 0.1-inch-diameter donuts at every lead access hole, 0.125-inch-diameter donuts at all terminal pads, and 0.025-inch-wide conductor paths for the wiring traces. The donuts and the conductors will also be drawn on separate layers (layers 1 and 3, respectively). It should be noted that AutoCAD automatically arranges Layer names in alphabetical order.

**Donuts.** We begin to generate the conductor pattern artwork by first making a NEW layer called Donuts (layer 1) current. Layers 0, 2, and 4 are left on. Next enter the DRAW menu and highlight DONUT:

**DONUT:** <u>\<Enter\></u>

Then at the command prompt for inside diameter, enter zero, since we will be using solid donuts:

```
Inside diameter <0.5000>: 0 <Enter>
```

At the outside diameter prompt, enter 0.1 inch, the diameter selected for this project:

```
Outside diameter <1.0000>: 0.1 <Enter>
```

Then AutoCAD will prompt for the center of the donut. Using the mouse, locate the cursor at the center of one of the lead access holes and push the pick button to draw a solid, filled donut. Repeat this process until all of the lead access holes are filled. Then change the donut size to 0.125 inch for filling the terminal pads by again entering the Draw menu and selecting Donut to be prompted for donut size desired. Figure 7.16a shows all of the donuts located. The donuts are often placed on a separate layer in case a solder mask is to be made (see Sec. 7.8).

**Conductors.** To draw the conductors, a NEW layer called Paths (layer 3) is made Current. Layers 1 and 4 are turned on as a guide for drawing the paths to the correct donuts. Conductor patterns are created using the POLY line or PLINE function. This function is used to draw lines with any defined width. To draw a conductor path, select DRAW from the screen menu. Now pick NEXT near the bottom of the menu. The continuation of the DRAW menu is displayed. Finally, select **PLINE:** at the top of this menu:

```
PLINE:<Enter>
```

The software command line then prompts for the starting point. From Point: Use the mouse to locate the center and select one of the donuts requiring a conductor path. You will be prompted with the following command line:

```
Arc/Close/Halfwidth/Length/Undo/Width/<Endpoint of line>:
```

Since we want to specify the width of the poly line (conductor path), type <W> followed by <Enter>. When prompted for the starting and ending width, type 0.025 inches:

```
Starting width <0.0500> 0.025<Enter> Ending width
                 <0.0250> <Enter>
```

At this point you are again presented with the command line shown above, where <Endpoint of line> information is required. Use the mouse to locate the transition point of the line or the center of a donut. Selecting this location with the pick button will generate the line. Repeat this process until all conductor paths are drawn. Figure 7.16b shows the completed conductor pattern artwork for the amplifier circuit designed in the previous chapter.

It should be noted that any lettering that is to be produced in copper on the circuit side of the pc board must appear in backward lettering. This is due to the fact that the circuit side is viewed through the component side layer when the conductor pattern layout is generated. A reversal in orientation of this view is required when a photo is made of the conductor pattern artwork. The backward lettering ensures that the correct view is used when processing the board (see Fig. 7.16b).

To form backward lettering, first select the **DRAW** pull-down menu from the top of the screen. Select Text in this menu and then select Set Style from the smaller dialogue box that appears beside the DRAW menu. The Select Text Font dialogue box now appears on the screen. Roman Duplex and the OK box are picked with the arrow pointer to

(a)

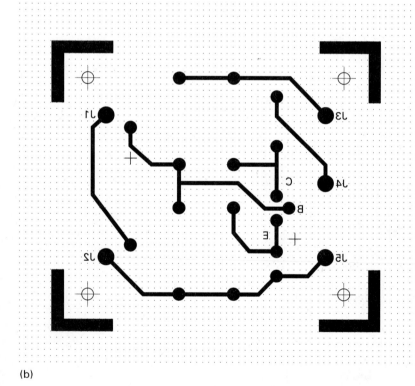

(b)

**Figure 7.16** Conductor pattern layout:
(a) donut layer with layers 0, 2, and 4
turned on; (b) conductor path layer 3
with donut layer 1 on and all other layers
turned off.

return to the drawing screen. The following are the continuous sequence of AutoCAD prompts and the response to each one after the <defaults>:

```
Command: '_style Text style name (or ?) <ROMAND>: romand
Existing style.
Font file <romand>: romand Height <0.0000>: .15 < Enter>
 Width factor <1.0000>: <Enter>

Obliquing angle <0>: <Enter>
Backwards? <N> y <Enter>
Upside-down? <N> <Enter>
Vertical? <N> <Enter>

ROMAND is now the current text style.
Regenerating drawing.
```

After entering Draw and DText from the menu, all characters will appear backward-reading.

**Corner Brackets.**  The sequence of steps outlined in the previous section to draw conductor paths is also used to draw the corner brackets. However, the width should be set to 0.1 in. for this detail (see Fig. 7.16b).

**Conductor Pattern Arrangements.**  When positioning conductor paths, several considerations need to be noted to avoid improper conductor pattern arrangements:

1. A separate terminal pad must be provided, with its center on grid, for each component lead to be soldered to the conductor patten. In cases in which large conductor areas are used instead of individual conductor pads, each component lead must still be provided with its own access hole.

2. Whenever possible, conductor paths should be along the vertical and horizontal lines of the grid system. Irregular paths are acceptable only if they simplify the conductor pattern.

3. Attention should be focused on the conductor width and spacing. The cross-sectional area of the conductor path is determined in part by the amount of current it will be expected to handle. As discussed in Chapter 5, once the foil thickness has been selected, the minimum conductor width allowable is determined as outlined in Sec. 5.3 with the use of Table 5.7. Once this width has been established, the minimum spacing between conductors must next be determined. Table 7.1 lists recommended conductor spacings for boards that will not be encapsulated (sealed) when completed. These spacings are *minimum* for a wide range of ac or dc voltages and take into consideration the *dielectric breakdown* voltage characteristics. These characteristics refer to that voltage at which flashover (arc) among conductor paths occurs. For optimum circuit reliability, the components should be positioned in such a manner that the conductor width and spacing are not less than the minimums specified throughout the entire conductor pattern. In this phase of layout, the design must not be compromised. Once these minimum values have been determined, any reduction in either one for the purpose of preferred component repositioning for convenience or for the sake of appearance is unacceptable.

An example that illustrates the use of Tables 5.7 and 7.1 for the purpose of selecting an appropriate conductor width and spacing follows. Assume the following specifications:

*Maximum continuous current:* 1.6 amperes
*Maximum voltage:* 36 volts

**TABLE 7.1**

Recommended Minimum Conductor Spacing

| Voltage Between Conductors Dc or Ac Peak (V) | Minimum Spacing |
|---|---|
| *Sea Level to 10,000 Feet* | |
| 0–150 | 0.025 inch |
| 151–300 | 0.050 inch |
| 301–500 | 0.100 inch |
| Greater than 500 | 0.0002 (inch per volt) |
| *Over 10,000 Feet* | |
| 0–50 | 0.025 inch |
| 51–100 | 0.060 inch |
| 101–170 | 0.125 inch |
| 171–250 | 0.250 inch |
| 251–500 | 0.500 inch |
| Greater than 500 | 0.001 (inch per volt) |

Table 5.7 shows that a conductor width of 0.030 inch will safely handle 1.9 amperes using 1-ounce copper foil. This width allows an acceptable margin of safety. The spacing for the specified voltage requirements can be determined from Table 7.1, which shows that the minimum spacing for a 0- to 150-volt range is 0.025 inch. Therefore, no conductor path on the finished board can be closer to any other path than this minimum. The solution to this problem, therefore, is

*Conductor width:* 0.031 inch

*Conductor spacing:* 0.025 inch

Notice that these specifications apply to the *finished* pc board at a 1:1 scale.

4. Finally, terminal area shapes and conductor path configurations must be evaluated so that only recommended artwork forms will be employed. Figure 7.17 shows various terminal shapes and path layouts with the preferred pattern configurations, along with those that should be avoided. The use of those configurations that are not recommended may introduce problems such as undesirable voltage distribution along ground paths, nonuniform solder flow onto the conductor pattern (especially at terminal areas), and weakening of the foil bond.

## 7.7 PRINTING THE CONDUCTOR PATTERN ARTWORK

To print the conductor pattern artwork layer, all other layers must be turned off. The PLOT command is entered at the command line:

```
Command: Plot <Enter>
```

The Plot dialogue box now appears on the screen as shown in Fig. 7.18a. Before attempting to plot the conductor pattern artwork, the following entries are established to obtain 1:1 scale plots on 8½- by 11-inch paper.

In the Paper Size and Orientation block, Inches is picked using the arrow pointer. Next the Size box is picked and a Paper Size subdialogue box is displayed on top of the Plot dialogue box (see Fig. 7.17b). Paper size A is selected followed by picking the OK box to return to the Plot dialogue box. Now pick the Rotation and Origin box, and their subdialogue box (Fig. 7.17c) appears. Pick 0 for rotation and 0.000 for both X and Y origins followed by clicking on OK. To obtain a 1:1 scale print, type the number 1 in both the Plotted Inches and Drawing Units boxes. By picking Full and then Preview, the posi-

| Recommended | | Not Recommended | |
|---|---|---|---|
| Pattern Design | Comments | Pattern Design | Comments |
| | Uniform distribution of solder around lead access hole | | Non-symmetrical pad area around lead access hole will result in a non-uniform solder flow |
| | Double filleted entries result in symmetrical solder fillets | | Non-uniform pad area will result in a non-uniform solder fillet |
| | Allows symmetrical solder fillets | | Uniform solder flow is prevented around outside edges of pad |
| | Pattern allows for symmetrical solder fillets | | Non-uniformity in solder flow due to large center area of pattern |
| | Proper path entry | | Inferior entry results in sharp conductor path edges detracting from appearance and hampering good solder flow |
| | 45° elbow maintains good foil bond | | Weakened foil bond due to sharp exterior edge of elbow which could create problems during etching |
| | For symmetrical soldering an individual access hole for each component lead is required | | Excessive pad area is removed to accommodate multiple leads resulting in foil bond weakening |
| | Shortest possible path length | | Excessive pad length and sharp angle at entries |
| | Lower voltage drops due to heavy buss | | High voltage drops along thin conductor paths |

**Figure 7.17**   Recommended conductor pattern configurations.

tion of the artwork will be shown. If the position needs to be changed, the origin values selected are simply changed. The Plot is now accomplished by making an <Enter>. Plotting commands automatically appear at the command prompts (see Fig. 7.19). However, another <Enter> has to be made after the instruction to Position Paper in Plotter.

## 7.8  SILK-SCREEN SOLDER AND MARKING MASKS

The procedures and techniques thus far discussed for designing component layouts and conductor pattern drawings represent the minimum requirements for producing pc boards. To provide for more efficient soldering, a *solder mask* may be employed. Solder masks consist of corner brackets and duplicate-size donuts located on the solder mask layer as

**Figure 7.18** Dialogue boxes: (a) Plot dialogue box; (b) Paper Size subdialogue box; (c) Rotation and Origin subdialogue box.

**Figure 7.19** Plotting commands that appear during printing of the conductor pattern artwork.

they appear on the conductor pattern layer. These donuts represent the terminal pads on the processed board to which component leads are soldered. The solder mask artwork is referred to as a *positive pad master*.

To form the solder mask over the conductor pattern on a processed pc board, a *silk screen* is necessary. This silk screen is prepared from a 1:1 scale photographic negative of just the donuts and their positions as they appear on the conductor pattern layout. Therefore, using AutoCAD, a donuts layer positive artwork is generated in addition to the component and conductor layers. The silk screen is produced by using the techniques considered in detail in Chapter 10. All of the areas seen as white on the positive solder mask artwork will be clear of film gelatin when the screen is processed. The silk screen solder mask becomes a negative artwork screening tool and must be of the same size as the donut layer. When the conductor side of the board is positioned under the silk screen, the application of an appropriate solder resist through the screen will coat all surface areas of the board, including the conductor paths with the solder stop-off film but will leave the terminal pads to be soldered exposed. A typical solder mask is shown in Fig. 7.20.

The solder mask will protect the conductor side of the board during and after the solder process since it will not be removed. This mask is applied *after* the board has been etched as described in Chapters 10 and 13. Solder will alloy only with the exposed bare copper terminal pads and provide sound electrical connections for component leads. A solder mask is used almost exclusively for automatic soldering methods such as wave soldering (described in Chapter 17). Since only the terminal pads are exposed, less solder is required, thereby reducing weight and cost. Without the mask, the entire conductor pattern would be alloyed with the solder.

A *marking mask* is employed to aid in component assembly. This mask shows component outlines, without terminal pads, their position, and identification numbers. When processed into a silk screen, the mask will allow this detail to be printed onto the component side of the pc board. The component outlines are drawn using AutoCAD techniques previously discussed in this chapter. Component identifications such as $R_1$, $R_2$, $C_1$, $C_2$,

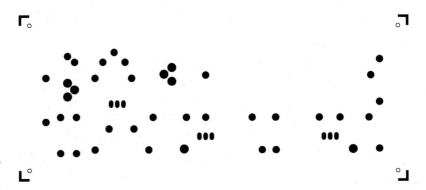

**Figure 7.20** Positive artwork solder mask.

**Figure 7.21** Positive artwork marking mask.

and $Q_1$, should be in accordance with the schematic diagram. Component outlines on the marking mask layer must also be the same size as those appearing on the component layer.

The marking mask is processed into a silk screen with the screen becoming a *negative* marking mask tool. The application of a suitable epoxy paint will produce the desired detail on the component side of the board. A completed marking mask for a pc board is shown in Fig. 7.21.

At this point, a differentiation should be noted between the lettering detail shown on the marking mask and that shown on the conductor pattern layer. Backward letter orientation, as previously mentioned, must be used on the conductor pattern layer. Forward-oriented lettering must be used on the marking mask layer to obtain the correct letter orientation when the component side of the board is screened through the marking mask. The application of letter symbols on both sides of the board aids in identifying components and rapidly locating test points when these needs arise.

The 1:1 scale artworks generated by the techniques discussed in this chapter are camera ready when printed and are reproduced into photographic negatives and positives by processes discussed in Chapter 9.

To exit AutoCAD, simply type QUIT at the command prompt.

## EXERCISES

### A. Questions

7.1    Explain how to access the main AutoCAD menu.

7.2    List the four settings that are made before starting a new drawing.

7.3    What keys are used to turn on or off Grid and Snap?

7.4    Describe the information that is displayed on the status line.

7.5    Outline the procedure for assigning a filename to a NEW drawing.

7.6 What are typical Grid and Snap settings for drawing a 1:1 scale layout on Auto-CAD?

7.7 How are individual drawing layers established?

7.8 Discuss the procedures for entering forward and backward lettering on a layer.

7.9 Describe the detail that normally appears on the component conductor, marking mask, and solder mask layers.

7.10 Discuss when work should be saved and how this is accomplished.


## B. True or False

Circle T if the statement is true, or F if any part of the statement is false.

| | | | |
|---|---|---|---|
| 7.1 | To begin a new drawing in AutoCAD, the Grid, Snap, Units, and Limits are set first. | **T** | **F** |
| 7.2 | No more than eight characters can be used for a filename. | **T** | **F** |
| 7.3 | Decimal dimensioning is used for pc board layouts. | **T** | **F** |
| 7.4 | Grid and Snap settings should be set to the same value. | **T** | **F** |
| 7.5 | Corner brackets are drawn inside the screen X and Y Limits setting. | **T** | **F** |
| 7.6 | Component outlines are drawn by selecting Pline from the DRAW menu. | **T** | **F** |
| 7.7 | Work is saved both periodically as the drawing is developed and when all layers are completed. | **T** | **F** |
| 7.8 | The inside diameter selected for donuts used to generate a conductor pattern and a solder mask is 0.010 inch. | **T** | **F** |
| 7.9 | A terminal pad must be provided for each component and device lead. | **T** | **F** |
| 7.10 | Marking masks reduce the amount of solder required on a pc board. | **T** | **F** |


## C. Multiple Choice

Circle the correct answer for each statement.

7.1 The status line is found at the (*top, bottom*) of the graphics screen.

7.2 When selecting the UNITS report, (*Decimal, Engineering*) format is chosen.

7.3 To turn the Grid on or off, the (*F6, F7*) key is depressed.

7.4 Drawing corner brackets is aided by using (*Zoom All, Zoom and Window*) from the View pull-down menu.

7.5 Components such as (*disk capacitors, resistors*) are constructed with their leads having axial orientation to the component body.

7.6 It is important to locate (*component bodies, terminal pads*) on grid when drawing the component layout.

7.7 DText is used when labeling on the conductor layer to obtain (*backward, forward*) letter orientation.

7.8 Drawing entities will be produced only on the (*Current, New*) layer.

7.9     The amount of (*current, voltage*) determines the minimum spacing between conductor paths.

7.10    The solder mask layout is a (*negative, positive*) artwork.


## D. Matching Columns

Match each item in column A to the most appropriate item in column B.

| COLUMN A | COLUMN B |
|---|---|
| 1. Settings | a. Backwards |
| 2. ZOOM | b. Solder mask |
| 3. DRAW menu | c. Limits |
| 4. Current | d. Save as |
| 5. DTEXT | e. Layer |
| 6. Voltage | f. Window |
| 7. PLOT | g. Resistor |
| 8. Donuts | h. Spacing |
| 9. Prototype | i. Pline |
| 10. Axial | j. Preview |


## E. Problems

7.1     Measure the body size, lead size, and lead spacing for several different components and devices. Convert your measurements to make 2:1 scale drafting aids using AutoCAD. Print a drafting aid for each part drawn.

**Figure 7.22**   Fourteen-pin dual-in-line test fixture.

7.2     Draw 1:1 scale component and conductor pattern layouts using AutoCAD of the hand sketch made for the tachometer in Problem 6.4. Make individual layers and print finished drawings.

7.3     Using AutoCAD, draw and print a 1:1 scale marking mask, including corner brackets, for the tachometer component layout generated in Problem 7.2

7.4     Draw a 1:1 scale donut layer for the tachometer using the conductor layer generated in Problem 7.2 and print the solder mask with corner brackets for the design.

7.5     Referring to the 14-pin dual-in-line test jig configuration shown in Fig. 7.22, draw a 1:1 scale conductor pattern layout of the device terminal pad layout and paths 1 through 14 with the donuts shown.

# 8 Double-Sided Plated-Through-Hole Printed Circuit Artwork Design

## LEARNING OBJECTIVES

*Upon completion of this chapter on artwork design and layout of double-sided plated-through-hole printed circuit boards, the student should be able to*

- Understand the basic differences between single-sided boards and double-sided plated-through-hole pc boards.
- Make interconnections between both sides of a double-sided pc board by using plated-through holes.
- Better understand how to use component bodies to bridge paths for greater flexibility.
- Create a circuit schematic using the EAGLE schematic module.
- Perform a computer-aided electrical rules check (ERC).
- Translate a schematic into a pc board layout using the EAGLE board layout editor.
- Create conductor paths using the EAGLE autorouter module.
- Understand the color code system adopted by industry to distinguish conductor patterns and other detail on both sides of a double-sided pc board layout.
- Perform a computer-aided design rules check (DRC).
- Output finished layouts or schematic.
- Become familiar with basic conductor pattern layout considerations for plated pc boards.
- Understand the purpose of including a thief area on a double-sided pc board layout for plated boards.

## 8.0 INTRODUCTION

As electronic systems have become more complex, the requirements of higher-density packaging have partially been met with the widespread use of double-sided printed circuit boards and the rapidly growing importance of multilayer and surface-mount pc boards (see Chapters 14 and 18). Double-sided board fabrication requires the preparation of at least two layouts, one for each side of the pc board. This, of course, adds to the time and costs of producing a double-sided board. Precise alignment of all terminal pads between both sides of the board is required for each lead access hole. This alignment problem is not a consideration in the preparation of single-sided boards since there is only one conductor pattern on one side of the board. In addition, double-sided boards require many more steps in fabrication, which increases their cost as compared to single-sided boards. Even though they are more costly and time-consuming to fabricate, double-sided pc boards allow for more components to be packaged in less space than single-sided boards and thus meet the demands for high-density electronic packaging.

## 8.1 BASIC DOUBLE-SIDED PRINTED CIRCUIT BOARDS

A double-sided pc board differs from a single-sided board in that both sides are processed with conductor patterns and terminal pads. The arrangement of conductor paths and pairs of terminal pads to form a double-sided pc board is shown in Fig. 8.1. As with single-sided boards, all components are mounted only on one side. This side is called the *component side* or *top*. Component leads pass through previously drilled and plated lead-access holes that have been processed through each pair of terminal pads on either side of the board. The component lead is *clinched* or bent over onto the terminal pad that is on the *circuit side* or *bottom* of the board. Note in Fig. 8.1 that the lead access hole drilled for the component lead also passes through a terminal pad on the component side of the board. Unlike single-sided designs, conductor paths may extend away from terminal pads on both sides of the board. Electrical interconnections between conductors on both sides of the board are made with *plated-through holes* (PTH) which become an integral part of the overall conductor pattern in the fabrication process.

When it is required only to make an interconnection between both sides of a double-

**Figure 8.1**  Double-sided PTH pc board conductor path and terminal pad arrangement.

sided pc board, a *via hole* is used. Also referred to as feed-throughs, via holes are not intended to accept component leads. For this reason, the terminal pads and drill hole sizes are small. Typically, a via hole pad size has a finished diameter of 0.050 inch with a drill hole size of 0.030 inch. The top and side view of a via hole would appear identical to that shown in Fig. 8.1 without the component lead and in Fig. 14.3 for a multilayer board.

The electrical interconnections between both sides of a pc board are formed by first drilling all lead-access holes in the bulk stock. Through a series of processes in chemical baths, all these holes are then plated with copper to form a continuous electrical connection between the copper surfaces of both sides of the board. Each copper surface is then imaged with the intended conductor pattern design. Further chemical processes result in additional copper and then solder to be plated on all the drilled hole walls. These processes, as well as those required to complete a double-sided plated-through-hole pc board, are discussed in detail in Chapters 11, 12, and 13.

The circuit side of the board can be considered a conventional single-sided layout. The conductors and terminal pads on the component side serve to extend the amount of conductor path routing. With both sides of the board used to route electrical interconnections, less space is required for conductor paths on the bottom side, thus allowing for greater component density.

One of the considerations involved in the fabrication of a double-sided pc board is the component bodies that must rest on the conductor paths which are on the top of the board. Most component and device cases, such as resistors, capacitors, plastic-cased transistors, and integrated circuits, are electrically insulated from their leads. The bodies of these components may safely contact a conductor path without electrical interference. However, some metal-cased integrated circuits and transistors have one lead attached to the case. These packages must not be allowed to come into contact with any of the conductor paths on the top side of the pc board. Electrical contact is avoided by mounting these components slightly above the board surface to create an insulating air gap.

## 8.2 DOUBLE-SIDED PC BOARD LAYOUT

The generation of a 1:1 scale component and conductor pattern layout for a double-sided board is accomplished by an entirely different manner than those procedures described in Chapters 6 and 7. The major differences are the elimination of using layout dolls, the elimination of making layout sketches by hand, and the substitution of a software drawing package by CadSoft Computer, Inc. called EAGLE 2.6 in place of AutoCAD.

EAGLE 2.6 is capable of producing 1:1 scale layouts for extremely complex single-sided, double-sided, and multilayer pc boards. However, a basic relaxation oscillator circuit, shown in Fig. 8.2, will be used as the vehicle to allow a greater focus on the manipulation of this software drawing package while gaining an in-depth understanding of layout concepts.

## 8.3 EAGLE 2.6 SOFTWARE FOR PC BOARD LAYOUT

EAGLE 2.6 is a powerful software drawing package optimized for printed circuit board layout and design. The complete package supplied by CadSoft Computer, Inc. contains a *schematic module,* board *layout editor,* and an *autorouter module.* Also available as part of the software is a large selection of components and devices, stored in a series of libraries as either *package macros* for pc board layouts or *schematic symbols* for schematic drawings. EAGLE 2.6 also supports up to 255 drawing layers. Layers below 100 have been predefined by the software manufacturer. For example, layer 1 represents the conductor paths called *tracts* on the *top* (component) side and layer 16 the *tracts* on the *bottom* (circuit) side of a pc layout. Some layers are assigned functions such as, *silk screen*

**Figure 8.2** Hand sketch of a basic relaxation oscillator circuit.

*top side* (layer 21), *silk screen bottom side* (layer 22), *pads* (layer 17), and *vias* (layer 18). Layers above 100 can be defined by the user with the **LAYER** command.

### 8.3.1 The EAGLE 2.6 Main Menu

Figure 8.3 illustrates EAGLE's *main menu,* containing a group of some of the available commands. Any action in EAGLE is started by calling one of these commands. Commands can be entered via the keyboard or selected with a mouse by first highlighting the command and then pressing the select (left) button. In certain EAGLE commands, pop-up menus are activated from which parameters for the initial command can be selected.

Twenty-four of the most common commands are displayed on the right side of the screen when the main menu is in view and the semicolon (;), highlighted in red, is the main terminator within the EAGLE software program. A main command line, indicating both dimensional units (*inch*) and drawing environment (*Board: UNNAMED*), appears at the top of the screen, and a cross (+) identifies the lower left corner of the drawing.

### 8.3.2 Creating a Circuit Schematic Using the Eagle Schematic Module

The relaxation oscillator shown in Figure 8.2 will be used to illustrate how a circuit schematic is generated in EAGLE and then how pc board component and conductor pattern layouts are created from the schematic. Create a new schematic by first highlighting the **EDIT** command within the main menu and selecting it with the mouse. From the pop-up menu, select first **SCHEMATIC** and then **NEW.** At the prompt, enter 555 Timer as the name for this new file, followed by a carriage return <CR>:

**enter name:** <u>555 Timer <CR></u>

The command line of the main menu will be updated to

**Sheet: 555 TIME.1/1**

indicating the first page of a schematic with 555 TIME as the file name.

To place schematic symbols for a schematic on the drawing, you must first choose the appropriate library in which the symbol resides. There are some 37 individual libraries

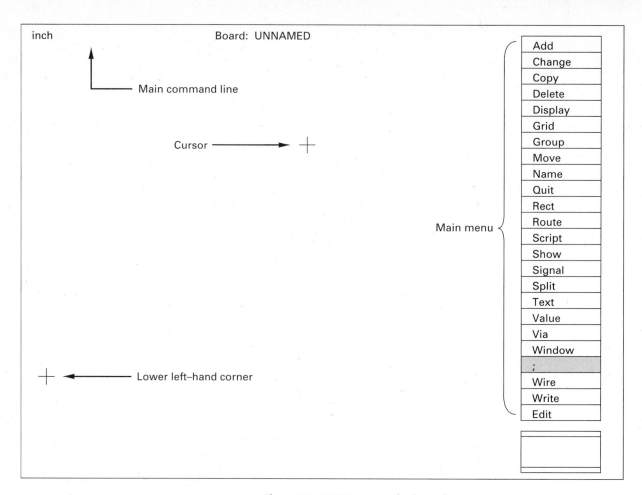

**Figure 8.3**  EAGLE start-up display with main menu.

in EAGLE 2.6. Each library is grouped according to function (ECL, MEMORY, 68000, etc.) or manufacturer (Maxim, Zilog, etc.). For the relaxation oscillator of Fig. 8.2 we find the 555 timer symbol in the **DEMO** library and the three resistors and two capacitor symbols in the **DISCRETE** library listed as **RESUS-7,5** and **CAP-7,5,** respectively. The **+VCC** and ground (**GNDA**) symbols are found in the **SUPPLY** library.

To select and place a resistor symbol, such as R1, first type **USE** to display the pop-up menu of libraries. Select **DISCRETE** to load this library and then highlight the **ADD** command from the main menu and choose **RESUS-7,5** with the mouse. As you move the mouse onto the screen you will drag the symbol for R1 along. If desired, rotate the symbol 90 degrees counterclockwise by pressing the right mouse button once, 180 degrees by pressing twice, etc. When the resistor is positioned correctly, press the select (left) mouse button. Repeat this procedure of library selection (**USE**), symbol selection (**ADD**), and symbol positioning (**MOVE, ROTATE**) until all elements are located on the drawing.

To label the value of each element, highlight and select the **VALUE** command. Then use the left mouse button to select the symbol and at the pop-up prompt, type the component value. For example, the value of R1 can be set to 10K by

```
R1: RESUS-7,5 10K <CR>
```

This procedure can be repeated to rename elements using the **NAME** command.

Electrically connecting the symbolic elements with the wiring (nets of airwire) requires the use of the **NET** and **JUNCTION** commands. Using the mouse, first highlight and then select the **NET** command from the main menu. Click on the lead end of an element and drag the mouse until it is positioned where desired and then double-click to terminate. Repeat until all elements are wired electrically as desired. To form the nets and component leads into connections, highlight the **JUNCTION** command. At the intersection of each net and component lead (pins), click the mouse select button to establish an electrical connection. Figure 8.4 illustrates the final schematic layout of the relaxation oscillator generated with the EAGLE 2.6 schematic module.

The final step in developing a circuit schematic is to perform an electrical rules check (ERC) to look for open pins and connections, etc. By typing **ERC,** EAGLE will check and display errors and warnings. The results of an ERC test are displayed below the main command line. A typical report might read as follows:

```
inch                        Sheet: 555 TIME.1/1
ERC: finished. 0 errors, 1 warning (C:\EAGLE\555 TIME.ERC)
```

**8.3.3 Translating a Schematic Into a PC Board Layout**    The circuit schematic developed in the last section can be converted into a pc board layout using EAGLE's **BOARD** command and the layout rules discussed in Section 8.2. Begin the process by first typing the **BOARD** command. All the components, along with their airwire connections (**RATS-NEST**), will appear beside a blank pc board outline (see Fig. 8.5a).

Then highlight the **MOVE** command and with the mouse, select each component

**Figure 8.4**    Finalized schematic of relaxation oscillator.

**Figure 8.5** Translating schematic into board: (a) EAGLE board command start-up display; (b) optimized parts placement with RATSNEST.

individually and move it into the pc board outline. When each component is approximately positioned, click on the select button. Repeat this process until all parts are within the pc board outline.

To obtain a more detailed image of all the parts and their relative location, select **WINDOW** from the main menu. Use the mouse to locate (left button) the lower left and upper right corners of the whole image containing all the parts. Then selecting or typing the semicolon (;) will window the image to best highlight all the parts (see Fig. 8.5b). Use the **MOVE** and **ROTATE** commands and employ the central, schematic, and peripheral viewpoints discussed in Chapter 6 to optimize the parts placement within the pc border. Once parts placement is optimized, the net positions are optimized by using the **RATS-NEST** command (see Fig. 8.5b).

### 8.3.4 Creating the Conductor Paths with an Autorouter

The autorouter module supplied with the EAGLE 2.6 software employs a ripup and retry algorithm that tends to create PCB layouts that are 100% routed without additional commands or any other manual intervention. This router can be configured to minimize the number of vias used on a double-sided board and also smooth the traces.

The autorouter is accessed by typing **AUTO;** then from the **Autorouter Setup** menu, select **Start.** Figure 8.6 shows the complete autorouted layout for the relaxation os-

**Figure 8.6** Composite view of double-sided pc board layout of relaxation oscillator: (a) top (red layer) and bottom (blue layer) track patterns; (b) legend for interpreting layer information.

cillator. Although individual layers can be turned on or off, the screen display of the autorouted pc board appears in Fig. 8.6a as a composite view with all of the layers turned on. A multicolor coding system has been adopted by industry to avoid confusion and misinterpretation of track patterns on different layers. The interpretation when observing the composite view is as follows: The layout is a top view with the first layer being that of the components. The tracks under the parts on the component side are shown with their pad connections and are presented in the color red, forming the next layer of the composite view. Still viewing the board from the top, the track pattern and pad connections on the bottom of the pc board would be "seen" as though one were looking through the board. Red and blue tracks may "cross" each other since each color represents conductor paths on different layers separated by the insulating board material. Because of the absence of colored illustrations in this book, a legend is given in Fig. 8.6b to aid in the interpretation of the autorouted layout shown in Fig. 8.6a. Blue tracks are shown as solid black paths and red tracks are shown as dotted paths.

Additional features such as corner brackets and mounting holes would be added. Nomenclature on the top side of the board is shown with forward reading orientation. Some pc boards may require labeling on the bottom side of the board. If any nomenclature is added to the bottom side of the board, it must be added in backward reading orientation on that layer to obtain forward orientation when viewing the actual bottom side of the pc board.

The final layout should be checked for spacing errors, short circuits, track widths, pad diameters, etc., using the design rule check (**DRC**) command. The results of a DRC test are displayed below the main command line. A typical report might read as follows:

```
inch                           Sheet: 555 TIME
DRC: Top: finished. no errors!
DRC: Bottom: finished. no errors!
```

**8.3.5 Creating Outputs with XPLOTS**   A program called **XPLOT** is used to output finished layouts or schematics. **XPLOT** allows the plotting of any drawing layer individually or in combination with other layers. **XPLOT** is invoked only from the DOS prompt. Therefore use the **WRITE** command to save the layout and **QUIT** to the DOS command prompt. Then type

```
XPLOT options filename layer
```

It should be noted that the information presented here on EAGLE's 2.6 software is the barest minimum information necessary to illustrate its use on a simple pc board layout. This software is quite powerful and can best be learned by reference to the CadSoft Company, Inc. user manuals. Information pertaining to requirements regarding conductor layers and patterns for multilayer pc boards and boards employing surface mount devices (SMT) is presented in Chapters 14 and 18, respectively.

## 8.4 LAYOUT CONSIDERATIONS FOR PLATING

Chapter 13 describes the process of pattern-plating both copper and 60/40 tin–lead alloy onto the conductor path areas of both sides of the pc board. Many of the problems associated with an improper balance of plated metals can be attributed to the component and conductor pattern layouts. If a double-sided pc board is to be electroplated to a uniform thickness and proper alloy, it must meet the following specifications:

1. The total conductor pattern areas to be plated on each side of the board must be as close to equal as possible.
2. The conductor pattern on a given side of the board must be uniformly and evenly distributed throughout.

Strict adherence to the specifications above is essential if the correct 60/40 tin–lead alloy is to be deposited during the plating process. An imbalance of conductor pattern areas between the two sides results in the side with the less area to be tin-rich-plated and the side with more area to be lead-rich-plated. If the ratio of alloys plated on either side is significantly different from the optimum 60/40 ratio, proper soldering of the board will not be realized.

If the conductor pattern is not uniformly and evenly distributed across either side of the board, the area with the heavy concentration of copper conductors will be low in tin-content plating, while the area with the light concentration of copper conductors will be plated with an alloy having high tin content. This will again result in soldering difficulties.

One final provision that must be included in the conductor pattern layout phase for boards that will be electroplated is an area of solid copper just outside the corner brackets of the pc board. This area is called the *thief* or *robber* area and its purpose is to direct some of the high current density at the edge of the board toward itself during the plating process. This will result in a more consistent 60/40 alloy on both sides of the pc board.

The thief area is included in the conductor pattern layouts and appears as a wide border just outside the corner brackets. At 1:1 scale, the thief area on the CAD artworks should be from half to 1 inch wide. Figure 8.7 shows the conductor patterns for both sides of the pc board with the inclusion of the thief area.

With the completion of printing the track patterns and addition of the thief areas, we can describe the sequential processing steps that are performed to fabricate a double-sided, plated-through pc board. Because these processes are involved and require a great deal of detailed explanation, they will be spread over four chapters. Chapter 9 begins the process by converting the printed track layers into silver halide film negatives and positives and diazo film positive phototools. Chapter 11 presents the drilling of the bulk stock and its preparation for copper deposition followed by a flash-plating process. Chapter 12 discusses the conductor pattern imaging process using dry film photo resist and the diazo phototools. Chapter 13 completes the fabrication of the double-sided plated-through-hole pc board by considering copper pattern plating, solder plating, photo resist stripping, etching, and infrared solder reflow.

The material in this unit provided the information necessary to design both single-

**Figure 8.7** Printed track patterns with the addition of thief areas.

sided and double-sided printed circuit boards, and two new software packages—Auto-CAD, Release 12, and EAGLE, Version 2.6—were presented to develop the conductor pattern artwork masters needed for fabrication. Next, in Unit 3, information is given on board fabrication. Single-sided, double-sided, and multilayer printed circuit boards are covered, as well as various accepted inspection techniques to assure quality.

## EXERCISES

### A. Questions

8.1     Describe specifically how interconnections are made between the circuit side and the component side of a double-sided plated-through-hole board.

8.2     What is the color-coding scheme for double-sided conductor pattern layouts?

8.3     When placing literal and numerical information on double-sided layouts, what is the convention in regard to foreward- and backward-reading directions?

8.4     What command is used to have an automatic electrical check made of the schematic diagram drawn on the screen?

8.5     To what does the term NET refer?

8.6     After the schematic is translated into an optimized component layout, what command optimizes the net positions?

8.7     When autorouting is completed, what command obtains an automatic test of all mechanical and electrical requirements being met?

8.8     To achieve uniform plating, what two layout specifications must be adhered to in circuit board design?

8.9     What is the purpose of the thief area?

8.10    What might be a possible cause of tin-rich plating?

### B. True or False

Circle T if the statement is true, or F if any part of the statement is false.

8.1     The bottom side of a double-sided pc board is also referred to as the circuit side.      **T    F**

8.2     A PTH forms a continuous metal path between copper on the top and bottom sides of a double-sided pc board.      **T    F**

8.3     Bodies of components, such as resistors and integrated circuits, must not contact conductor paths on a pc board.      **T    F**

8.4     On a composite view of both sides of a double-sided pc board, conductor paths must not be allowed to cross each other.      **T    F**

8.5     When using CAD, component and conductor pattern layouts must be drawn to a 2:1 scale.      **T    F**

8.6     In the layer color coding system for double-sided pc boards, RED represents the tracks on the component side.      **T    F**

8.7     Nets are optimized using the electrical rules check after autorouting is completed.      **T    F**

8.8     The purpose of the thief area is to direct some of the high current density at the edge of the board toward itself during the plating process.      **T    F**

## C. Multiple Choice

Circle the correct answer for each statement.

8.1    All the literal and numerical information should be (*foreward-*, *backward-*) reading on the component side and (*foreward-*, *backward-*) reading on the circuit side of a double-sided pc board.

8.2    The circuit side of a double-sided pc board is the (*top, bottom*) side.

8.3    Airwire positions are optimized by using the command (*ERC, RATSNEST*).

8.4    Tracks having the color (*blue, red*) represent conductor paths on the circuit side of the board.

8.5    An area of a pc board having a heavy concentration of conductor paths will receive a (*high, low*) tin content when solder plated.

## D. Matching Columns

Match each item in column A to the most appropriate item in column B.

| COLUMN A | COLUMN B |
|---|---|
| 1.  Thief area | a.  Output |
| 2.  Airwire | b.  PTH |
| 3.  Circuit side | c.  Track |
| 4.  Via | d.  Junction |
| 5.  Autoroute | e.  Short circuit |
| 6.  XPLOT | f.  Plating |
| 7.  DRC | g.  Blue |

# Unit
# THREE

# Printed Circuit Board Fabrication and Inspection

# 9 PCB Photographic Processes

## LEARNING OBJECTIVES

*Upon completion of this chapter on photographic processing, the student should be able to*

- Become familiar with the operation of a typical copy camera for photographically reducing printed circuit artworks into phototools.
- Expose and process photographic film.
- Use the 21-step exposure table in developing film.
- Use the technique of contact printing to generate positives from reduced negatives.
- Produce diazo phototools for proper double-sided registration.
- Use a densitometer on processed film.

## 9.0 INTRODUCTION

Until recently, 2:1 and 4:1 scaled *taped artwork masters* were produced by manual means on clear acetate sheets with opaque tape. Today most new PCB designs are generated using CAD software packages such as those illustrated in Chapters 7 and 8 of Unit 2. The *artwork master* that results from these new software tools is almost always a 1:1 scaled conductor pattern. Most CAD software is capable of producing negative phototools directly. If a Gerber-type photoplotter is available, all software can produce a high-quality 1:1 scaled positive print (black conductor pattern on white paper) with a laser printer.

The best method to photographically process old 2:1 or 4:1 scaled taped artwork masters or a new 1:1 scaled CAD-generated artwork master into *working tools* to be used to fabricate the finished printed circuit board is with a copy camera. The resulting positive or negative photographic films are commonly known as *phototools*.

This chapter begins with a discussion of the setup, use, and adjustments of a typical copy camera for photographically reducing old taped artworks or processing new CAD artworks into phototools. This is followed by a presentation of films, chemicals, and bath preparations as well as film processing. Also included is the technique of 1:1 scaled contact printing to obtain image reversals. The chapter concludes with information on diazo phototools for double-sided printed circuit board fabrication.

The accuracy and quality of the phototools described in this chapter will determine the quality of the resulting printed circuit board. For this reason, this chapter is sufficiently detailed so as to emphasize careful attention to each of the processes described. Properly prepared phototools can avoid pc board fabrication problems.

***Caution:*** *A chemical hazard exists in the materials presented in this chapter. Approved face shields (goggles)* and *chemical-resistant gloves, clothing, and footware must be worn prior to performing any of these operations.* Also, the section on safety outlined at the beginning of the book should be reviewed for more specific information.

## 9.1 PHOTOGRAPHIC IMAGING OF THE ARTWORK MASTERS

Commercial photocopy cameras are capable of accepting artwork masters in excess of 48 by 60 inches. These masters may be reduced into positives or negatives by a factor of 5:1, 4:1, 2:1, or 1:1. This reduction is dependent on the type of lens available on the camera. One-to-one reversals of CAD-generated artwork masters can also be made with the copy camera.

Artwork reductions made on a copy camera can be produced to a tolerance of ±0.002 inch, although ±0.005 inch is the tolerance typically specified. The results of artwork exposure with a copy camera have high-quality line resolution, high contrast between light and dark areas, and negligible imperfections such as pinholes (voids) in the opaque areas of the processed film.

A typical commercial copy camera used for artwork reduction in the printed circuit board industry is shown in Fig. 9.1. This is a nuArc Copy Camera Model SST1418, which is capable of handling a film image size of up to 14 by 18 inches and artworks as large as 21 by 25 inches if front-lighted and 15 by 19 inches for back-lighted work. When used with the appropriate film, this camera will provide excellent-quality exposures that are ideally suited for small- to medium-size pc boards.

The nuArc Model SST1418 copy camera will be described here to illustrate photographic procedures and typical camera operation. Other cameras used for printed circuit photography will involve similar operating procedures but may have distinctive differences so as to require the operator to consult the manufacturer's instructions for any operating variations. The nuArc camera has eight basic components: (1) transparent glass copyboard with cover glass and translucent rear panel for backlighting; (2) four 800-watt quartz iodide lamps; (3) lens with diaphram control, electric shutter, and cover; (4) bellows; (5) vacuum-frame film holder with vacuum pump and foot switch; (6) focusing plate; (7) two scale-adjustment control wheels; and (8) control console with focus, scale illumination, and master switches and lamp timer.

To operate this camera for reducing old taped artwork masters, the procedure is as follows. First, the lens cover is removed. The copyboard is then rotated to the horizontal position for artwork loading. This is done by pulling out the frame release handle and gently lowering the frame. The cover glass is then unlatched and raised until the internal lock in the cover support arm engages to hold the cover open. The artwork to be processed is placed onto the transparent base of the copyboard and centered (see Fig. 9.2). To aid in obtaining consistent uniformly developed exposures, a transparent-to-opaque *photographic step exposure table* is placed over a clear area of the artwork but outside the corner brackets so that is does not become a part of the resultant phototool. This is also shown in Fig. 9.2. (A detailed discussion of the use of a step table is given in Sec. 9.2, on processing film.)

With the artwork, together with the step table, positioned on the copyboard, the cover glass is lowered and locked into place. Raising the copyboard to its locked vertical position will establish the correct orientation for exposure. For maximum contrast between light and dark areas of the artwork, backlighting is the preferred method. This is shown in Fig. 9.3. Note that the lamps are directed at approximately 45-degree angles to the rear translucent light-diffusing panel of the copyboard. With this arrangement, the artwork is uniformly and brilliantly illuminated without *hot spots* or reflections during film exposure. For new CAD-generated artwork masters, the lamps are directed at approximately 45-degree angles to the front of the copyboard. This is necessary since the artwork master is usually printed on white paper.

All adjustments for photoreduction are done at the rear of the camera. Power to the control console is first applied by depressing the tops of the *master power* rocker switch and the *scale tape light* switch. The scale reduction settings are then made. Also located at the rear of the camera are two control wheels for adjusting both the reduction scale and the focus of the artwork image. One control wheel is labeled *lensboard* and the other is la-

**Figure 9.1** Horizontal copy camera. *Courtesy of nuArc Company, Incorporated.*

beled *copyboard*. Scale adjustment using these control wheels is made in conjunction with scale markings on two stainless steel calibration tapes that are viewed through 2 × magnifying lenses located at the center of the control console above the wheels. To adjust for a 2:1 scale reduction, the *copyboard* control wheel is rotated until the accompanying calibration tape reads *50*. Clockwise rotation of this wheel moves the copyboard *away* from the lens while a counterclockwise rotation moves it *toward* the lens. The *lensboard* control wheel is then adjusted so that its calibration tape also reads *50*. Clockwise rotation of this wheel moves the lens *toward* the copyboard and counterclockwise rotation moves it *away* from the copyboard. With both calibration tapes set at *50*, the camera is set for a 2:1 artwork reduction. The controls used for setting the camera to the desired reduction scale are shown in Fig. 9.4. Note that new CAD-generated artwork masters, printed at a 1:1 scale, require both tapes to be set at 100.

To check the accuracy of the tape setting as well as the focus, the vacuum frame film holder is first opened by releasing the latch. The frame is then lowered to the horizontal position. The ground-glass focusing plate, which is stored above the film holder, is next unlatched and swung downward until it locks into the viewing position (see Fig. 9.5). Under safe light conditions, a visual check of the image may now be made. By depressing

*PCB Photographic Processes*

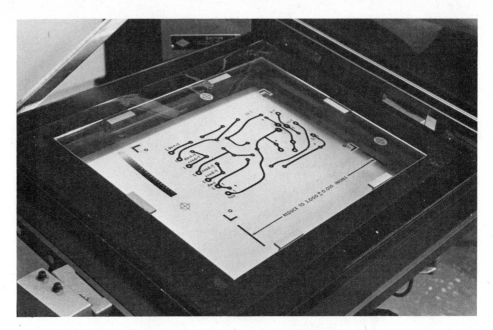

**Figure 9.2** Loading taped artwork master and step table onto camera's copyboard.

the top of the *timer/focus* rocker switch, the camera lights will turn on and the artwork image will be seen on the focusing plate, which has horizontal as well as vertical scale markings for making general measurements. However, best results can be achieved in checking the accuracy of the setting on the reduction scales by using a *precision transparent polyester scale,* such as the Bishop Graphics Accuscale, placed flat on the focusing glass in line with the reduction marks. Any fine adjustment required is then made only with the lensboard wheel by rotating it slightly clockwise or counterclockwise until exact

**Figure 9.3** Correct position of quartz lamps for backlighting of copyboard. *Courtesy of nuArc Company, Incorporated.*

**Figure 9.4** Detail of the control console and adjustment control wheels. *Courtesy of nuArc Company, Incorporated.*

**Figure 9.5** Focusing plate positioned to inspect artwork image.

**Figure 9.6** Lens with diaphram percentage control. *Courtesy of nuArc Company, Incorporated.*

alignment of the projected artwork scale reduction marks and the precision scale markings results on the focusing glass.

After the accuracy of the artwork reduction image has been verified, the focusing glass may be raised and locked into its vertical storage position above the film holder. The *timer/focus* switch should then be positioned to *timer* and the *tape* switch moved to *off* by depressing the bottoms of these rocker switches. In this mode, the exposure lamps will be off as well as the scale illumination lamps.

Proper film exposure is a function of the *amount* of light that strikes the film and the *length* of exposure time. The amount of light is set by the aperture opening within the lens system, the size of which is designated by *f/stops.* The maximum amount of light is transmitted through the lens using the widest aperture opening (lowest f/stop number). The nuArc Model SST1418 camera has an f/stop range of f/10 to f/22. It has been determined empirically that an aperture setting of f/16 will result in high-quality artwork phototools. This opening is set with the pointer using the upper scale shown in Fig. 9.6.

Exposure time is controlled by an electrical timer that is synchronized with the shutter mechanism of the lens. Refer again to Fig. 9.4. The timer pointer is set to the desired number of seconds of exposure and is activated by depressing the center button (see Fig. 9.7). Immediately, the lights are illuminated and the lens shutter is opened. The film is exposed for the preset time, at the end of which the timer activates the switching that turns off the lamps and closes the shutter. After each exposure, the timer will recycle to the preset position in preparation for another time exposure. With the optimum aperture opening of f/16, a timer setting of 10 seconds results in high-quality negatives ideally suited to pc artwork processing. A flow chart of the photographic reduction procedures using the nuArc SST1418 copy camera is shown in Fig. 9.8.

**Figure 9.7** Time of film exposure controlled with an electrical timer. *Courtesy of nuArc Company, Incorporated.*

Turn off ——————→
Room Lights,
Turn On
Safe Lights

**Figure 9.8** Flow diagram for photographic reduction of artwork master.

The nuArc SST1418 copy camera or other similar unit used for photographic processing of artwork masters requires the use of suitable film. This film must possess high line contrast, reasonably fast processing time, and excellent dimensional stability. Several manufacturers have produced sheet film with these qualities, among them Kodak Kodalith Ortho Type 3 and Dupont Cronalith film. For purposes of illustration, the Kodak Ortho 3 film and its associated chemicals are discussed. If other film is used, the specific manufacturer's literature on film processing must be consulted.

The Kodak negative film is processed in the following sequence: *exposure, develop, stop, fix, wash,* and *dry*. Each of these steps will be described in detail.

To load the camera, a sheet of Kodak negative film is first removed from its light-tight box under *safe-light* conditions (Wratten 1-A or equivalent). Under normal lighting, this film would appear pale *red* but appears as light gray under the safe lamp. The *emulsion* side of the film (as opposed to the 0.004-inch-thick triacetate support backing side) is discernible by its lighter color. (If extremely high-dimensional stability and durability are required, 0.007-inch film is available at slightly higher cost.) The film is centered onto the vacuum frame film holder so that its emulsion side is facing upward toward the operator. This is shown in Fig. 9.9. The vacuum pump is then turned on by a foot switch. With the vacuum holding the film firmly in place, the film holder is raised to its vertical position and latched. The artwork to be reduced is positioned on the copyboard so that it is centered over the translucent backing. An exposure step-table having 21 transparent-to-opaque sequential blocks is taped onto the artwork outside the corner brackets. With the setup in place, the glass cover is secured and the copyboard is raised to its vertical position. Before making the exposure, the operator should check the f/stop setting, timer setting, and reduction scale tapes as well as the position of the lamps.

The film is exposed by depressing the timer *start* button shown in Fig. 9.7. After exposure, the film is removed from the vacuum frame and is ready to be developed.

Kodak film is developed by the use of two concentrated chemicals: *Kodalith Liquid Developer Solution A* and *Kodalith Liquid Developer Solution B*. Since these chemicals

**Figure 9.9** Vacuum frame holds film in position for exposure. *Courtesy of nuArc Company, Incorporated.*

are in liquid form, they require no special mixing equipment. For best results, they should be mixed with water that is at a temperature of between 65 and 70°F (18 to 21°C). (This water temperature is recommended for mixing all film baths.) The following five steps outline the preparation of the developer.

1. Determine the volume of developer required.
2. Divide this total in half to determine the amount of diluted developer A and diluted developer B required.
3. Divide this figure by 4 to find the amounts of both concentrates A and B needed.
4. Add three equal parts of water to concentrate A as determined in step 3; repeat for concentrate B.
5. Mix diluted A to diluted B to yield the total volume as determined in step 1.

The concentrates A and B are never added together. They must be diluted before mixing.

An example for the preparation of the developer will aid in understanding the various steps. In this example, it is required that *1 quart of Kodalith Developer be prepared.* Following are the five steps just presented:

1. Total volume required—1 quart (32 ounces).
2. Subvolume—32/2 = 16 ounces of diluted A and 16 ounces of diluted B.
3. 16A/4 = 4 ounces of A concentrate; 16B/4 = 4 ounces of B concentrate.
4. Four times 3 = 12 ounces of water. Four ounces of A plus 12 ounces of water = 16 ounces of diluted A solution; 4 ounces of B plus 12 ounces of water = 16 ounces of diluted B solution.
5. Sixteen ounces of A diluted solution added to 16 ounces of B diluted solution yields 32 ounces (1 quart) of working developer.

The prepared solution is placed in a plastic developing tray. Under safe-light conditions, the exposed film with the emulsion side *up* is placed in the tray and constantly agitated until the film is developed completely. At a developing solution temperature of 68°F (20°C), the development time is between 2 and $2\frac{1}{2}$ minutes, the time decreasing as the temperature increases. At 74°F (23°C), the developing time is $1\frac{1}{4}$ to 2 min and at 80°F (27°C) reduces to 1 to $1\frac{1}{2}$ minutes.

| Bath Temperature | Developing Time |
|---|---|
| 68°F | 2 min. 15 sec. |
| 69°F | 2 min. 10 sec. |
| 70°F | 2 min. 7 sec. |
| 71°F | 2 min. 3 sec. |
| 72°F | 1 min. 58 sec. |
| 73°F | 1 min. 52 sec. |
| 74°F | 1 min. 49 sec. |
| 75°F | 1 min. 45 sec. |
| 76°F | 1 min. 40 sec. |
| 77°F | 1 min. 37 sec. |
| 78°F | 1 min. 32 sec. |
| 79°F | 1 min. 26 sec. |
| 80°F | 1 min. 22 sec. |
| 81°F | 1 min. 18 sec. |
| 82°F | 1 min. 15 sec. |

**Figure 9.10** Developing time as a function of bath temperature.

A chart showing approximate developing time in minutes for specific bath temperatures is shown in Fig. 9.10. Proper developing time is critical. Depending on how long the film is allowed to remain in the developing bath, it can be under- or overdeveloped. An explanation of the developing process will aid in understanding the reason for paying close attention to the required time. When the film is initially exposed in the camera, certain portions remain underexposed because the black conductor pattern on the copyboard absorbs light, whereas the surrounding white area reflects light. As a result, the underexposed film portions form a *latent* image of the conductor pattern. When the film is developed, it yields a white conductor pattern image on a black background. During the developing process, the total emulsion surface is softened and feels oily to the touch. For this reason, care needs to be exercised to avoid scratching this surface. If the film is allowed to remain in the developer for an excessive amount of time, the entire emulsion surface will become totally black. The film negative should be agitated in the developer solution long enough for a clear-cut image to appear with the desired contrast. Optimum contrast can be gauged by continual visual inspection of the device outline configuration. Attention should be focused on the smallest lead holes to be developed. When they appear as distinct sharp black dots, the developing process should be terminated *immediately* by quickly placing the film into a *stop bath*. If lead holes are not present, continual comparison of the depth of color in the black areas of the developing film between *both* sides is made. As soon as both sides appear to have equal degrees of blackness, the film is ready to be placed into the stop bath.

Determining proper developing time by the methods just described is appropriate for experienced technicians. For the beginner, however, a more exacting method may be used. This method takes into account the light source and exposure time, lens setting and reduction scale, developing solution bath temperature, and time in the developing bath. It compensates automatically for any variations in these details to result in a correctly developed film. A Stouffer chart, shown in Appendix VII, is placed outside the corner brackets of the artwork on the copyboard of the camera. As the developing process progresses, the blocks in the step table will begin to blacken, starting with the transparent block, step 1. By monitoring the step chart, the film is removed from the developing solution and placed into the bath at the point where the *sixth* step of the table has completely blackened with the seventh, and perhaps the eighth and ninth steps, partially blackened. This method of film developing assures consistent results regardless of any variations in the exposure setting. The appearance of a properly developed step table is shown in Fig. 9.11.

Whichever technique of film developing is employed, care should be exercised to avoid the possibility of under- or overdeveloping. Immediately after the film is properly developed, it is quickly placed into a *stop bath*. This stop bath is made up of 1 part of *glacial acetic acid* to 32 parts of water. To serve as an example of this mixture, to prepare 330 milliliters of stop bath, 10 milliliters of glacial acetic acid is mixed with 320 milliliters of water.

The developer is completely neutralized in the stop bath when the oily characteristics of the film emulsion no longer exist. This neutralization process will take approximately 1 minute. There is no visible change in the image. The negative is gently agitated in this stop bath for approximately 1 minute and then is placed in a tray containing *Kodak Rapid Fixer (Solution A)* with a *hardener (Kodak Solution B bath for fixer)* at 68°F (20°C) for approximately 1 to 2 minutes. The black areas of the film are insoluble in the fixer, but the white emulsion of the conducting pattern image will dissolve in the solution. The negative must be agitated in the fixer solution for a sufficient period of time for all the emulsion in the conducting pattern area to be dissolved, exposing the underlying transparent backing. A black tray should be used because it permits the complete dissolving of the white emulsion to be more easily detected. The visual impression is a totally black film. The hardener is included in the fixer solution to stabilize the remaining emulsion.

To prepare 1 gallon of fixer bath, 1 quart of solution A is added to 2 quarts of water

**Figure 9.11** Properly developed 21-step table for Kodalith Ortho Type 3 film (settings: f/16, 10 seconds, 2:1 reduction).

**Figure 9.12**  Developing sequence for reduced negative.

at 65 to 70°F (18 to 21°C). Three-and-one-half fluid ounces of solution B (hardener) is stirred into the solution. Finally, sufficient water is added to make the total solution equal to 1 gallon. A negative film being removed from the fixer solution is shown in Fig. 9.12.

After the fixing process, the film negative is washed in a spray of water at approximately 70°F (21°C) for several minutes to remove all traces of the fixing solution (Fig. 9.13). Once the film has been washed, it can be allowed to dry by simply suspending it with a clamp attached to an overhead line. This method is not recommended, however, since the surfaces of the negative tend to waterspot. To prevent this spotting a squeegee or photographic sponge may first be used to remove the excess water and then the negative should be allowed to dry. The film is now completely processed into a negative of the printed circuit pattern. The negative should be handled carefully because it will scratch easily. Because of the involved nature of the photographic process, a flow diagram is provided in Fig. 9.14 to guide the technician through the individual steps.

**Figure 9.13**  Final film rinse.

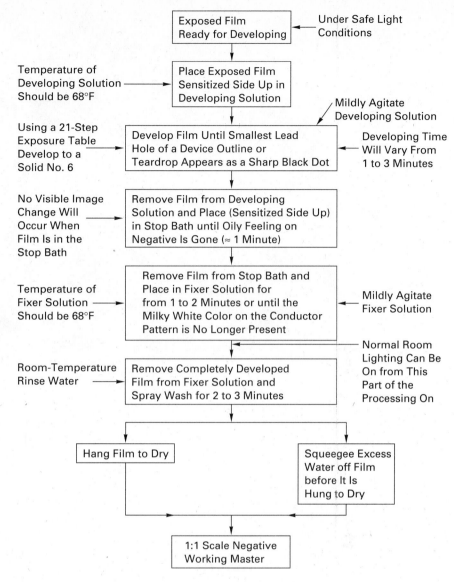

**Figure 9.14** Flow chart for developing negative film.

## 9.3 CONTACT PRINTING

When the artwork masters have been photographically reduced into negative transparencies, the required 1:1 scale *positives* of the marking mask negatives can now be processed. If the film used in the initial reduction process yields a negative transparency, one more operation will generate the positive transparencies. To produce a positive from a reduced artwork negative is in fact to obtain a positive transparency of the original artwork master that is reduced to the proper 1:1 size scale for processing a pc board marking mask. (The process of converting a positive transparency of a marking mask into a silk screen is discussed in detail in Sec. 10.5.)

The processing of a 1:1 scaled positive from the same-size negative is best accomplished by *contact printing* using the same Kodak Kodalith Ortho Type 3 film used for the reduction of the original artworks. An arrangement showing a marking mask being contact-printed is shown in Fig. 9.15. This system consists of a vacuum frame, pump, lamp and adjustable intensity, and exposure timer controls.

**Figure 9.15**  Negative and film being exposed with contact printer.

For contact printing, a sheet of Ortho 3 film is removed from its light-tight box under safe-light conditions. It is placed, emulsion side up, on the black nonreflective mat in the vacuum frame. The reduced negative is then placed on top of the film and the glass cover of the vacuum frame is lowered and locked into place. The vacuum pump is switched on to maintain pressure of the negative against the film. Good contact is achieved when 25 in. Hg or greater is registered on the vacuum gauge. The contact printer lamp is positioned approximately 3 feet directly above the film to be exposed. For exposure, the intensity control of the light source is switched to *intensity level 2* and the timer is set at 10 seconds. The *start* button is then depressed to activate the light and the timer.

After exposure, the vacuum pump is switched off and the film is removed from the frame. Still under safe-light conditions, the film is processed through the same developer, stop bath, fixer, and water rinse procedures as described in the preceding section. The flow diagram of Fig. 9.14 used to process negative film is also applicable for contact print processing.

## 9.4  DIAZO PHOTOTOOL FOR DOUBLE-SIDED PRINTED CIRCUIT BOARDS

For processing a double-sided pc board with plated-through holes, two 1:1 scale positives are required. These *silver halide* positives are generated from contact-printing the negatives, as described in the previous section. Silver halide positives, however, are not suitable for proper registration, as will be explained here. Correct registration of the 1:1 phototools on both sides of a previously drilled panel is accomplished by aligning the solid terminal pads of the phototools over the holes. The aim is to obtain uniform annular rings about all holes. Because the conductor pattern on silver halide positives is 100% opaque, the drilled holes cannot be viewed through the pads on the positive to obtain proper registration. For this reason, phototools must be made from the 1:1 silver halide positives, which have a lower visual density (light-transparent color) of the opaque conductor pattern, yet be opaque to ultraviolet radiation. These positives will allow the drilled holes to be viewed through the pads for proper registration.

One type of film available for producing this special phototool is a light-sensitive *diazo emulsion* film on a polyester base. Diazo film, whose emulsion is sensitive to ammonia gas and is commonly used in the pc board industry, makes direct image transfer (without reversal) when contact printed with standard silver halide positives. The contact printing is accomplished under gold safe-light conditions using the same ultraviolet light-exposure units used for photo-resist exposure (see Chapter 12). The film is then dry-developed in hot ammonia gas.

To illustrate the process of producing two diazo phototools (one for the component side and one for the circuit side, we will use *Dynachem Type G-2 Trans Opaque* film. This diazo film is supplied on a dimensionally stable 0.007-inch polyester base. It should be stored at 50°F (10°C) or below when not in use in order to extend its shelf life. It is always processed in a gold safe-light environment such as that found in the photo-resist exposure area of a pc board fabrication facility.

To make a 1:1 contact-printed diazo positive from a silver halide positive, a sheet of G-2 film is first placed in the vacuum frame of an exposure unit (similar to the one shown in Fig. 9.16) on a black nonreflective surface. The film is positioned with the emulsion side facing *upward*. In this position, the notch on the sheet of film will be in the upper right corner. The silver halide positive is then placed on top of the diazo film with its emulsion side *down*. The frame cover is closed and the vacuum pump is turned on, which will bring both films in tight contact with each other before the frame is slid under the exposure lamps. Exposure time is determined empirically. It has been found that with the 400-watt ultraviolet light source shown in Fig. 9.16, proper exposure is completed in under 2 minutes. Those areas on the diazo film that are exposed to the light through the transparent areas of the silver halide positive will remain clear when developed in ammonia gas. Those areas under the opaque conductor pattern of the silver halide positive are unexposed to the light. In the diazo developing process, these areas became an amber color, which is transparent to visible light but opaque to ultraviolet light.

After the diazo film has been exposed, it is removed from the vacuum frame and developed in a gold-lighted area using a *Dynachem Diazo G-2 Film Processor,* which is shown in Fig. 9.17. This unit is a simple conveyor having hard-rubber rollers that trans-

**Figure 9.16**   Exposure unit for double-sided pc boards.

**Figure 9.17** Exposed film being developed with the Dynachem diazo developing unit.

port the film slowly through an environment of ammonia gas heated to 200 to 220°F (93 to 104°C). By simply introducing the film into the unit, total developing time is approximately 1 minute.

When the diazo film has been developed, the amber conductor pattern image will become visible (see Fig. 9.18). Since it is now possible to see through the solid terminal pads, proper registration of the film over a predrilled board can be effectively accomplished.

Since the processed films will be used as phototools to image pc boards, they must have areas that will effectively block ultraviolet light and other areas that will transmit UV energy. This is true for both silver film and diazo film. With diazo film, the amber areas must block UV light while the clear portions transmit UV energy. Accurate measurements of the ability of the film areas to block or transmit UV light can be obtained easily and quickly with the use of a Macbeth Transmission Densitometer. This is shown in Fig. 9.19. To use the instrument, the two areas of film are alternately placed over the sampling area and readings are shown on a digital display. The opaque area that is to block UV light must show a reading of 4.00 or greater. The clear portion must produce a reading of 0.6 to 0.7 for diazo film and 0.02 to 0.06 for silver films to ensure effective transmission of UV energy.

The procedure for using the densitometer is as follows:

**Figure 9.18** Diazo film phototools with double-sided masking tape.

**Figure 9.19** The densitometer measures the UV filtering and transmission characteristics of phototools. *Courtesy of Macbeth, a division of Kollmorgen Corporation.*

**Step 1.** The filter wheel is rotated to the UV position. This places the ultraviolet filter at the sampling aperture.

**Step 2.** The rocker switch located on the rear panel is turned on. Three figure Cs will appear on the digital display and the light in the density button will appear.

**Step 3.** The sampling aperture arm and the zero button are pressed simultaneously, which causes 0.00 to appear on the display.

**Step 4.** The calibration film strip is placed under the sampling aperture and depressed. This should result in a reading of 3.00. If a different reading is displayed, the CAL TRANS potentiometer on the rear panel is adjusted for a reading of 3.00.

**Step 5.** The sampling aperture arm is released and the calibration film is removed.

**Step 6.** The diazo film's amber area (black for silver halide) is placed under the sampling aperture and depressed. A reading of 4.00 or greater should be displayed to indicate a properly exposed and developed film.

**Step 7.** The sampling aperture arm is released and the film is shifted to a clear area and the aperture is again depressed. For effective UV transmission, readings of 0.02 to 0.06 should be displayed for silver film and 0.6 to 0.7 for diazo film.

The densitometer is an important instrument in assuring the effectiveness of the phototools to filter and transmit UV energy.

Final preparation of the phototools requires that the two sheets of diazo film first be cut to size. Two pieces of double-sided Scotch or masking tape are then placed on each phototool on the side that will contact the photosensitized pc board during exposure (Chapter 12). This is the side where the literal information appears backward-reading. The tape ensures that registration will be maintained between the phototool and the pc board throughout the exposure process.

With the completion of the diazo phototools, we will continue with the discussion of the sequential processes in the fabrication of both single- and double-sided pc boards. Because these processes are involved and require a great deal of detailed explanation, they will be spread over the next four chapters. Chapter 10 is devoted to the fabrication of single-sided pc boards using a simple print-and-etch technique and the negative photo-tools discussed in this chapter. Chapter 11 presents the drilling of the bulk stock and its preparation for copper deposition, followed by a flash-plating process. Chapter 12 discusses the conductor pattern imaging process using dry film photo resist and the diazo phototools described in this chapter. Chapter 13 completes the fabrication of a double-sided plated-through-hole pc board by considering copper pattern plating, solder plating, photo-resist stripping, etching, and infrared solder reflow.

## EXERCISES

### A. Questions

9.1    What type of camera is recommended for photoreducing artworks for printed circuits?

9.2    How should the step table be positioned on the copyboard with the artwork?

9.3    Which of the following f/stops will allow the transmission of the least amount of light: f/5.6, f/11, f/16, or f/22?

9.4    (a) What is the recommended temperature of the developing solution?
(b) If the developing solution temperature is 78°F, what is the estimated time for developing exposed film?

9.5    Explain the term *latent image*.

9.6    Briefly describe the three methods for determining proper developing time.

9.7    What is the purpose of the stop bath in the developing process?

9.8    What areas of the developed film dissolve in the fixer solution?

9.9    How does the developed film, after contact printing, compare to the artwork through which it was exposed?

9.10   Define the term *phototool*.

9.11   How are hot spots or reflections eliminated during film exposure?

9.12   What are the six processing steps for Kodak negative film?

9.13   What determines when the film should be removed from the fixer tray?

9.14   Why is diazo film chosen over silver halide film for phototools used to image double-sided pc boards?

### B. True or False

Circle T if the statement is true, or F if any part of the statement is false.

9.1    When producing a phototool, a 1:1 scaled image reversal occurs.          T    F

9.2    To obtain properly developed exposures, the step exposure table is placed over a clear area of the artwork inside the corner brackets.          T    F

9.3    Maximum contrast between light and dark areas of the artwork results when back-lighting is used on the camera.          T    F

9.4    Under safe-light conditions, the emulsion side of Kodalith film appears lighter than the triacetate backing side.          T    F

9.5    Using an f/stop setting of 16, Kodalith film is exposed for 2 to $2\frac{1}{2}$ minutes.          T    F

9.6 The emulsion around the conductor pattern dissolves when the film is placed into the fixer bath. **T F**

9.7 A stop bath is used before the fixer bath to prevent overdeveloping. **T F**

9.8 When contact printing, the film to be exposed is placed in the vacuum frame with the emulsion side facing up. **T F**

9.9 Reversal of the image does not result when contact printing with Kodalith film as it does when photoreducing on a camera. **T F**

9.10 Ultraviolet light will not pass through the areas of silver halide film but will pass through the amber-colored areas of diazo film. **T F**

## C. Multiple Choice

Circle the correct answer for each statement.

9.1 Using a copy camera for artwork master reduction, a phototool can be produced to a tolerance of (*±0.01*, *±0.002*) inch.

9.2 Maximum contrast between light and dark areas of the artwork results with (*front-*, *back-*) lighting on the copy camera.

9.3 The recommended exposure setting for a copy camera is (*f/16 at 10 seconds*, *f/8 at 30 seconds*).

9.4 The emulsion side of film is placed facing (*up*, *down*) in the developing tray.

9.5 Developing time (*is*, *is not*) a function of developing solution temperature.

9.6 Neutralizing time in the stop bath is approximately (*1 second*, *1 minute*).

9.7 The exposed film becomes a transparency in the (*fixer*, *stop*) bath.

9.8 Contact printing is used to obtain 1:1 scale (*negatives*, *positives*) of the marking mask.

9.9 (*Diazo*, *Silver halide*) film uses an emulsion that is developed in a dry ammonia gas.

9.10 Diazo film is exposed using (*ultraviolet*, *infrared*) sources.

## D. Matching Columns

Match each item in column A to the most appropriate item in column B.

| COLUMN A | COLUMN B |
|---|---|
| 1. Fixer solution | a. Aperture opening |
| 2. Contact printing | b. 2 to $2\frac{1}{2}$ minutes |
| 3. Developing process | c. Glacial acetic acid |
| 4. f/16 | d. 1:1 scale reversal |
| 5. Diazo | e. Transparent film |
| 6. Emulsion side of film | f. Lighter color |
| 7. Support backing of film | g. Darker color |
| 8. Stop bath | h. Ammonia gas |

# 10
# Single-Sided PCB Processing: Print-and-Etch Technique

## LEARNING OBJECTIVES

*Upon completion of this chapter on the print-and-etch technique for processing single-sided circuit boards, the student should be able to*

- Apply photo resist to copper foil.
- Expose the photo resist.
- Develop the pc board and apply dye.
- Prepare etchant solution and etch a pc board.
- Use a wide-shank carbide-tip drill bit to drill the required holes in an etched pc board.
- Strip and completely clean an etched board.
- Process contact films for silk-screen printing.
- Be familiar with the various available types of silk screens and printing inks and paints and select the appropriate type for a given application.
- Prepare and use a silk screen for printing on pc boards.

## 10.0 INTRODUCTION

Of the many methods available for the processing of single-sided pc boards by the simple print-and-etch technique, the use of liquid photo resist (wet film) is perhaps the most economically feasible for prototype work. This technique requires inexpensive materials and equipment, thus making it ideally suited for this purpose. The topics discussed in this chapter include detailed discussions on circuit board cleaning, photosensitive resist application, etching, drilling, and the silk-screen process for preparing marking masks.

In the wet-film technique of circuit board processing, certain methods are more suitable for prototype work, whereas others lend themselves more readily to mass production applications. The selection of a specific process will depend on such factors as costs, quantity of boards to be fabricated, and labeling and solder mask requirements. The use of the various chemicals and materials for processing are included to aid in the selection of the most appropriate method. Also included is information on simple construction of some of the equipment necessary for the various phases of board fabrication.

***Caution:*** *A chemical hazard exists in the materials presented in this chapter. Approved face shields (goggles), chemical-resistant gloves, clothing, footware, and a respirator must be worn prior to performing any of these operations.* Also, the section on safety outlined at the beginning of the book should be reviewed for more specific information.

For developing the conductor pattern onto the foil side of the copperclad pc board, a negative of the original conductor pattern artwork master is required. This negative becomes one of the basic elements used to process either one board for prototype work or hundreds of boards for mass-production requirements.

The surface of the foil must first be treated so that selected areas of the copper can be protected from the etching solution, which will remove all the copper except the conductor pattern. The copper may be treated by several methods to accomplish selective foil pattern protection. The method to be considered here is *photosensitizing*. In this process, the foil is treated with a *photosensitive resist,* exposed through a 1:1 negative of the conducting pattern, and developed. The result will be a 1:1 positive conductor pattern produced on the surface of the foil, which is etchant resistant, whereas the undesired areas of the foil are unprotected and will be removed by the etching process.

Several chemicals are readily available to implement this photosensitized technique. The chemicals and procedures discussed in this section are those of the *Eastman Kodak Company*. Chemicals of other manufacturers may require processing and exposure times that are different from those discussed.

The process begins by cutting the necessary-size copper-clad board from sheet stock. (See Chapter 5 for information on shearing pc boards.) The work must then be thoroughly cleaned before the photo resist is applied. This cleaning is essential in order to remove all contaminants, such as grease or copper-oxide film, which readily forms on exposed copper. Unless these contaminants are removed, they will adversely effect the adhesion of the resist with the copper surface.

The copper is first chemically cleaned in a 5 to 10% by volume bath of hydrochloric acid and water. This bath will remove copper oxide and grease film for the improvement of photo resist adhesion. To prepare this bath, 75 milliliters of hydrochloric acid is slowly added to 925 milliliters of cold tap water in a container of Pyrex or acid-resistant plastic such as polyethylene or polypropylene. (**Caution:** Never add water to acid. This causes violent eruptions to occur.)

After the board is removed from the acid bath, the copper foil is then scrubbed with a 3M Scotch-Brite 96 pad while water-spray rinsing. This technique is shown in Fig. 10.1. Firm pressure should be applied with small overlapping circular motions. If the recommended cleaning pad is not readily available, any gentle abrasive household cleanser that contains no bleach is acceptable. Coarse abrasives should not be used because they would deeply scratch the soft copper foil. Proper cleaning removes contaminants without causing surface imperfections in the foil. Cleaning is sufficient when water flushed over the surface of the copper "sheets" and no water "beads" remain. Remaining grease contaminents will appear as dry spots from which the water tends to pull away. To complete the cleaning process, the foil is thoroughly rinsed with water and swabbed with a lint-free cloth to remove any remaining cleaner deposits. Once the copper is completely cleaned, all surface moisture must be removed. This is accomplished by first patting dry with a paper towel. Since moisture severely affects resist adhesion a final drying operation is important. Total drying will be achieved if the board is placed in a forced-air oven for approximately 10 minutes at a temperature of 120°F (49°C). After the board is removed from the oven and allowed to cool to room temperature, the application of the photo resist can begin. The board must be handled carefully because fingerprints will contaminate the copper. The cleaning and drying process appears time-consuming and tedious, but it is only through a thoroughly clean and dry surface that a quality printed circuit will result.

The application of liquid photo resist is done by *spraying, dipping, roller coating, low-speed whirling,* or *flow coating.* The first four methods are common to industrial applications and result in a consistently uniform layer of resist. These methods require expensive specialized equipment and, for this reason, are not intended for prototype pc

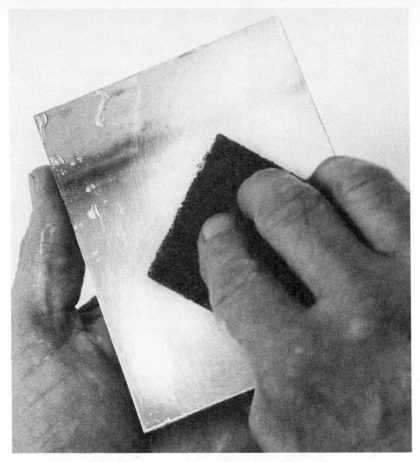

**Figure 10.1** Printed circuit board material contaminant removal is best accomplished with a Scotch-Brite pad and cold water flush.

board fabrication. Flow coating requires no special equipment, and even though it does not produce a consistently uniform resist thickness, the results are acceptable.

Presensitized photo resist pc boards ready for imaging are commercially available in a variety of sizes. These are maintained in light-tight plastic bags.

Since the photo resist is sensitive to ultraviolet light, safe-light conditions are required in its use. Lighting should be provided by red safe light (Wratten 1-A or equivalent) or fluorescent tubes covered with gold or amber plastic sleeves with the ends sealed with black vinyl electrical tape to prevent light leaks.

Kodak KPR liquid resist lends itself well to flow-coating applications. It is clear in color and is applied directly to the pc board without thinning. For ease of application, a convenient amount of liquid resist may be transferred into a small bottle or other dispenser. Figure 10.2 shows the application of Kodak resist with the use of a dropper. The resist is initially applied onto the center of the pc board. The board is then rotated and tilted in a circular motion so as to allow the resist to flow outward toward all four edges, thereby covering the entire surface of the copper. With this process completed, one corner of the board is tilted and placed into the mouth of the resist container to drain off all excess resist. The board is then placed in a horizontal position to dry at room temperature in a dust-free area. In approximately 10 minutes, the board is tack-free.

To ensure successful imaging with proper resist adhesion to the copper surface, a baking cycle is required. The baking removes any residual resist solvent that may remain after being air-dried. Trapped solvent will prevent the conductor pattern image from being fully protected during the etching process.

Baking can be accomplished by either one of two methods. The first method is to place the board into an oven for 20 minutes at a temperature of 176°F (80°C) or a maxi-

**Figure 10.2**  Flow-coating a pc board.

mum of 10 minutes at 250°F (120°C). The second method uses a 275-watt ruby-faced infrared heating lamp positioned 12 to 14 inches above the board. This process will require 20 to 30 minutes to eliminate resist solvent. In any type of baking cycle, extreme care must be exercised to avoid heat or light fog affecting the film. This would degrade the film and cause later processing problems. The resist surface temperature should not exceed 250°F (120°C). If using a lamp, reflectors should be opaque with no ultraviolet leaks.

After the photo resist hardens in the baking cycle, it becomes sensitive to light. For this reason, care must be exercised to protect it from exposure to any intense light when removing the board from the oven or from under the lamp.

The board should be allowed to cool to room temperature prior to initiating the imaging process. Large-diameter 16-mm film cans provide a convenient means of storing sensitized boards for short periods of time.

Kodak KPR photo resist can be exposed with a variety of sources rich in ultraviolet light, such as carbon-arc lamps, pulsed-xenon lamps, and unfiltered ultraviolet fluorescent lamps. A commercial exposure unit, shown in Fig. 10.3, uses high-pressure mercury vapor lamps and is ideally suited for photo resist imaging.

Exposure time will depend on such factors as light intensity, type of light source, distance of the source from the resist surface, and thickness of the resist layer. Optimum exposure can best be determined by trial and error.

It is essential that the light source be provided with a shield that directs the light onto the work and at the same time protects the eyes from exposure to the light rays. Light in the ultraviolet range is dangerous to the eyes and prolonged exposure must be avoided. If a light shield is not available, protective glasses should be worn. The pc board is placed

**Figure 10.3** Exposure unit for photo resist imaging.

onto the copyboard with the sensitized side *up* under safe-light conditions. The reduced negative of the conductor pattern is positioned directly onto the sensitized surface. It is held in direct contact with and flat against the pc board by the weight of the glass. The border delineation marks on the negative must be aligned with the corners of the board. In addition, the negative is positioned so that all literal and numerical designations are readable. Care must be exercised to ensure that the sensitized surface will be exposed through the correct side of the negative. If an orientation error is made and the board is later etched, it will be impossible to correct the conductor pattern.

Those photo resist areas exposed to the light through the clear conductor pattern are polymerized and will become insoluble to the developer and thus remain on the copper. Those areas under the opaque portions of the negative are not exposed and remain soluble to the developer. This photo resist will wash away, leaving a photo resist mask on the surface of the copper that duplicates the desired positive conductor pattern.

The exposed board is developed with the use of a Pyrex or stainless steel tray that contains at least 1 inch of Kodak KPR developer. The board is placed into the tray with the imaged copper side facing up to prevent damage to the photosensitized emulsion. The amount of time that the board remains in the developer solution is critical. Normally, approximately 2 minutes at room temperature is sufficient to completely remove all of the resist areas that have not been polymerized by the light sources. The tray should be gently rocked for the full developing time. During the entire developing process, the remaining resist remains clear but a faint image of the conductor pattern becomes visible when the board is viewed at an angle under normal lighting conditions.

After the board is developed, it is removed from the tray, being handled only by the edges since the resist is soft and can be easily damaged. Both board surfaces are then flushed with cold tap water at high volume and low pressure for at least 30 seconds on each side.

The developed resist image is next inspected to determine if there are any imperfections, such as pinholes, bridges, or open paths. This inspection process makes use of Kodak KPR dye. The developed and rinsed board is dipped into a bath of this blue-colored dye for approximately 30 seconds. It is then removed and immediately flushed with cold tap water. This is shown in Fig. 10.4. The dye will adhere only to the photo resist remaining after the developing process.

**Figure 10.4** Conductor pattern inspection is aided with the use of dye.

The board is next dried in an oven for 10 minutes at 250°F (120°C) or under a lamp at 6 to 8 inches for 10 minutes. In no case should the resist surface temperature exceed 250°F. After the drying cycle, the board is allowed to cool to room temperature.

The board can now be inspected and any errors or pattern imperfections that may be present can be corrected by using a sharp, pointed knife. If there are any breaks in the photo resist covering of the desired conductor pattern at this point, they can be repaired by the application of *etchant ink resist* (non-light-sensitive). This resist is carefully applied with a fine artist's brush to close the break. It must overlap the photo resist on either side of the break while maintaining width. This technique is shown in Fig. 10.5. There are a number of asphaltum-based materials available from graphic arts suppliers that may be used for this purpose. Some may require dilution to a workable consistency. The solvent used should be in the turpentine family. Touch-up pens are also available to close breaks in conductor paths.

**Figure 10.5** Application of etchant resist to repair defective pattern.

Figure 10.6  Flow diagram for sensitizing printed circuit boards with photo resist.

After the conductor pattern touch-up has been completed, another baking cycle is recommended to remove the solvent from the resist. A 20-min cycle at approximately 200°F (93°C) is sufficient to remove all solvents.

The photo sensitizing procedure just discussed is an effective method of preparing a pc board for etching in prototype work. A flow diagram to aid the technician in following the many steps in this process is shown in Fig. 10.6.

## 10.2 CIRCUIT BOARD ETCHING

*Etching* is the process of attacking and removing the unprotected copper from the pc board to yield the desired conductor pattern. Several commercial etchants and techniques are available for processing pc boards. These materials and methods attack the unprotected copper yet do not affect the adhesive, supporting laminate, and photo resist. The most common chemicals used by industry as an etchant are *ferric chloride, ammonium persulfate, chromic acid,* and *cupric chloride.* Of these, ferric chloride is commonly used since it is less expensive and potentially the least dangerous.

Methods of etching include *tray rocking, tank etching,* and *spray etching.* Tray rocking is the simplest system, consisting of a tray of etchant attached to a powered rocking table. Tank etching involves dipping the work into a vat containing the etchant, which is maintained at an optimum temperature. Spray etching, although a much faster process, calls for etchant to be pumped under pressure onto the surface of the pc board. This method requires an elaborate and expensive system. These three methods will be described using ferric chloride as an etchant.

Ferric chloride ($FeCl_3$) is an etchant that will react with the unprotected areas of the copper surface and produce copper ions, which are soluble in solution. This chemical is available premixed to the correct concentration. Ferric chloride crystals are also available if liquid storage is a problem but require premixing and concentration testing.

During the etching process, the concentration weakens because the soluble cupric and ferric ions precipitate out of solution in the form of a sludge that tends to settle on the bottom of the etching vat. When the etching time becomes excessively long, the used etchant should be discarded and replaced with a fresh solution.

Ideal etching conditions require that the etchant be heated to a temperature of between 100 and 130°F (38 to 54°C). Temperatures above this range should be avoided because of the fumes that are generated. Although etchant temperature is not critical, the activity of the etching process is increased when the solution is heated above room temperature. With the temperature maintained reasonably constant, the length of time required for etching will be consistent. However, because of the weakening of the solution, etching time will progressively increase.

In addition to heating, efficient etching requires that the etchant be continuously agitated to allow "fresh" solution to flow over the copper. A simple means of achieving this agitation is to use a Pyrex tray filled with etchant and placed on top of a *rocker table.* Rocking motion is imparted to the table by means of a small *motor, cam, connecting arm,* and off-center *pivot* arrangement. The pc board is placed copper side *up* in the tray. Only one board should be etched at one time since, because of the rocking motion, they tend to come in contact with each other. This could cause some of the etchant resist to be scraped off of a surface. As the table is rocked, a wave is generated in the etching solution that travels back and forth across the tray and board surface, keeping the major portion of any sludge formed off the copper surface. There are two disadvantages in using this arrangement. First, no method is provided for heating the etchant unless an infrared heat lamp is used, in which case etchant temperature is difficult to control. Second, some sludge is continually carried over the surface of the board, which reduces some of the effectiveness of the etchant. However, outweighing these disadvantages is the fact that this system is relatively simple and inexpensive and suitable for prototype work. A 5- by 5-inch square of 1-ounce single- or double-sided pc board can be completely etched in approximately 30 minutes.

A more elaborate system for agitating the solution is shown in Fig. 10.7, consisting of a *vat* (Pyrex or other suitable nonmetallic material), *heating tape, variable transformer, air feed,* and *thermometer.* The temperature of the etchant contained in the vat is controlled by varying the voltage level delivered to the heating tape by the transformer. This temperature is monitored by a thermometer placed in the solution in proximity to the boards being etched. The etchant is circulated (agitated) by means of air pumped into the vat under low pressure and escaping through a perforated plastic tube placed along one side of the vat. The sludge (copper and ferric ions) settles to the bottom of the vat, allowing relatively sludge-free etchant to continually circulate across the surface of the work. The boards to be etched are suspended in the solution from noncorrosive rods and clamps with their copper side facing the source of the air bubbles. This method, although more expensive, can etch a 5- by 5-inch 1-ounce board in approximately 15 minutes.

Forced–air Input

Rubber Hose

Etchant

Glass Tube

Rubber Stopper

Plastic Clamp

Epoxy

Plastic Tube

Plastic PC Board
Clamp Assembly

Hard–rubber Tank

**Figure 10.7**   Forced-air etching system.

Spray etching involves jet spraying the surface of the board with pressurized nozzles. Systems such as that shown in Fig. 10.8 also are available with a heating element to decrease etching time. With such a system, etching times of between 1 and 2 minutes can be realized without difficulty. The etching cycle is controlled by a mechanical timer. Several precautions should be pointed out. The thermostat setting should not exceed the manufacturer's recommended maximum limits to avoid damage to the equipment. In addition, spray etchers of this type may be hazardous if the cover is not properly seated prior to operation to completely contain the etchant within the system.

When the etching is completed, the board is rinsed under water for approximately 30 seconds and patted dry. This washing process will not remove the photo resist, which is allowed to remain on the board to protect the copper during drilling and punching operations. Aprons should be worn to protect clothing and tongs used when transferring etched boards from the etchant to the initial rinse. After this washing and drying process, the boards are now prepared for drilling and punching.

## 10.3 PRINTED CIRCUIT BOARD DRILLING AND PUNCHING

For drilling printed circuit boards, a specially constructed drill bit is available. This bit, shown in Fig. 10.9, consists of a short length of twist bit with an enlarged shank of approximately $\frac{1}{8}$ inch. Printed circuit bits used for drilling through epoxy-resin boards are made of tungsten carbide. Although carbide is a brittle material, it does not dull as quickly as HS drills. Carbide withstands the heat better at speeds of 15,000 to 80,000 rpm, which are typical in CNC (computer numerical control) drilling. Automated drilling is used on double-sided printed circuit boards for the plated-through-hole (PTH) process discussed in Chapter 11.

**Figure 10.8**  Industrial etching system.
*Courtesy of Chemcut Corporation.*

Drilling is always performed from the *copper* side of the processed board for the following reasons: (1) to use the small etched center holes in terminal pads, which aid in bit alignment (it is extremely difficult to locate hole positions for drilling from the reverse side of the board, especially if the insulating base material is opaque), and (2) to minimize the possibility of pulling the copper foil away from the base (Fig. 10.10). The maximum lead diameter for some common components and devices together with the recommended drill sizes for lead access holes are shown in Table 10.1.

**Figure 10.9**  Special carbide-tip printed circuit twist drills.

**Figure 10.10** Printed circuit board drilling with wide-shank carbide-tip drill bit.

Punching should also be performed from the copper side for the same reasons just mentioned. If necessary, the grade of board used may require elevated temperatures for punching (see Sec. 5.2). Where necessary, an oven should be used for heating just prior to the punching operation. The punch and die should be sharp for best results. Before additional processing can be undertaken, the board must be completely cleaned of all photo resist, dye, and etchant impurities.

## 10.4 CONDUCTOR PATTERN SURFACE CLEANING

To prepare the etched copper surface for additional processes, the photo resist and the dye are removed by dipping the board in a bath of Kepro PRSK photo resist stripper for 1 minute. The board is then removed from the bath and scrubbed with a 3M Scotch-Brite 96

**TABLE 10.1**

Lead Hole Drilling Reference for Printed Circuits

| Component or Device | Lead Diameter (in.) | Drill Size | Decimal Equivalent of Drill (in.) |
|---|---|---|---|
| 1/8-watt resistor | 0.016 | #75 | 0.0210 |
| 1/4-watt resistor | 0.019 | #72 | 0.0250 |
| 1/2-watt resistor | 0.027 | #66 | 0.0330 |
| 1-watt resistor | 0.041 | 3/64 | 0.0469 |
| 2-watt resistor | 0.045 | #55 | 0.0520 |
| Disk capacitor | 0.030 | #64 | 0.0360 |
| TO-5 case style | 0.019 | #72 | 0.0250 |
| TO-18 case style | 0.019 | #72 | 0.0250 |
| DO-14 case style | 0.022 | #70 | 0.0280 |
| 77-02 plastic power transistor | 0.026 | #67 | 0.0320 |
| TO-99 (8-pin IC) | 0.019 | #72 | 0.0250 |
| TO-116 (14-pin DIP) | 0.023 | #69 | 0.0292 |

pad while water-spray rinsing. This scrubbing is essential to remove any photo resist residue. If not removed, this residue can cause problems in soldering leads to terminal pads. The board is given a final water rinse for about 30 seconds.

Etching in a ferric chloride solution produces contaminants on the copper foil, which will adversely affect soldering or plating. These salts of ferric chloride require an additional cleaning process. The following procedure for their removal is recommended. The contaminating salts remaining on the copper are made water soluble by dipping and mildly agitating in a bath of Lonco Copperbrite No. 48 HT. A final water rinse for 1 minute will remove all residues from the copper surface. After this final rinse, the board is wiped dry with a lint-free cloth.

## 10.5 SILK-SCREEN PRINTING

A relatively inexpensive method of reproducing identical pc boards is by the silk-screen printing technique. This process utilizes ink resist that is applied through the stencil of the desired pattern that has been affixed to a fine-mesh silk screen. (Nylon or wire mesh can also be used.) Although the silk-screen method is extensively used for processing pc conductor patterns, the illustrative descriptions provided will involve preparing marking and solder masks. The end result will be stencils that will yield acceptable tolerances of line definition. Artwork for silk-screen processing can be generated by CAD techniques as discussed in Chapter 7.

The necessary materials to produce a silk-screen stencil are a silk or nylon screen mounted on a printing frame with hinged clamps, stencil film, film developer blackout, rubber squeegee, paint or solder resist, and the artwork master processed into a 1:1 scale positive or negative transparency. These materials and the procedures for producing and using a silk screen for the marking mask shown in Fig. 10.11 will be discussed as an illustrative example.

The contact films used in the development of silk-screen stencils are normally extremely slow speed types with respect to exposure time and, as a result, do not require darkroom facilities. The film selected for this example is *Ulano Super Prep*. Even though the film speed is slow, it is recommended that this film be used in subdued light avoiding direct sunlight or high-output fluorescent lamps. The 1:1 scale positive transparency, as shown in Fig. 10.11, is used to implement this film.

A piece of plain white paper, larger than the artwork, is first placed flat to serve as a background. The Super Prep film, allowing at least a 1-inch border on all sides, is placed on top of the white sheet with the sensitized (emulsion) side of the film positioned *against* the white paper. (The sensitized side of the film is determined by observing the *dull* finish

**Figure 10.11** Positive 1:1 scale marking mask

as opposed to the glossy side of the polyester backing.) The exposure unit shown in Fig. 10.3 will be used for this example. The 1:1 scale positive marking mask transparency is then placed on *top* of the film and held firmly in place with the glass printing frame. Exposure times will vary from 5 to 15 minutes, depending on such factors as film type and light source. Optimum exposure times can be experimentally determined and once established, should not vary appreciably. Those areas of the film emulsion exposed to the light source will become insoluble in warm water after the developer bath. Those areas protected from the light by the opaque portions of the positive will dissolve in warm water after the developer bath. After exposure, the film is placed in a tray of developer with the emulsion side *up* for between 1 and 2 minutes. One end of the tray should be slowly raised and lowered to allow the solution to agitate gently over the surface of the film.

The developer for Ulano Super Prep film is available in powder form and packaged in premeasured packets labeled *A* and *B*. To prepare the developing solution, one packet of each of the powders is dissolved together in 16 ounces of water at a temperature of approximately 65 to 70°F (18 to 21°C). This developer is light sensitive and should be prepared in subdued light and stored in an amber-colored bottle. The solution has an effective lifetime of approximately 24 hours, after which time it begins to deteriorate. One indication of deterioration is the appearance of wrinkles on the polyester support of the film during the developing process. When this occurs, the developer should be discarded and replaced with a fresh mixture.

After the film has been in the developer for the necessary amount of time, it is removed and immediately placed in a tray of water at a temperature of approximately 100°F (38°C). The emulsion side of the film is again facing upward to prevent any smearing of the softened emulsion from the bottom of the tray. Gentle rocking of the water bath will remove the soluble portions of the emulsion (those areas protected during exposure). Approximately 1 minute is all the time required for the water bath to process the film. The film being processed in the water is shown in Fig. 10.12.

The film is next removed from the warm-water bath and rinsed with cold water for about 30 seconds to remove any remaining emulsion from those areas of the negative film that must be completely transparent. (***Caution:*** Extreme care must be exercised when handling the film after it is taken from the warm-water bath since the remaining emulsion is soft and tacky and is easily smeared. The cold-water rinse is best applied by placing the film on a sheet of glass with the polyester support backing against the glass. The emulsion side can then be gently rinsed under a cold-water tap.)

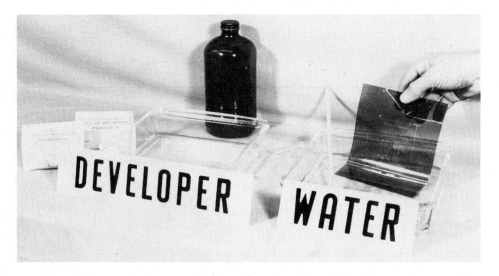

**Figure 10.12** Warm water used to process film after developing.

**Figure 10.13**    Removing excess moisture from both sides of a silk screen.

The silk-screen negative stencil of the original artwork master can now be obtained by firmly pressing the soft emulsion of the film against the surface of the silk screen. After the entire emulsion side has been pressed against the silk screen, both sides of the screen are gently blotted with soft absorbent paper towels to remove any excess water (Fig. 10.13). A block-out material, such as *Ulano No. 60 Water Soluble Fill-in,* is then spread with a small brush over the unused areas of the screen bounding the stencil. The block-out confines all excess paint or resist to the upper surface of the screen during printing (Fig. 10.14).

The screen is then allowed to dry thoroughly. If a small fan is available, drying time will require only approximately 30 minutes. After the screen has dried, the transparent backing of the film is peeled from the screen, which is then carefully inspected by tilting it back and forth while holding it up to an intense light. No residual film should remain on those areas of the screen that are not blocked and appear as translucent. This residual film will appear to glisten and, if not removed, will impair smooth transfer of the paint or resist through the screen stencil during its application. This film is removed with standard mineral-oil spirits and a stiff brush. The emulsions forming the stencil or the block-out are resistant to these solvents and will not be affected by this cleaning process. The use of water, however, should be avoided because these emulsions are water soluble.

The stencil just developed is now ready for use. Before proceeding, however, some consideration needs to be given to the type of screen to select for a particular application. In this section, the concern centers around screens for marking and solder masks. Therefore, a moderate degree of control must be maintained on positioning paint or resist flow through the screen in order to effect acceptable line definition.

Screens are classified in two categories, *mesh number* and *durability.* Mesh numbers for silk screens, generally running from No. 1 to No. 25, refer to the number of filaments per inch. A No. 1 mesh screen has approximately 30 filaments per inch and is considered very coarse. Mesh No. 25 represents a fine screen of approximately 196 filaments per inch. The more the filaments per inch, the greater the control of paint or resist flow through the screen and consequently the sharper the resulting pattern. For extremely critical line definition, *monofilament* nylon mesh screens with as many as 283 filaments per inch are also available. For applying epoxy paint for marking masks and solder resist for

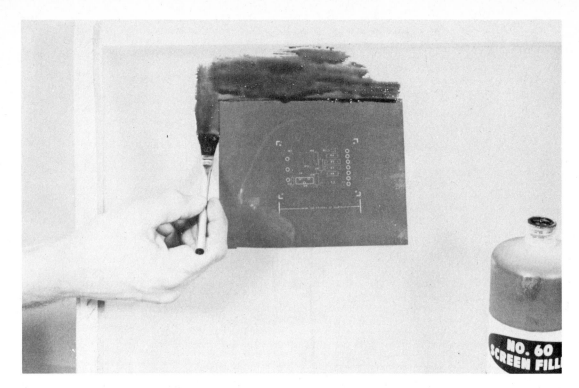

**Figure 10.14**  Block-out applied to unused areas of screen.

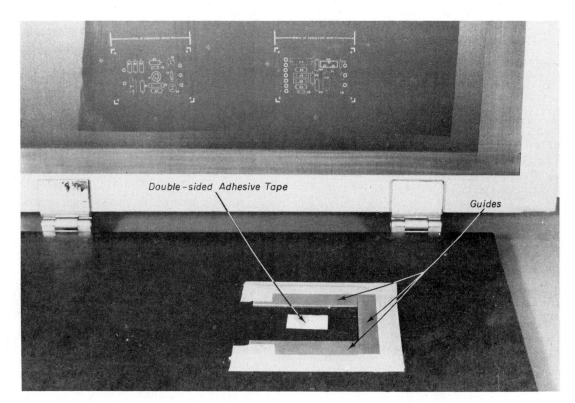

*Double-sided Adhesive Tape*

*Guides*

**Figure 10.15**  Register guides to aid in board alignment.

solder masks, a screen with a mesh number of 20 will produce adequate results. This mesh is also suitable for conductor pattern stencils.

The durability of the screen is a measure of its ability to withstand continued reuse. Durability classifications are *X*, *XX*, and *XXX*, with the latter used only if many reproductions are to be made. For prototype work, screens with a durability classification of *X* are adequate. However, if the screen is to be continually stripped clean and reprocessed for other stencils, *XX*, *XXX*, or wire mesh screens should be considered.

To facilitate the silk-screen labeling process, it is first necessary to provide a base and a set of clamp hinges to accommodate the frame and also to provide positive position-

(a)

(b)

**Figure 10.16** Use of silk screen for board marking: (a) a single stroke of the squeegee labels the board; (b) results of the screening process.

ing of the surface to be labeled. These aids are shown in Fig. 10.15. The frame is clamped to the printing base by means of the clamp hinges in such a manner that the stencil is closest to the base as the frame is swung down into the printing position. With the frame properly clamped, the pc board is placed on the printing base under the frame and its corners are aligned with the border delineation marks on the stencil. Once the desired alignment is achieved, three small *register guides* cut from either cardboard or metal, no thicker than the board, are positioned against any three board edges. This procedure is also shown in Fig. 10.15. The register guides are secured in place with masking tape and help hold the work in position during printing. They also allow for rapid positioning of identical workpieces when many boards are to be processed.

Once the work and stencil are properly aligned, a liberal amount of epoxy-type paint is poured across the left-hand margin of the marking mask stencil from top to bottom. A hard-rubber squeegee, preferably one that will span the entire stencil, is brought down onto the screen behind the pool of paint. With a sweeping motion, the squeegee is drawn across the stencil, spreading the paint as it moves. This technique is shown in Fig. 10.16a. Familiarity with this technique will enable the technician to completely label the entire work surface with a single sweeping motion.

After the stenciling is completed, the screen is raised from the work (Fig. 10.16b). Small boards may tend to cling to the stencil when it is raised. Usually, they will slide slightly before they fall, thus causing the labeling to smear. This problem can be avoided by using double-sided adhesive tape placed on the underside of the work when it is first positioned on the base.

When labeling has been completed, the paint must be allowed to dry thoroughly before handling. Depending on the type of paint used, drying time will vary from 2 hours to a complete curing time of 4 days. Drying time can be reduced by baking according to

**Figure 10.17**   Solder mask with boards screened.

**Figure 10.18** Flow diagram for developing silk-screen stencils.

paint manufacturers' specifications. Certain lacquers and vinyls used in the silk-screen process for labeling will air dry in 5 to 15 minutes. No difficulty will be encountered in obtaining any desired labeling characteristics in terms of color, durability, or finish because there are hundreds of different types of paints, enamels, and lacquers available from graphical suppliers.

When the labeling process has been completed, the excess paint can be removed from the margins of the stencil with a small spatula and stored. The remaining paint is then washed off with an appropriate solvent. If the existing stencil is no longer to be used, the screen can be reclaimed for further stencil use by washing it in warm water. This will dissolve the water-soluble emulsion and blockout. Difficult areas can be gently scrubbed with a soft-bristle brush.

The discussions above have been limited to silk-screen printing for producing marking masks. This process is also used for processing solder masks. For this purpose, the procedures for developing the stencil are identical except that a 1:1 *negative* is used initially to generate the silk-screen film. Also, instead of using an epoxy paint, a solder resist, such as *Lambert No. 184-10-V green solder resist,* is applied through the screen.

Silk-screen stencils and finished boards with solder masks applied are shown in Fig. 10.17. Because of the complexity of the silk-screen process, a flow diagram is provided in Fig. 10.18 to help the technician follow the various steps.

After the pc boards have been processed, the components and hardware must be mounted and leads soldered securely to the conductor pattern. The procedures and equipment necessary for this phase of construction are discussed in the succeeding chapters.

## EXERCISES

### A. Questions

10.1 What is the purpose of photo resist?

10.2 How should pc board copper surfaces be cleaned prior to the application of photo resist?

10.3 List the various methods of applying photo resist.

10.4 Define the term *etching.*

10.5 What would be the result if a positive phototool was used to contact print the photosensitized board instead of a negative phototool?

10.6 What is the correct orientation for the negative when contact printing the photosensitized board?

10.7 Define the term *polymerize.*

10.8 What precaution must be observed when developing exposed photo resist?

10.9 After the application of liquid photo resist, why are the pc boards dried in an oven?

10.10 Determine the required drill sizes for $\frac{1}{4}$-watt resistors, disk capacitors, and DO-14 case leads. Which component requires the largest hole size?

### B. True or False

Circle T if the statement is true, or F if any part of the statement is false.

10.1 Etching efficiency increases with agitation and decreased temperature.     **T   F**

| | | |
|---|---|---|
| 10.2 | Before drilling a pc board, each terminal pad is center punched. | **T  F** |
| 10.3 | The flow coating of photo resist requires no special equipment. | **T  F** |
| 10.4 | Ferric chloride is commonly used as an etchant. | **T  F** |
| 10.5 | Drilling of a pc board is performed from the opposite side of the copper conductor pattern. | **T  F** |
| 10.6 | The contact films used in the development of silk-screen stencils do not require that they be used under safe-light conditions. | **T  F** |
| 10.7 | Some grades of pc board require elevated temperatures prior to punching. | **T  F** |
| 10.8 | Etchant resist is removed by scrubbing the board with a Scotch-Brite pad. | **T  F** |
| 10.9 | Developer solution for Ulano Super Prep film is light-sensitive. | **T  F** |
| 10.10 | A silk screen with a mesh number of 2 is ideally suited for conductor pattern stencils. | **T  F** |

## C. Multiple Choice

Circle the correct answer for each statement.

10.1   Light-sensitive photo resist is exposed with a (*black, infrared*) light.

10.2   When preparing a bath of hydrochloric acid cleaner the (*water, acid*) is slowly added to the (*water, acid*).

10.3   The drilling of a processed pc board is performed from the (*insulation, copper*) side.

10.4   Kodak KPR liquid resist is (*clear, amber*) in color.

10.5   To expose photo resist, the negative is positioned so that literal and numerical designations are (*forward-, reverse-*) reading.

10.6   An exposed board to be developed is placed into the developing tray with the imaged copper side facing (*up, down*).

10.7   After the touch-up process has been completed, the board is baked to remove (*gases, solvents*) from the resist.

## D. Matching Columns

Match each item in column A to the most appropriate item in column B.

| COLUMN A | COLUMN B |
|---|---|
| 1.  Copperbrite | a.  Clean copper foil |
| 2.  Sun lamp | b.  Rubber squeegee |
| 3.  Water "sheeting" | c.  2 minutes |
| 4.  Etchant | d.  Contamination remover |
| 5.  Photo resist developing | e.  3600 angstroms |
| 6.  Silk-screening | f.  30 filaments/inch |
| 7.  Mesh number | g.  Ferric chloride |

## E. Problems

10.1 Prepare silk screens for the marking and solder mask layouts designed for the tachometer circuit of Problem 6.4. (The screens are used in Problem 10.4.) Obtain the meter movement specified for the circuit. Using a commercially available tachometer as a calibration reference, redesign the meter scale and construct a silk screen for the new scale.

10.2 Fabricate a pc board from the artwork master of the seven-fixed-interval UJT timing circuit designed in Problem 6.2 from $\frac{1}{16}$-inch FR-4 board with 1-ounce copper foil. Components are assembled in Problem 16.2.

10.3 Construct a pc board from the artwork master for the series feedback regulator circuit designed in Problem 6.3 from $\frac{1}{16}$-inch FR-4 board with 1-ounce copper foil. Components are assembled and soldered in Problems 16.3 and 17.3, respectively.

10.4 Fabricate a pc board from the artwork master for the tachometer circuit designed in Problem 6.4 from $\frac{1}{16}$-inch *XXXP* board with 1-ounce copper foil. Using the silk screens fabricated in Problem 10.1, construct the meter scale, mark the pc board, and mask the conductor pattern for soldering. Components are assembled to the pc board in Problem 16.4.

# Double-Sided PCB Processing: Electroless Copper Deposition and Copper Flash Plating

## LEARNING OBJECTIVES

*Upon completion of this chapter on double-sided printed circuit board copper deposition and flash plating, the student should be able to*

- Stack and shear pc boards to their correct size.
- Understand the major components and performance characteristics of production CNC drill/router machines.
- Understand the equipment and processes involved in an electroless copper deposition line.
- Leach all tanks and accessories in an electroless copper deposition line.
- Properly mix, maintain, and control all the solutions of the electroless copper deposition line.
- Successfully operate an electroless copper deposition line.
- Understand the equipment and processes involved in copper flash plating.
- Properly mix, maintain, and control the solution of the flash plating bath.
- Flash-plate copper onto pc boards.

## 11.0 INTRODUCTION

In Chapter 7, a number of design requirements for the generation of artwork suitable for double-sided pc boards were presented. Recall that major emphasis was placed on electrically interconnecting circuitry on the component side of the board to that on the circuit side. For prototype applications, where only a few boards of any specific type are processed, this side-to-side interconnection may be accomplished by first wave-soldering the processed board on the circuit side and then hand-soldering the leads on the component side to its circuitry. Although this method is time-consuming, it is adequate for obtaining the desired results.

Volume production manufacturing requires that side-to-side interconnections be made an integral part of the pc board in the fabrication process. This interconnection technique, adopted by high-volume pc board manufacturers, is the *plated-through-hole* (PTH) process, which lends itself to mass-production applications and results in a much higher degree of finished board control. Because it is an extensive multistepped process, the discussion of double-sided plated-through-hole pc board fabrication will extend into the following two chapters. The following chapter descriptions provide an overview of this complex process.

This chapter begins with a detailed discussion of shearing pc board stock to size and then drilling all of the required holes. Through a series of chemical baths, a very thin layer of copper is then deposited electrolessly over the entire board surface covering all exposed areas, including the edges and the walls of the drilled holes. This layer of copper results in the initial electrical interconnection between the copper foil on both sides of the board through the plating of the walls of each drilled hole. To increase the thickness of the copper deposited on the walls of the holes, the board is further treated in a *copper flash plating* process, which is a standard electrolytic method employing a copper plating bath and electric current.

In Chapter 12 the fabrication process continues with the application of dry-film photo resist to image the copper foil on both sides of the board with the desired conductor patterns. (Double-sided conductor pattern artwork is presented in Chapter 7.) The board is completed in Chapter 13 by first additionally electroplating the exposed conductor pattern and hole walls with copper followed by a solder electroplating process. This solder plate serves as the etching resist and ensures a highly solderable pc board. After all of the plating is completed, the dry-film resist is stripped from the board followed by etching, reflowing, and finally routing to size.

As can be seen from this introduction, the fabrication of a double-sided plated-through-hole pc board requires many sequential processes, some of which demand exacting procedures. For this reason, each of the processes are described in detail.

*Caution: A chemical hazard exists in the materials described in this chapter. Approved face shields (goggles), chemical-resistant gloves, clothing, footware, and a respirator must be worn prior to performing any of these operations.* Also, the section on safety outlined at the beginning of the book should be reviewed for more specific information.

## 11.1 ELECTROLESS COPPER DEPOSITION: GENERAL INFORMATION

The discussion of double-sided plated-through-hole fabrication begins with the selection of the appropriate pc board stock. For our purposes, we have selected grade FR-4 double-sided board stock having a thickness of 0.059 inch and a copper thickness of 1 ounce per square foot ($oz/ft^2$) on each side. This thickness is most often specified, although thinner copper thicknesses are becoming more popular for fine-line applications (i.e., less than 10-mil path widths and spaces). The stock is first sheared into the correct panel size to accommodate the required conductor pattern artwork (see Chapter 7).

Two precisely located tooling holes are then punched into each panel as well as into a layer of backup material and entry material. All these layers are then stacked and pinned together for drilling. A pinned lay-up of panels for drilling is shown in Fig. 11.1. The bot-

**Figure 11.1** Stacking and pinning setup used on CNC machines for multipanel drilling.

tom backup panel is a piece of $\frac{1}{8}$-inch XP hardboard material. Three double-sided copper clad pc boards are shown stacked on top of the bottom panel. The top drill-entry material is a thin sheet of aluminum. The purposes of the entry and backup panels are to (1) prevent burring of the drilled holes, (2) clean and cool the drill bit to prevent tearing hole walls, and (3) protect the underlying surface of the drilling machine worktable. The layers are held firmly together with steel stacking pins especially made for this purpose. The pins are pressed into the previously drilled tooling holes.

A typical multispindle CNC (computer numerical control) drilling machine, used to automatically drill separate stacks of panels simultaneously, is shown in Fig. 11.2. These machines are also used to route individual pc boards to size and shape. The major components of a CNC drill/router machine are (1) computer controller; (2) granite slab base; (3) moving worktable, servomotor drive, and feedback system; and (4) directly programmable 15,000- to 80,000-rpm spindles. The massive base is required to resist the immense inertial forces of the rapidly moving worktable as well as to support the entire machine structure.

Typical drilling performance characteristics of CNC drilling machines are (1) $x$ and $y$ table positional speeds of up to 500 inches per minute, (2) positioning accuracies of ±0.0002 inch with repeatability of ±0.0001 inch, (3) drill hole accuracy of ±0.001 inch, (4) production speeds of 400 "hits" (drilled holes) per minute, and (5) automatic drill size changes in 10 seconds. The CNC machine is programmed to drill multiboard panels in a step-and-repeat mode and to take optimum paths between holes to further reduce drilling time.

After the boards have been drilled, they are unstacked and mechanically deburred on both sides using 400-grit sand paper. These boards are now ready for the *electroless copper deposition* process. This process will deposit a layer of copper approximately 20 to 30 millionths of an inch ($\mu$inches) thick on all board surfaces, that is, the copper foil on both sides, the board edges, and most important, around the inner glass/epoxy surface of each drilled hole. To accomplish this, the drilled boards are first secured in specially designed racks and then processed through 16 chemical baths and water rinses. In this plating line, the chemical makeup of baths, the solution temperatures, the immersion time rate of board agitation, the air agitation of solution, and the quality of water rinses are rigidly

**Figure 11.2** Automatic driller/router machine. *Courtesy of Excellon Automation, Division of Excellon Industries, an Esterline Company.*

**Figure 11.3** Cross-sectional view of a double-sided plated-through-hole board after copper deposition.

Insulating Material

Electroless Copper Deposited on Surface of a Drilled Hole

Copper-to-Copper Interconnection

Laminate Copper Foil

≈25 µin

Laminate Copper Foil (Typically 1oz/ft²)

controlled. After the boards have been processed through the 16-bath copper deposition line, they will exhibit a slightly brownish copper appearance throughout, that is, on the copper foil surfaces, the edges, and walls of the drilled holes. A cross-sectional view of a board after it has been processed with the copper deposition baths in shown in Fig. 11.3.

## 11.2 ELECTROLESS COPPER DEPOSITION LINE: MECHANICAL CONSIDERATIONS

A typical electroless copper deposition line is shown in Fig. 11.4a. It consists of 16 stations, or bays, all of which are the same size. In the line shown, each bay is designed to accommodate a 10-gallon (37.85-liter) bath. Thus their physical dimensions are approximately 6 inches wide, 24 inches deep, and 24 inches high. The bays are constructed of a welded rigid plastic material such as stress-relieved polypropylene with PVC plumbing. Each bay is fitted with a PVC ballcock drain valve leading into a central drain manifold to facilitate solution removal. A high-torque low-rpm motor is mounted to the left side of the deposition line. The motor is coupled through a cam and a 1-inch-square fiberglass drive bar to a pair of 1-inch-square rails. These rails are positioned on Teflon guides and run the length of the line along the top outside edges. The racked boards are lowered into each of the baths with the arms of the rack resting on the rails. A typical rack is shown in Fig. 11.4b. The motor-driven rails provide a transverse motion of the racked boards which forces the solution through the holes as the boards travel back and forth in the baths. The rate of travel should be relatively slow, not to exceed 12 complete cycles per minute with a 2- to 4-inch stroke cycle.

The deposition line can be divided into two major bay groupings: (1) water rinse stations and (2) chemical treatment stations. There are eight of each of these groupings in the line. Refer to the deposition line shown in Fig. 11.4a. The first bay on the left is the only one required to be operated above room temperature. It is fitted with a 1000-watt stainless steel heater that is sufficient to raise the temperature of 10 gallons of solution to the required 150°F (66°C) in a reasonable amount of time. A temperature control with a probe is used to maintain this temperature and prevent overheating of the solution. Bay 14 in Fig. 11.4a is the electroless copper bath and is fitted with a length of PVC tubing connected to an aquarium pump. The tubing extends to the bottom of the bay. The bubbles generated from the pump aid in stabilizing the electroless copper solution.

Each of the rinse stations is equipped with an overflow drain near the top rim of the bay. This is shown in Fig. 11.4a. Rinse water is supplied from a mixing valve that regu-

**Figure 11.4** Standard 16-bay electroless copper deposition line: (a) detailed front-sectional view; (b) 316 stainless steel rack for copper deposition.

**189**

lates the temperature to a central manifold water system (PVC) that feeds each rinse station. From this system, a $\frac{1}{2}$-inch PVC water pipe is routed through the rear wall of each rinse bay, down the back of the tank, and runs along the bottom center of the tank. Several small equally spaced holes are drilled into the bottom section of the pipe. In this way, clean water enters continuously from the bottom of the tank and moves upward, finally leaving the bay through the drain at the top. This results in contaminants, which are generated from rinsing the racked boards, being forced upward and out into the drain. For optimum results, the rinse-water temperature should be maintained between 60 and 70°F (16 and 21°C). Sufficient volume of water to have a bath turnover of from 5 to 10 times the capacity of the tank per hour should be supplied. This will maintain the level of rinse-water contamination, even for a constantly working line, within acceptable limits. For the relatively small deposition line shown in Fig. 11.4a, which has eight 10-gallon rinse bays, a turnover of 10 times per hour for each bay would require a water supply of 800 gallons per hour.

To reduce the large volume of water required and to result in more effective board rinsings than that provided by rack movement alone, a PVC air sparger is installed in each of the rinse stations. These $\frac{1}{2}$-inch-diameter pipes are L-shaped and enter the top of the bay through a hole at the rear of the tanks. They run downward and along the bottom of the tanks. Each sparger has equally spaced holes drilled along the bottom of its total horizontal length. This pipe is returned to a central manifold which in turn is connected to an oil- and dust-free blower system equipped with airflow control valves. Air entering each rinse bay at the rate of $1\frac{1}{2}$ to 2 cubic feet per minute for each square foot of tank surface will result in greatly increasing the solution movement. This provides more effective rinsing with minimum water input required.

The rack used with the copper deposition line is made from 316-type stainless steel and is designed so that it will rinse freely and entrap no chemicals. The rack shown in Fig. 11.4b has a two-tier arrangement of slotted spacers. The slots are approximately $\frac{1}{2}$ inch apart, $\frac{1}{4}$ inch deep and 20 thousandths of an inch wider than the thickness of the pc boards to be processed. Boards mounted onto the rack are therefore slightly off vertical, which results in good solution-to-hole movement and more effective rinsing. The rack is fitted with front and back extensions so that it may be positioned onto the work rod agitator.

In this section we have described the mechanical aspects of a typical low-production, hand-operated deposition line common to many printed circuit fabrication shops. To aid in efficiency, several improvements can be made to this basic line. For example, the addition of a hoist to raise and lower the rack into the baths would increase productivity. Also, the installation of conductivity meters in the rinse stations to monitor the level of contamination would reduce production time by sounding an alarm when levels of contamination exceed preset limits.

In the next section we describe the procedures for starting up a new deposition line and also the chemical baths involved in the copper deposition process.

## 11.3 ELECTROLESS COPPER DEPOSITION: CHEMICAL PROCESSES

After the construction of a new copper deposition line, all bays, racks, and hardware must be thoroughly *leached* (chemically cleaned) to remove all contaminants prior to filling any of the tanks with chemical baths or rinse water. All heaters, plumbing, and spargers should be installed so that they may be leached at the same time. The leaching process is required only for startup purposes, that is, before the chemicals are added to the tanks for the first time. A detailed leaching process is given in Appendix IX. The student should refer to this appendix before proceeding to the next section. It should be emphasized that even though a new tank may look clean, it is probably the single largest source of bath contaminations. Those contaminations, such as mold release film, are typically the cause of many copper deposition failures. For this reason it is absolutely essential that all parts which will come into contact with any of the chemical baths be properly leached.

The basic 16-bay deposition line as described in Sec. 11.2. and shown in Fig. 11.4a can be divided into three major categories, described by the functions of each bath. Viewing the line from left to right, these categories are (1) bays 1 through 7 for *cleaning*, (2) bays 8 through 13 for *activating*, and (3) bays 14 through 16 for *copper deposition* and *holding* (see Fig. 11.5). During the cleaning cycle, the copper surfaces of the pc board as well as the hole walls and board edges are cleaned and conditioned to accept activation and subsequent copper deposition. The *activation* process "seeds" (deposits) particles of precious metal (palladium) onto the surfaces of the nonconductive hole walls in order to provide a site for subsequent copper deposition. In the *copper deposition* process, these sites are deposited with 20 to 30 millionths of an inch (20 to 30 μinches) of copper resulting in a completely metalized board capable of being *flash* electroplated with copper (Sec. 11.5) to a thickness of approximately 0.000125 inch (125 μinches).

To describe the complete deposition process in detail, a standard *low-speed electroless copper process* developed by the *Shipley Company, Inc.* is shown in flow-chart form in Fig. 11.5 and it will be discussed in detail. This is called the *Cuposit 328Q* process and it will be described using the 16-bay deposition line discussed in Sec. 11.2 and shown in Fig. 11.4a. Recall that each of the bays has a capacity of 10 gallons.

As shown in Fig. 11.5, the boards are first racked in accordance with the discussion of Sec. 11.2. Remember that during this entire process, the rails supporting the plating

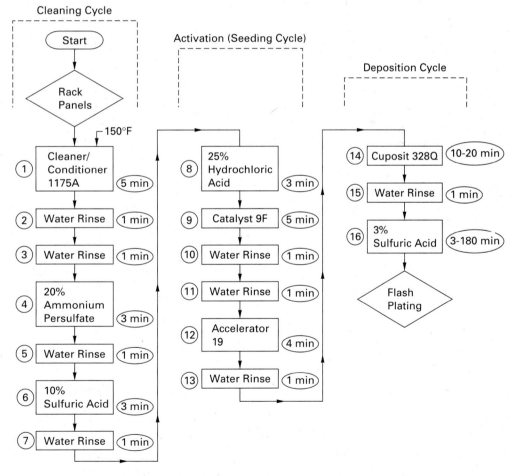

**Figure 11.5** Flow chart depicting the Shipley copper deposition 328Q process.

rack are driven back and forth to ensure the proper movement of solution through the holes in the board.

Each of the baths shown in Fig. 11.5 will now be described.

**Bay 1: Cleaner/Conditioner 1175A.** This is a strongly alkaline bath designed to (1) clean oils, stains, and oxides from the copper surfaces, and (2) condition the glass/cloth hole wall surface to accept the activation (seeding) layer. This 2.5% by volume bath is prepared by adding 0.25 gallon (946 milliliters) of 1175A concentrate to 9.75 gallons (36.9 liters) of tap water to result in 10 gallons of working solution. With the bath maintained at an operating temperature of 150°F to a maximum of 180°F (66 to 82°C), the racked boards are loaded into the solution for 3 to 5 minutes.

**Bay 2: Water Rinse.** The rack is removed from bay 1, allowed to drain, and then rinsed in this room-temperature (≈70°F) water bath for 1 minute. Both air and water spargers are used in this and subsequent rinse baths to prevent chemical drag-out from one bath to the next.

**Bay 3: Water Rinse.** The rack is removed from bay 2, allowed to drain, and then again rinsed for 1 minute in water at room temperature.

**Bay 4: 20% Ammonium Persulfate.** To ensure a good copper-to-copper bond (electroless to foil), a light etch is required on the copper foil to coursen the surface and also to remove any contamination. This *micro-etch* bath is prepared by adding 15 pounds of ammonium persulfate crystals to 10 gallons of tap water. (This is a ratio of $1\frac{1}{2}$ pounds of AP per gallon of water.) The solution is stirred with a leached paddle until all the crystals are dissolved. Finally, 10 ounces (300 milliliters) of reagent or chemically pure (CP) grade sulfuric acid ($H_2SO_4$) is added to complete the working solution.

The rack is removed from bay 3, allowed to drain, and then placed into the ammonium persulfate micro-etch bath, which is maintained at room temperature, for 3 minutes. After this process, the boards should exhibit a uniform mat salmon color over all copper surfaces.

**Bay 5: Water Rinse.** The rack is removed from bay 4, allowed to drain, and rinsed in room temperature water for 1 minute.

**Bay 6: 10% Sulfuric Acid.** This bath is necessary to ensure that all complex ammonium persulfate crystals embedded into the copper surface from the micro-etch bath are removed. To make 10 gallons of 10% solution of sulfuric acid, 1 gallon (3.785 liters) of reagent-grade sulfuric acid is slowly added to 9 gallons (34 liters) of water.

The rack is removed from bay 5, allowed to drain, and then loaded into the 10% sulfuric acid bath, maintained at room temperature, for 2 to 3 minutes.

**Bay 7: Water Rinse.** The rack is removed from bay 6, allowed to drain, and then rinsed in room-temperature water for 1 minute.

**Bay 8: 25% Hydrochloric Acid.** This bath will further prevent the next bath (bay 9) from being contaminated with any of the previously used chemicals. To make 10 gallons of 25% hydrochloric acid solution, 2.5 gallons (9.46 liters) of reagent-grade hydrochloric acid is slowly added to 7.5 gallons (28.39 liters) of distilled water.

The rack is removed from bay 7, allowed to drain, and loaded into the 25% hydrochloric acid solution for 1 to 5 minutes. This bath is also operated at room temperature. No rinse is required after this process since this solution is compatible with that of bay 9.

**Bay 9: Catalyst 9F.** This is a proprietary colloidal solution of palladium–tin in hydrochloric acid. In this solution, the palladium is surrounded and protected by stannic tin molecules. The purpose of this bath is to "seed" (implant) palladium onto the nonconductive glass/epoxy wall surfaces of the drilled holes. This is necessary so that these surfaces will accept the subsequent electroless copper deposition process. To prepare a 10-gallon bath of Catalyst 9F, 4 gallons (15.14 liters) of either reagent-grade or CP-grade hydrochloric acid is slowly added to 4 gallons of distilled water and stirred to mix. To this solution, 2 gallons (7.57 liters) of Catalyst 9F concentrate is added and again stirred.

The rack is removed from bay 8, allowed to drain, and then loaded into the Catalyst 9F bath for 3 to 10 minutes. This bath is operated at room temperature.

**Bay 10: Water Rinse.** The rack is removed from bay 9, allowed to drain, and then rinsed for 1 min in room-temperature water.

**Bay 11: Water Rinse.** The rack is removed from bay 10, allowed to drain, and again rinsed for 1 min in room-temperature water.

**Bay 12: Accelerator 19.** Accelerator 19 is a proprietary solution designed to dissolve the tin molecules that were deposited in bay 9. This exposes the palladium sites to achieve uniform copper deposition. The Accelerator 19 bath also increases the life of the subsequent copper bath by minimizing drag-in of Catalyst 9F. To prepare a 10-gallon bath of Accelerator 19, 1.666 gallons (6.306 liters) of concentrate is added to 8.333 gallons (31.54 liters) of distilled water.

The rack is removed from bay 11, allowed to drain, and then loaded into the Accelerator 19 bath, operated at room temperature, for 4 to 8 minutes.

**Bay 13: Water Rinse.** The rack is removed from bay 12, allowed to drain, and then rinsed in room-temperature water for 1 minute.

**Bay 14: Cuposit 328Q Electroless Copper Bath.** Cuposit 328Q is also a proprietary solution designed to deposit approximately 25 $\mu$inches of copper onto properly *activated* surfaces. To prepare 10 gallons of Cuposit 328Q solution, add 1.25 gallons (4.73 liters) of Shipley Copper Mix 328A to 7.5 gallons (28.39 liters) of distilled water and stir thoroughly. To this mixture, 1.25 gallons of Copper Mix 328Q is added and again stirred well. Finally, 0.25 gallon (946 milliliters) of Copper Mix 328C is added and stirred. This bath is more stable if gentle air bubbling is allowed to agitate the solution to reduce the formation of cuprous oxide.

The rack is removed from bay 13, allowed to drain, and then loaded into the Cuposit 328Q bath for a period of time ranging from 8 to 20 minutes, depending on the bath temperature and solution concentration. These time and concentration relationships are shown in Fig. 11.6. The typical range of bath operating parameters are shown on the three-dimensional surface response curve. The plating time varies from a low of 8 minutes if the bath is at its optimum copper concentration of 100% and at its highest recommended temperature of 78°F (point *a*) to as much as 20 minutes if the bath is operated at its low limit of 65°F with a copper concentration of 80% (point *h*). The shaded area of the response curve bounded by points, *a, d, h,* and *e* shows the effects that both temperature and copper concentration have on plating time.

Plating time can additionally be determined from either of the simplified two-dimensional curves also shown in Fig. 11.6. The time-versus-percent copper concentration graph relates plating time as a function of % copper for each of four constant temperatures. The time-versus-bath temperature graph relates plating time as a function of temperature for two constant copper concentrations (80 and 100%).

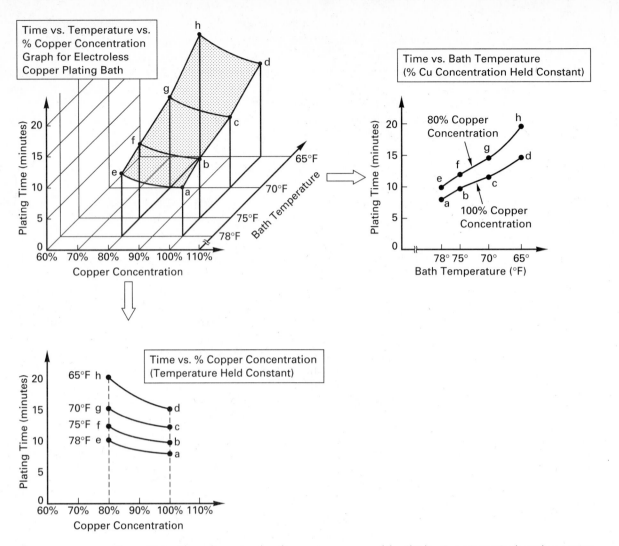

**Figure 11.6** Three-dimensional surface response curve of the Shipley Cuposit 328Q electroless copper deposition bath. This graph demonstrates the effect on plating time as both bath temperature and percent copper concentration are varied. All data presented are to obtain approximately 25 μinches of electroless copper.

**Bay 15: Water Rinse.** The rack is removed from bay 14, allowed to drain, and then rinsed in room-temperature water for 1 minute.

**Bay 16: 3% Sulfuric Acid.** This bath is intended as an alkaline neutralizer and also serves as a storage tank to prevent the electroless deposited copper on the processed boards from rapidly oxidizing which would occur if they were exposed to air. To prepare 10 gallons of 3% sulfuric acid, 0.3 gallon (1.46 liters) of reagent- or CP-grade sulfuric acid is slowly added to 9.7 gallons (36.71 liters) of water.

The rack is removed from bay 15, allowed to drain, and then loaded into the 3% sulfuric acid bath for a period of time from 3 minutes up to 3 hours. This bath is operated at room temperature.

After the boards have been processed through the 16-bay electroless copper deposition line, they are completely coated with approximately 25 μinches of copper. The next process is to immediately panel plate or flash plate the boards to build up the copper thickness on the hole walls an additional 0.001 inch (100 μinches). This will ensure electrical integrity between the copper foil surfaces on opposite sides of the board through the plated holes.

Before introducing the copper flash plating process, we complete the discussion of electroless copper deposition by considering visual inspections for bath monitoring and bath replenishments necessary to maintain quality control.

## 11.4 ELECTROLESS COPPER DEPOSITION: PROCESS MONITORING AND CONTROL

To maintain the deposition line in good chemical balance, several monitoring operations must be undertaken on a routine basis. Most often, monitoring is performed by visual interpretation of bath color or by the amount of dwell time that the boards are processed through any given bay. For the Shipley low-speed electroless copper deposition line described in the preceding section, each of the baths is considered in terms of their monitoring and process control.

**Water Rinse.** Rinse water should always be maintained at temperatures above 60°F with a rate of flow from 5 to 10 bath volumes per hour. If air agitation is provided, the lower rate of flow may be used with good results. To maximize the effectiveness of the rinse water, a *solution conductivity* or *resistivity* meter should be used. A specially designed test probe leading to the meter is placed directly into the rinse water. (The readings of either one of these types of meters is simply the reciprocal of the other.) These meters measure the contamination in the rinse water by monitoring the increase of conductivity. As the level of contamination rises, the purity of the water is degraded resulting in higher values of conductivity. Conductivity is measured in units of microsiemens per centimeter (μS/cm). Ideally, the conductivity meter may be set to alert the operator when contaminants have degraded the rinse water to a conductivity reading of approximately 3500 μS/cm. For this reading, the rinse water should be discarded and replaced with fresh water.

**Cleaner/Conditioner 1175A.** Tap water is used to replenish the solution due to evaporation loss caused by the high operating temperature of this bath. The strength of the bath can be checked visually. If *light* oxides are not removed from the copper surface after a dwell time in the bath of 1 minute, the solution should be discarded and replaced with a fresh batch. Another method of monitoring the effectiveness of this bath is to keep account of the number of square feet of pc boards that have been processed. The bath should be replaced after 200 square feet of surface area per gallon of solution are processed. For our 10-gallon tank, this would translate into 2000 square feet of board surface.

**20% Ammonium Persulfate.** This solution should be replaced when its color becomes medium blue. In addition, the solution strength can be monitored by visual inspection of the board surfaces. If a uniform pink or salmon appearance is not displayed over the entire copper surfaces on both sides of the board after the proper dwell time in the bath, the solution should be replaced.

**10% Sulfuric Acid.** This solution should be replaced when its color becomes light blue.

**25% Hydrochloric Acid.** This solution should be replaced when its color becomes pale green.

**Catalyst 9F.** This solution will exhibit a dark brown-to-black color at room temperature when freshly prepared. A crude check on this bath is to observe if the color turns to a pale amber. This is an indication that the solution has not been properly maintained and should be replaced. Another check on this solution is to observe the boards after they have been loaded for the proper dwell time. A uniform tan color should appear on all insulation surface edges for FR-4 and G-10 grade boards. This bath should be maintained at a strength of at least 10% by periodic replenishments made according to the schedule shown in Fig. 11.7. The procedure for determining the concentration of Catalyst 9F is as follows:

1. Maintain the bath level by additions of concentrated hydrochloric acid (reagent or CP grade).
2. Remove a 20-milliliter sample of the bath. In a separate beaker, mix 45 milliliters of distilled water and 15 milliliters of reagent-grade hydrochloric acid. This constitutes 60 milliliters of a 25% by volume solution of hydrochloric acid.
3. Mix the 20-milliliter sample and the 60 milliliters of 25% hydrochloric acid solution into a bottle that is the same size and shape of the color standard bottles that are available from the manufacturer and, in our case, from the Shipley Co., Inc. The mixture is covered and the solution shaken well. The color of the sample is then compared with the color standards that indicate various bath strengths. This comparison is made by first spacing the color standard bottles approximately 2 inches apart in good lighting (near a window works well). The prepared sample bottle is then placed between adjacent standard bottles and moved until its color most closely matches one of the standards. The percent concentration of the sample is equal to the value written on the cap of the matching standard bottle.
4. Replenish the bath by first removing a bath volume equal to the amount of Catalyst 9F to be added according to the replenishment schedule shown in Fig. 11.7.

**Accelerator 19.** The level of this bath should be maintained with distilled water only. A crude check on the strength of the solution is to observe when it turns cloudy or blue. At that time it should be immediately replaced. The amount of board processing can also be used as a measure of solution control. After 300 square feet of pc board stock per gallon of solution have been processed, the bath should be replaced.

**Cuposit 328Q.** Shipley's low-speed electroless copper bath is deep blue in color. The copper concentration is controlled by using a set of blue color standards, which are

| Catalyst 9F Replenishment Schedule for a 10-Gallon Bath | |
|---|---|
| Bath Strength Determined by Comparison Against Color Standards | Additions* of Catalyst 9F Concentrate |
| 100% | None |
| 90% | 750 ml |
| 80% | 1500 ml |
| 70% | 2250 ml |
| 60% | 3000 ml |
| 50% | 3750 ml |

Maintain Concentration Above 70%

*Note: Remove an equal volume of bath before making additions. Mix bath completely after additions are made.

**Figure 11.7** Recommended replenishment schedule for the catalyst 9F bath.

obtained from the manufacturer. The color-matching procedure is similar to that described for the Catalyst 9F solution. The procedure for determining the copper concentration is as follows:

1. Measure out 89 milliliters of Cuposit copper mix color indicator, which is obtained from the manufacturer.

2. Add 11 milliliters of electroless copper bath to the above. The solution is thoroughly mixed and placed in the same size and shape bottle as the color standards. The prepared sample is then compared against the blue color standards to determine the copper concentration of the bath.

3. Replenish the bath by first removing a bath volume equal to the amount of 328A and 328Q concentrates according to the replenish schedule shown in Fig. 11.8.

It should be emphasized that the bath should be replenished to the 100% level for each morning startup. During the day, sufficient work should be processed through the bath so that it is depleted to 70% copper concentration or below for overnight storage.

The three primary elements in the Cuposit 328Q solution are *copper sulfate, sodium hydroxide,* and *formaldehyde*. The copper sulfate provides the copper metal for deposition, the sodium hydroxide controls the pH (alkalinity) of the solution and the formaldehyde acts as a reducing agent during the plating process. Even though there are other elements in the solution, these three are mainly expended during the plating process and must be replenished. As was previously discussed, the copper concentration is monitored and controlled through the use of color standards. Both the sodium hydroxide and the formaldehyde can be monitored and controlled through chemical analysis with the use of a pH meter and a titration procedure. A less exacting but effective monitoring and control method is by visual inspection of the processed boards according to the following procedure. Over a weekend of bath inactivity, the copper concentrate of the solution is first checked and replenished to operating strength using the color standards as previously described. A pc board is processed in the bath and it is inspected to see if the copper appearance is *pink* in color. If not, sodium hydroxide and formaldehyde must be added. The Shipley Company supplies sodium hydroxide as a concentrate which is labeled Cuposit Z and formaldehyde concentrate labeled Cuposit Y. These concentrates are added to the solution in 200-milliliter steps. First, 200 milliliters of Cuposit Z is added followed by 200 milliliters of Cuposit Y. Again, a board is processed and inspected. If it is pink in color,

| Cuposit 328Q Replenishment Schedule for a 10-Gallon Bath | | |
|---|---|---|
| Copper Concentration of Bath Determined by Color Standards | 328A Additions* | 328Q Additions* |
| 100% | None | None |
| 90% | 475 ml | 475 ml |
| 80% | 950 ml | 950 ml |
| 70% | 1425 ml | 1425 ml |
| 60% | 1900 ml | 1900 ml |
| 50% | 2375 ml | 2375 ml |

Maintain Bath Above 70% for Daily Operation

*Notes: 1. Remove an equal volume of bath before making additions.
2. Replenish bath by adding the correct amount of 328A concentrate <u>first</u>, mixing the bath completely and then adding the correct amount of 328Q concentrate. Again stir bath completely to mix solution.

**Figure 11.8** Recommended replenishment schedule for the Cuposit 328Q electroless copper bath.

no further additives are required. If it does not appear as pink, additional Cuposit Z and Cuposit Y are added in 200 milliliter steps and another board is processed and inspected. This procedure continues until a processed board appears as pink.

**3% Sulfuric Acid.** This bath must be in a pH range of from 1 to 2. The solution should be replaced when the pH level falls outside this range.

Monitoring and controlling of the copper deposition line may first appear tedious and difficult. However, with a short period of experience, you will find that it will become quite simple and will require only a modest amount of time each day.

## 11.5 FLASH PLATING: GENERAL INFORMATION

In the preceding section it was described how drilled pc boards were completely metalized with a thin ($\approx$25 μinches) deposit of electroless copper over all surfaces and the walls of each hole. These processed boards are allowed to remain in the 3% sulfuric acid solution in preparation of a copper *flash* plating which will build up the copper thickness to approximately 0.000125 inch (125 μinches). This plating is required to prevent the electroless copper deposited on the hole walls from oxidizing, which would result in voids, or nonconducting areas, after subsequent board processing.

Flash plating is accomplished in a plating tank containing a copper sulfate plating solution. The pc board is suspended in this solution and electrically connected to the negative side of a power supply. The positive side of the power supply is connected to a large copper *anode*, which is also suspended in the plating solution. An electric current is passed through the plating tank causing copper ions from the plating solution to be attracted to the board and be deposited onto all copper surfaces. These copper ions are supplied to the plating solution by the corrosion of the copper anode. A thickness of 125 μinches of copper can be electroplated by this method in a little more than 4 minutes. The result of flash plating is shown in Fig. 11.9. Note that the buildup of copper is continuous and uniform over the entire copper surfaces of the board in addition to the hole walls.

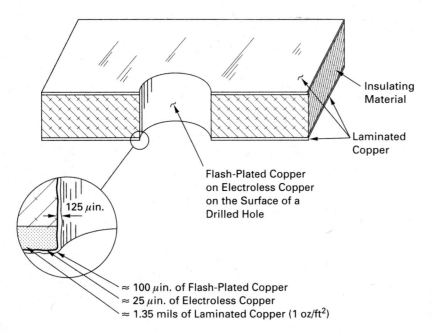

125 μin.

Flash-Plated Copper on Electroless Copper on the Surface of a Drilled Hole

Insulating Material

Laminated Copper

$\approx$ 100 μin. of Flash-Plated Copper
$\approx$ 25 μin. of Electroless Copper
$\approx$ 1.35 mils of Laminated Copper (1 oz/ft$^2$)

**Figure 11.9** Cross-sectional view of a double-sided, plated-through-hole board after flash plating.

## 11.6 COPPER PLATING TANK AND PLATING ACCESSORIES

This section describes the plating tank, plating rack, and the power supply for a typical prototype plating process. A rack for plating printed circuits holds either one or more boards in a fixed position in the plating solution. It must be capable of handling the large currents required in the plating process and not be easily affected by the plating solution. A basic plating tank setup is shown in Fig. 11.10a. The rack contacts the cathode work rod of the plating tank and is held securely with a wing nut or thumb screw. Tight contact between the thumb screw and cathode reduces resistance and holds the rack rigidly to the rod. The lower section of the rack connects to the copper foil of the pc board and is held firmly by another thumb screw or setscrew.

Plating racks are often fabricated from square copper rod stock, because this material is capable of handling 1000 amperes per square inch of cross-sectional area. A commercially fabricated plating rack capable of handling at least four boards at one time is shown in Fig. 11.10b. Because most of the rack assembly will be in contact with the plating solution, it is protected with a pestasol coating to prevent plating buildup. The contact points for the boards are fabricated of 300 series stainless steel. A smaller custom single board rack is shown in Fig. 11.10c. It is constructed of copper stock and stainless steel screws. Subsequent discussions on plating will make use of this rack.

Because the copper will be electroplated from a sulfate plating solution, the recommended material for the tank is stress-relieved polypropylene or polyethylene plastic.

A tank size of 12 inches long by 12 inches wide by 12 inches deep will be adequate for most prototype work. This tank will have a solution capacity of approximately 6 gallons (22.7 liters) when the plating solution is 9.75 inches from the bottom. The exact volume can be calculated by dividing the number of cubic inches of solution by 231 cubic inches per gallon. The following example will demonstrate this calculation.

*Example:* Determine the exact volume of plating solution for a 12 inches by 12 inches by 12 inches tank.

$$\text{Volume of solution (cubic inches)} = \text{length} \times \text{width} \times \text{depth of solution}$$
$$= 12 \text{ inches} \times 12 \text{ inches} \times 9.75 \text{ inches}$$
$$= 1404 \text{ cubic inches}$$

$$\text{Solution volume (gallons)} = \frac{\text{volume of solution (cubic inches)}}{231 \text{ cubic inches/gallon}}$$
$$= \frac{1404 \text{ cubic inches}}{231 \text{ cubic inches/gal}} = 6.08 \text{ gallons}$$

A 12-inch-deep tank with approximately 10 inches of plating solution (2 inches of clearance to the rim of the tank) will allow the insertion of the anodes and the racked pc board to be plated without causing the solution to spill over the top of the tank.

The tank is fitted with two anode work rods and one cathode work rod. These rods are made of $\frac{3}{8}$-inch-diameter solid copper rod stock. This diameter will easily support the weight of the copper anodes and the plating rack with pc boards as well as handle the large currents used in the plating process. The work rods are held securely in position with pairs of Lucite or plastic blocks fastened onto opposite ends of the tank rim. The anode work rods are positioned over the tank parallel to one another and 1 inch inward from the sides of the tank. The cathode work rod is positioned parallel to the anode rods and centered between them. The on-center spacing between either anode work rod and the cathode work rod is 5 inches (see Fig. 11.10a). These distances must remain constant between the surfaces of the board and the anodes for proper plating.

Thumb Screw

Plating Rack

Cathode Work Rod

Oil-Free Air Line

(+)

(−)

(+)

Anode Work Rods

Drilled and Electroless Copper-Plated Panel

Air Sparger

Copper Anode

Plating Tank

(a)

(b)

(c)

**Figure 11.10** Plating tank and accessories for electrolytic copper plating: (a) prototype copper plating tank; (b) commercial plating rack; (c) custom single-board rack.

For uniform plating, effective solution movement through the holes in the board is required. The amount of time required to electroplate a given thickness of copper can be reduced with more vigorous solution movement. To accomplish this, an air agitation system is built into the plating tank (see Fig. 11.10a). The source of air should be an oil-free, low-pressure blower delivering approximately 1.5 to 2 cubic feet per minute of air for each square foot of plating tank surface area. The air is conveyed to the plating solution through a $\frac{1}{2}$-inch PVC sparger system. Note in Fig. 11.10a that the sparger pipe is positioned directly below the pc board to be plated. A series of $\frac{1}{16}$-inch-diameter holes are drilled into the bottom of the sparger pipe at 45-degree angles to its center line. The sparger is positioned approximately 1 inch above the bottom of the tank. Air entering the pipe exits through these small holes and is deflected off the bottom of the tank. The upward motion of these expanding air bubbles causes vigorous solution movement across the outside surfaces of the board as well as through all the drilled holes.

The negative terminal of a plating power supply is electrically connected to the cathode work rod. Both anode work rods are electrically connected together (anode strap) and then connected to the positive terminal of the power supply (see Fig. 11.11). All electrical connections should be made with AWG No. 12 stranded wire protected with a chemical-resistant plastic insulation.

The plating power supply should have the following characteristics:

1. Continuous adjustment of output voltage from 0 to 6 volts dc.
2. Current capacity in excess of plating current requirements (a maximum of 15 amperes is acceptable for the plating tank illustrated).
3. Percent ripple at full load current should be equal to or less than 5%.
4. A meter to monitor the dc output voltage at the + and − terminals.
5. A meter to monitor the load current.

The circuit schematic for a typical plating power supply is shown in Fig. 11.12.

The types of material used for the anode, together with its shape and surface area, will effect the quality of the resulting plating onto the pc board surface. Only *phosphorized* copper anodes should be used, since they will corrode properly in an acid copper-plating solution. They should contain 0.04 to 0.06% phosphorus by weight. Copper anodes designated as OHFC or pure copper anodes should not be used.

Anodes are available in various shapes and in a wide range of sizes (Fig. 11.13). The standard shapes are *flat, oval, round, sawtooth,* and *gear tooth.* A flat anode is preferable in a small plating tank since it results in more uniform current density. The oval, sawtooth, and other special shapes are designed to increase the surface area of the anode. The criterion for selecting an anode is its surface area relative to the area of the pc board

**Figure 11.11** Typical prototype copper plating setup.

## Parts List

SW1–DPST switch
L1–neon lamp
T1–2-A 0- to 120-V ac 0.25-kVA autotransformer
F1–1.2-A 250-V, SLO-BLO fuse
T2–Stancore P-6433 5.0 V CT at 15-A transformer
CR1-CR2–IN3900, 20 A at 100 PIV
C1, C2–24,000 $\mu$F at 35-WV dc electrolytic capacitor
R1–1 k$\Omega$, 1/2 W, 10%
M1–0- to 25-A meter
M2–0- to 10-V meter

**Figure 11.12**  Schematic diagram of plating power supply.

to be plated. For the plating of copper, an anode-to-cathode surface area ratio of approximately 1:1 is recommended. This means that the surface area of the anode facing the pc board in the plating tank should have a surface area approximately equal to the one side of the pc board that it is facing. Uniform plating through the holes of double-sided pc boards will result if the overall exposed copper path area is equal on both sides of the pc board. Double-sided plating is accomplished with the use of two anodes, one positioned on each side of the pc board, both having identical surface areas. This will simplify and optimize the plating process.

The anodes must be provided with a hook whose length will allow the anode to extend from the anode work rod down into the plating solution yet allow the top of the anode to protrude slightly above ($\frac{1}{4}$ inch) the surface of the solution. The length of the anode should be such that it does not extend below the bottom of the pc board more than 1 inch.

To illustrate the calculation of the correct size of an anode for a specific application, we will assume the requirement of processing a 5- by 5-inch card. The area of copper that will be exposed to the plating solution is determined by multiplying the length $L$ of the board by its width $W$. This will yield the overall area in square inches of exposed copper surface for one side. For our example, this results in 5 inches by 5 inches, or 25 square inches. This figure is used to determine the anode area required for plating as well as the plating power supply current density (Sec. 11.8) to obtain optimum plating.

Flat    Oval    Round    Sawtooth    Gear-tooth

**Figure 11.13**  Standard anode shapes.

There are two restrictions involved in the selection of the anode size required for plating: (1) Its surface area must approximately equal that of the cathode, and (2) it cannot extend more than 1 inch below the pc board. Selecting an anode whose length is 6 inches (1 inch longer than the board) will satisfy the length criterion. From the range of standard available widths, a 5-inch-wide anode will result in an area of 30 square inches, which is slightly larger than that of the cathode (25 square inches). Since each side of the pc board has equal areas, two 5- by 6-inch anodes will meet all of the requirements for flash plating our example board even though the ratio of anode area to work area is slightly larger than the optimum 1:1 proportion.

Prior to the placement of the anodes into the plating tank, proper procedures must be followed to avoid contaminating the plating solution. All plating solutions are extremely susceptible to organic contamination. For this reason, the anodes must be properly cleaned and correctly *bagged*. Anode degreasing is accomplished by scrubbing the surface with a brass bristle brush and cold tap water. The anode is then rinsed thoroughly and dipped into a 10% by volume solution of sulfuric acid for approximately 10 minutes. After this acid dip, the anode is ready to be bagged.

The anode bags used should be made of Dacron or polypropylene (never cotton) and should be 2 inches wider and 2 inches longer than the anode. The bag must be *leached* (purified) of all sizing or organic contaminations that it was exposed to in its manufacturing process. Leaching is accomplished in three steps. First, the bag is boiled in water for at least 1 hour. It is then removed from the boiling water and rinsed in cold water. Finally, the bag is soaked for 24 hours in $\frac{1}{2}$% by volume of Copper Gleam PC solution (see Sec. 11.7). One gallon of this solution will be sufficient for two bags of the size described. To mix 1 gallon of this solution, 190 milliliters of Copper Gleam PC is added to 3594 milliliters of water. The treated bags are then tied around the cleaned anodes. This is shown in Fig. 11.14a. The anodes are now ready to be installed into the plating tank (Fig. 11.14b), which, together with all the hardware and plumbing, must be thoroughly leached prior to adding the plating solution and the anodes. The leaching process is the same as that described in Sec. 11.3 for the electroless copper deposition line. Again, leaching is required only when the plating solution is added for the *first* time. Recall that detailed information for leaching is given in Appendix IX and should be reviewed before proceeding to the next section.

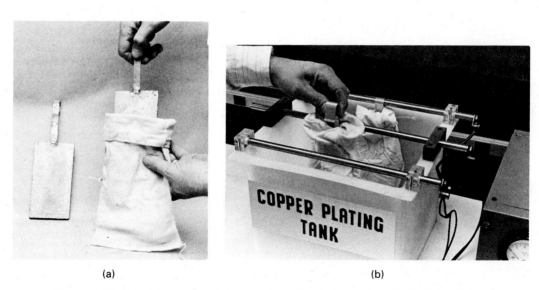

(a)                                    (b)

**Figure 11.14**  Anode bag preparation: (a) anode and leached anode bag; (b) anode installation into copper plating tank.

After the plating tank and the accessories have been leached, the plating solution is prepared. For our purposes we have selected a copper sulfate solution, since it requires only one principal additive and is thus easy to control. A key ingredient in this plating bath is manufactured by the *Lea Ronal Company* and is sold with the trade name of *Copper Gleam PC*. The use of this solution results in a bright copper deposit that is ideally suited for plating pc boards.

The copper sulfate plating solution is made up of the following:

1. Copper sulfate pentahydrate liquid
2. Sulfuric acid (reagent grade)
3. Hydrochloric acid (reagent grade)
4. Copper Gleam PC additive/brightener
5. Distilled water

Highly pure copper sulfate provides the source of copper in the plating solution. The most convenient use of this chemical is in liquid form. Sulfuric acid is added to the solution to improve its conductivity and also aids in anode corrosion. Reagent-grade sulfuric acid is used to prevent contamination from entering the plating cell. A small amount of hydrochloric acid is added since chlorides aid in brightening the plating. The Copper Gleam PC provides a bright ductile copper deposit over the entire board surface. The formulation of 10 gallons of Copper Gleam PC copper sulfate plating solution is given in Table 11.1, which shows the required amount of each ingredient and the order of mixing.

To prepare a plating bath for the 6-gallon tank described previously, Table 11.2 lists the ingredients and order of mixing. The amounts of each ingredient were derived by simply taking $\frac{6}{10}$ of those amounts listed in Table 11.1. The 6-gallon bath is prepared by first placing 3.72 gallons (14.08 liters) of distilled water into the leached tank. Next, 1.68 gallons (6.36 liters) of liquid copper sulfate pentahydrate, having a copper concentration of 2.25 pounds of copper per gallon, is added to the water. The solution is stirred slowly until it is thoroughly mixed. The 0.6 gallon (2.27 liters) of sulfuric acid is then added, stirred, and the solution is allowed to cool to room temperature. Finally, 2.7 milliliters of hydrochloric acid and 0.03 gallon (114 milliliters) of Copper Gleam PC are added and the solution is thoroughly mixed. The height of the solution should be marked on the outside of the tank to note any decrease in level due to evaporation. Any significant decrease of solution should be replaced so as to return it to the original level in the tank. The bath should be maintained at room temperature and loosely covered with sheet plastic when not in use to protect the solution from dust and other foreign matter.

All plating solutions require periodic chemical analysis. The techniques of chemical

**TABLE 11.1**

Formulation of 10 Gallons of Copper Plating Solution
With Copper Gleam PC Brightener

| Order of Mixing Ingredients | Compound | Amount | |
|---|---|---|---|
| 1 | Water (distilled or deionized) | 6.2 gal | 23.5 liters |
| 2 | Copper sulfate pentahydrate (2.25 lb copper/gal) $CuSO_4 \cdot 5H_2O$ | 2.8 gal | 10.6 liters |
| 3 | Sulfuric acid (reagent grade) | 1.0 gal | 3.8 liters |
| 4 | Hydrochloric acid (reagent grade) | — | 4.5 mL |
| 5 | Copper Gleam PC | 0.05 gal | 190 mL |
| | Total | ≈ 10.05 gal | ≈ 38.1 liters |

**TABLE 11.2**

Formulation of 6 Gallons of Copper Plating Solution
With Copper Gleam PC Brightener

| Order of Mixing Ingredients | Compound | Amount | |
|---|---|---|---|
| 1 | Water (distilled or deionized) | 3.72 gal | 14.08 liters |
| 2 | Copper sulfate pentahydrate (2.25 lb copper/gal) $CuSO_4 \cdot 5H_2O$ | 1.68 gal | 6.36 liters |
| 3 | Sulfuric acid (reagent grade) | 0.6 gal | 2.27 liters |
| 4 | Hydrochloric acid (reagent grade) | — | 2.7 mL |
| 5 | Copper Gleam PC | 0.03 gal | 114 mL |
| | Total | ≈ 6.03 gal | ≈ 22.83 liters |

analysis for the type of bath just described are beyond the scope of this book. However, since the amount of solution described in Table 11.2 is relatively small and not significantly expensive, the entire bath should be replaced if plating problems are encountered. A fresh batch of solution can be used successfully for a reasonable length of time if extreme care is exercised to minimize the introduction of organic contaminations into the bath. Detailed chemical analysis procedures for the bath described may be obtained from the Lea Ronal Company.

Before pc boards can be plated, the anodes must be conditioned (electrolyzed) after they have been cleaned, bagged, and hung into the plating bath. The conditioning process causes a film to be formed on the anodes and involves simply hanging sample copper-clad panels from the cathode rod into the plating solution. The power supply is adjusted to a value that will result in a current density of 10 amperes per square foot of surface area on the sample panels used. A suitable anode film is formed in approximately 4 hours.

## 11.8 COPPER SULFATE FLASH PLATING

With the pc boards that have been processed through the electroless deposition line stored in the 3% sulfuric acid holding tank and the plating tank fully prepared, the flash plating process can begin. The air supply is turned on and the rack of boards is removed from the holding tank. They are immediately placed into a rack, such as the ones shown in Fig. 11.10, and are ready for the electroplating process. The power supply is turned on and set just slightly above its zero reading. The racked boards are placed into the plating solution and swished back and forth for 15 seconds before electrical connection is made to the cathode rod. This 15-second dip will remove any minor oxides that may have formed, thereby reactivating the surfaces for electroplating. The rack holding the pc board is then lowered and hooked onto the cathode rod and secured with the thumb screw (refer to Fig. 11.10a).

The current density recommended by the manufacturer for flash plating is 30 amperes per square foot (≈ 208 milliamperes per square inch) of pc board surface. *However, it is important to note that this high-current density initially subjected to the 20 μinches of electroless copper deposited on the hole walls would cause its destruction.* It is therefore essential that one begin the plating process by applying one-fifth the recommended current density, that is, 6 amperes per square foot (≈ 42 milliamperes per square inch), for 2 minutes and then an additional 4 minutes at the recommended rate of 30 amperes per square foot.

The power supply is to deliver 30 amperes of current for every square foot of copper that is exposed to the plating solution. In Sec. 11.6 it was determined that our example panel has a surface area of 25 square inches. Since this is a double-sided board, the total surface area exposed to the plating solution is twice this amount or 50 square inches. We will now calculate the amount of current that the power supply is to deliver for the first 2 minutes (i.e., at one-fifth the recommended rate of 208 milliamperes per square inch) as follows:

initial current setting

$$= \frac{1}{5} \text{ recommended current density} \times \text{total surface area}$$

$$= \left(\frac{1}{5}\right) \frac{208 \text{ milliamperes}}{\text{square inch}} \times 50 \text{ square inches}$$

2080 milliamperes ($\approx$ 2 amperes)

When the power supply is set at 2 amperes, you will note that the output voltage is low (< 1 volt).

After 2 minutes at the 2-ampere rate, the power supply should be readjusted for the recommended current density for an additional 4 minutes. The final current setting is calculated as follows:

final current setting

$$= \text{recommended current density} \times \text{total surface area}$$

$$= \frac{208 \text{ milliamperes}}{\text{square inch}} \times 50 \text{ square inches}$$

$$= 10,400 \text{ milliamperes}$$

$$= 10.4 \text{ amperes} (\approx 10 \text{ amperes}).$$

Recall that the power supply, specified in Sec. 11.6, has a current capacity of 15 amperes, which is well in excess of the requirements of our plating application.

Thickness of the copper plated onto the surface of the pc board is dependent on (1) the current density and (2) the plating time. For the copper sulfate bath operated at 30 amperes per square foot, copper is deposited at the rate of 100 $\mu$inches every $3\frac{1}{2}$ to 4 minutes. To result in a thickness of 125 $\mu$inches, the required plating time would be slightly more than 4 minutes.

After the flash plating process has been completed, the power supply is turned off and the rack is loosened from the cathode work rod and immediately removed from the tank. The pc board is then thoroughly rinsed in cold tap water for at least 1 minute and then dried. Inspection of the plating should reveal a smooth and bright deposit of copper on all surfaces, including the walls of all the drilled holes.

The next phase in the processing of the pc board is to photosensitize both of its sides so that the conductor patterns may be imaged. Chapter 12 presents the procedures for photosensitizing and imaging in preparation for the final processing steps discussed in Chapter 13.

## EXERCISES

### A. Questions

11.1    The formula

$$\text{gallons} = \frac{L \times W \times H}{231}$$

(a)                    (b)                   (c)

**Figure 16.21**   Transistor and integrated circuit sockets: 3-, 4-, 8-, 10-pin sockets for TO-5 case style; (b) 14-pin DIP socket with wire-wrap terminals; (c) 24-pin socket for large-scale integration (LSI). *Courtesy of Augat, Inc.*

priate diameter mounting holes are first drilled into the pc board in accordance with the socket lead orientation. The socket terminals are then inserted into the board and swaged into place with an arbor press. If desired, the pins can later be soldered to the copper foil of the pc board. High-density *DIP test panels* using the socket terminals are shown in Fig. 16.22b.

The wiring and testing of new IC packages, such as the 52-pin PLCC shown in Fig.

(a)

(b)

**Figure 16.22**   Socket terminal nomenclature and DIP panels: (a) wire-wrap socket terminal; (b) high-density test panels using wire-wrap socket terminals. *Courtesy of Augat, Inc.*

*16.5 Device Sockets*                                                   **311**

**Figure 16.23** A modern 52-pin plastic lead chip carrier (PLCC).

16.23, require special sockets. These sockets accept the pins of the new high-density chips and provide wire-wrap capability for rapid prototyping.

## 16.6 DEVICE HEAT SINKING FOR PRINTED CIRCUIT BOARD APPLICATIONS

Heat sinking devices on pc boards is usually less difficult than heat sinking chassis-mounted devices. When devices and heat sinks are assembled on the pc board, insulators are not necessary because the board itself is an insulator. This is not the case with chassis-mounted heat sinks (as was discussed in Chapter 2).

Several commercially available heat sinks used in low- and medium-power applications and the case styles to which they are adaptable are shown in Fig. 16.24. These heat sinks are fabricated from aluminum or beryllium-copper alloys. They are secured to the device case by either a press fit resulting from the spring tension of the sink material or by mechanical clamping techniques. *Silicon grease* (thermal joint compound) used where the device case contacts the heat sink can reduce the thermal resistance between the two by as much as 100%. These sinks are finished in either a black anodized or natural unaltered metal surface. The black finish is more efficient in radiating heat away from the device case.

Techniques for heat sinking DIP-type ICs are shown in Fig. 16.25. The sink shown in Fig. 16.25a requires no hardware for mounting, since it is held by a friction fit between

(a)

(b)

**Figure 16.24** Printed circuit heat sinking for low- and medium-power devices: (a) sink for TO-5 and TO-18 case styles; (b) dual TO-5 sink. *Courtesy of International Electronic Research Corporation.*

(b)

(a)

**Figure 16.25**  Methods of heat sinking DIP integrated-circuit packages: (a) sinking for socket assembly, *courtesy of Astodyne, Inc., Wilmington, MA, subsidiary of Roanwell Corp.;* (b) assembly technique for dual surface heat sinking, *courtesy of Thermalloy.*

the base of the device and the top surface of the socket. For those DIP-type ICs that are to be mounted directly to the pc board without the use of a socket, the heat sink with its method of assembly as shown in Fig. 16.25b may be used. This arrangement allows heat to be removed from both the top and bottom surfaces of the device. The mounting used to secure the heat sink to the device is also used to secure the assembly to the pc board.

The selection of the proper heat sink is dictated by design factors discussed in Chapter 2. However, it is the responsibility of the technician to select the most appropriate type and style to meet the packaging requirements.

## EXERCISES

### A. Questions

16.1    What is the purpose of a polarizing key on connectors?

16.2    What is the difference between a full clinch and a dead head?

16.3    Why is it important to use pliers with smooth jaws when bending component leads?

16.4    What care should be taken when assembling disk capacitors to pc boards?

16.5    What is the advantage to staircasing device leads?

16.6    What are the advantages of using device sockets?

16.7    Why are heat sinks generally made with a black finish?

16.8    Why is heat sinking components on pc boards less difficult than heat sinking those that are chassis mounted?

16.9    Why are lead heat sinks used?

### B. True or False

Circle T if the statement is true, or F if any part of the statement is false.

16.1 Hand- or pressure staking is employed when installing turret terminals. **T F**

16.2 Both anvil and setting tools must be hollow when swaging double-ended terminals. **T F**

16.3 Axial lead components dissipating up to 2 watts may be flush mounted on the pc board. **T F**

16.4 Dead heads are used if it is expected that the component may be removed at a later time. **T F**

16.5 Socket terminals can be used in place of device sockets. **T F**

16.6 A heat sink may be applied to the device case while soldering the leads. **T F**

16.7 Some connectors are plated with a gold alloy to reduce wear. **T F**

## C. Multiple Choice

Circle the correct answer for each statement.

16.1 Solder terminals are secured to the pc board by (*welding, staking*).

16.2 For maximum rigidity, components should be installed with (*full-clinch bends, dead heads*).

16.3 A major advantage of using device sockets is (*service, system reliability*).

16.4 A component lead bend should begin no closer than ($\frac{1}{2}$, $\frac{1}{16}$) inch from the body end.

16.5 When a component is assembled into its mounting holes, the leads are bent (*toward, away from*) the conducting path at a (*60-, 30-*) degree angle.

16.6 When installing disk capacitors, their leads should be provided with (*full-clinch bends, dead heads*).

## D. Matching Columns

Match each item in column A to the most appropriate item in column B.

| COLUMN A | COLUMN B |
|---|---|
| 1. PCB connectors | a. Arbor press |
| 2. Solder terminal | b. Silicon grease |
| 3. Pressure staking | c. TO-92 |
| 4. Lead bend | d. Bifurcated contacts |
| 5. Transistor case style | e. Dead head |
| 6. Heat sink | f. Metal stud |

## E. Problems

16.1 Cut two $1\frac{1}{2}$-inch lengths of $\frac{1}{8}$-inch-diameter brass or copper rod. File a taper on one end and a flat side over the entire length of each piece. Remove the teeth from only the ends of the jaws of a small alligator clip and construct a lead heat sink similar to that shown in Fig. 16.26.

16.2 Assemble the components onto the pc board designed for the seven-fixed-inter-

**Figure 16.26**

val UJT timer of Problem 6.2 and fabricated in Problem 10.2. The assembled pc board is soldered in Problem 17.2. The indicating lamp ($L_1$), on–off switch ($S_8$), and battery clip are mounted to a chassis in Problem 26.5.

16.3   Mount the components on the pc board designed for the series feedback regulator in Problem 6.3 and constructed in Problem 10.3. The assembled board is soldered in Problem 17.3.

16.4   Install the components on the pc board designed for the tachometer in Problem 6.4 and fabricated and labeled in Problem 10.4. The assembled board is soldered in Problem 17.4. The tachometer pc board and meter are chassis- or panel-mounted in Problem 26.6.

# 17

# PCB Soldering

## LEARNING OBJECTIVES

*Upon completion of this chapter on printed circuit board soldering, the student should be able to*

- Understand the composition and temperature characteristics of soft solder.
- Know the major classifications of fluxes and their applications.
- Know the available soldering iron tip configurations and their temperature requirements for a variety of soldering applications.
- Use a soldering iron to solder pc board connections.
- Use a personal computer to obtain a thermal profile of a soldering iron.
- Be familiar with common solder joint deficiencies, their causes, and methods to correct them.
- Use de-soldering aids and tools.
- Understand the wave-soldering process and the basic components of a wave-soldering system.
- Know how to obtain a thermal profile of a wave soldering process.

## 17.0 INTRODUCTION

With the components properly assembled, the final process necessary to complete the construction of a pc board is to produce sound electrical connections between the leads and the foil pattern. These connections can be produced by several methods, the most common of which are *welding* and *soft soldering*. Welding requires the use of complex and expensive equipment and, as such, is rarely used in fabricating a prototype pc board. Soldering, on the other hand, can be performed quickly with less expensive equipment and results in excellent mechanical and electrical connections and protects the joint from oxidation. For these reasons, soldering will be discussed exclusively in this chapter.

Soldering is a metal solvent or chemical alloying action of the solder with the surfaces of the metal parts between which an electrical connection is formed. This completely metallic contact is produced by the application of soft solder with the heat of a soldering iron to the joint between the component lead and the terminal pad. The resulting connection is electrically sound, with the new alloy formed having different electrical and mechanical characteristics than either the solder or the metals joined. The soldering of pc boards requires the development of proper techniques if quality results are to be obtained. The two methods of soldering discussed in this chapter are *hand* and *wave* soldering. These methods are used extensively in pc board applications.

Specific information is also provided on the following topics: soft solder and its characteristics; the types of *flux* available and its application; the soldering iron, including the *tip, power rating,* and *tinning* processes; and hand-soldering techniques, with emphasis on the proper soldering of leads, swaged terminals, and conductor paths (tinning). Included is the technique of obtaining the thermal profile of a soldering iron with the use of a personal computer. A discussion of the characteristics of correct solder connections is presented to aid in visual inspection. *De-soldering* and the solvents used to remove the flux residues formed on soldered connections are also included. Finally, information on production-line methods of soldering, specifically wave soldering, is presented together with a method of obtaining a thermal profile of the wave soldering process.

*Caution: Prior to performing any of the operations described in this chapter, approved safety glasses must be worn.* Also, the section on safety outlined at the beginning of the book should be reviewed for more specific information.

## 17.1 SOFT SOLDER AND ITS CHARACTERISTICS

Soft solder used extensively in electronic equipment construction is an alloy principally of *tin* and *lead*. Soft solder is differentiated from hard solder by its tin content and lower melting point. The amount of tin contained in soft solder ranges from 50 to 70%. The tin–lead ratio determines the strength, hardness, and melting point of the solder.

Solder liquefies at temperatures between 361 and 621°F (183 to 327°C), the exact temperature depending on the tin–lead ratio. A metal such as copper, which has a melting point of 1981°F (1083°C), can be successfully alloyed with solder at temperatures well below this value because of the *solvent action* of solder when it is liquefied. At the melting point of solder, a thin film of metal is dissolved from the copper surface, forming an alloy and establishing an electrically continuous joint. The formation of this alloy between the metal and the solder has its unique physical properties, such as torsional, shear, and tensile strength, which are different from those of either the solder or the metal. These properties will vary widely and will depend on the depth of alloying into the metal surface.

Pure tin melts at 450°F (232°C) (point *B*, Fig. 17.1) and the melting point of pure lead is 621°F (327°C) (point *A*). When these two elements are combined, the melting

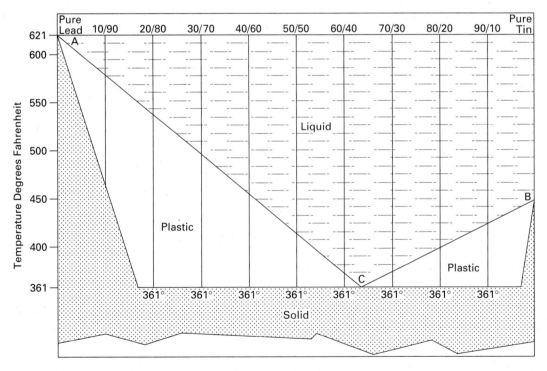

**Figure 17.1** Tin–lead fusion diagram. *Courtesy of Kester Solder, Division of Litton Systems, Inc.*

*17.1 Soft Solder and Its Characteristics*

point of the solder formed can be below that of either pure metal. A composition of 63% tin and 37% lead melts at 361°F (183°C), which represents the lowest melting point and most rapid transition from the solid to the liquid state of any other tin–lead ratio. This point, C, is also shown in Fig. 17.1. The 63/37 solder is termed the *eutectic* composition. All solder, with the exception of the eutectic composition, which melts sharply from solid to liquid, passes through a stage of softening between its solid and liquid stages known as the *plastic range*. A fusion diagram showing the temperature ranges for the various states of solder for all tin–lead combinations is shown in Fig. 17.1.

To determine the proper tin–lead ratio for a specific application, the function of the connection must be examined. The necessary properties that will dictate the composition of solder are mechanical resistance to fractures due to stress, ability to form a continuous metallic connection at low temperatures, and cost. These factors are discussed here.

Since tin is more expensive than lead, solder with a higher tin content is more costly. It has been shown empirically that the highest joint resistance to stress exists with a 63/37 tin–lead ratio (eutectic solder). This concentration, therefore, affords the best alloying qualities in addition to the lowest melting point.

For most hand-wiring and printed circuit applications, solder with a tin–lead ratio of 60/40 is commonly used, because of its excellent wetting action. *Wetting* is the term used to describe the ability of the solder to readily spread and alloy uniformly over the entire metal surfaces to be joined. Eutectic solder is sometimes selected to take advantage of the lowest and sharpest melting-point characteristics wherein maximum precautions are necessary to avoid component heat damage and upsetting the joint as it cools. If these two considerations are not critical, 60/40 solder is an excellent compromise.

Solder for electronic applications is available in bars, sheets, wire spools, and special forms, such as pellets, rings, and washers. For hand-wiring purposes, solder wire ranges from 0.030 to 0.090 inch in diameter. The larger sizes are used for general-purpose work and the smaller for delicate soldering applications such as pc boards.

Solder wire is also available with a core containing flux (see Sec. 17.2) in specific amounts to promote sound solder connections. For this reason, flux-core solder wire is used almost exclusively for electronic applications.

## 17.2 FLUX

The interaction of metal parts with the atmosphere forms a thin layer of oxide on their surfaces. This oxidation increases as the metal is heated and will severely interfere with the solvent action of solder, thus preventing alloying and the formation of an electrically continuous joint. Consequently, the oxide must be removed. *Fluxes* are used for this purpose. They are chemical agents that aid in soldering by removing thin films of oxide present on the metal surfaces to be soldered. When applied to the joint, the flux attacks the oxides and suspends them in solution, where they float to the surface during the soldering process. When the joint is heated, the presence of flux also prevents further oxidation in addition to lowering the surface tension of the metals, thereby increasing the wetting action. It is important to remember that flux is not a cleaning agent for removing grease or other contaminants. Its sole function is to remove the oxide film. For optimum soldering results, the parts must be thoroughly cleaned before the flux is applied.

The flux in no way becomes a part of the soldered connection but aids in the process. Upon completion of the soldered joint, a flux residue appears on its surface, which contains the captured oxides. This residue should be removed with an appropriate solvent.

The ability to rapidly remove oxide films from metal surfaces constitutes *activity* of the flux. It would appear that a highly active flux is ideally suited for electronic construction since it would afford rapid alloying and thereby reduce the possibility of heat damage to the components. This, however, is not the case. Highly active fluxes may be corrosive at room temperature and, if allowed to remain as a residue, will deteriorate the conductor surfaces or reduce the resistance of the insulation between soldered connections. Corrosive damage to components may also occur, since some of the active residues will gradually spread as they absorb moisture from the atmosphere. Even with suitable solvents, there is no assurance of complete flux residue removal, especially around closely spaced terminals, connectors, or conductors on pc boards where complete solvent flushing is not possible.

There are three major classifications of flux: (1) *chloride* (inorganic salts), (2) *organic* (acids and bases), and (3) *rosin* fluxes. The chloride types are the most active (highly corrosive) fluxes. They absorb moisture from the atmosphere and strongly react with acid even at room temperature. Organic fluxes are slightly less active than the chlorides and are used mainly for confined areas in which fast soldering time is important and corrosion problems are not critical. Many of the organic fluxes are converted to an inert residue after thermal decomposition. They do not absorb moisture and are difficult to remove. For the reasons given above, chloride- and organic-type fluxes are not recommended for use in electronic construction. The rosin-type fluxes are used almost exclusively because of their noncorrosive characteristics at room temperature. They are corrosive at temperatures near the melting point of solder. Consequently, they attack the oxide film during the heating cycle but are inactive when room temperature recurs. Rosin fluxes are available with activating agents that greatly improve their activity. These activated rosin fluxes are much more corrosive than pure rosin when heated and present the appearance of an instantaneous melting, wetting, and flowing action of the solder. They are essentially as noncorrosive at room temperature as the pure rosin types and are often preferable if a higher degree of flux activity is dictated such as in wave soldering.

Liquid flux may be applied by *wiping, dipping, spraying,* or *sponging.* Wiping or dipping methods are not extensively used. Uniform flux coatings are difficult to realize when wiping with a brush and thorough wetting of all surfaces may not result when dipping because of air pockets or cavities created during this process. Another disadvantage of these two methods is that application of an excessive amount of flux requires extensive removal of residue after the soldering has been completed.

Spraying flux involves applying a fine mist to the joints. The use of this method ensures that leads and terminals are more uniformly and completely coated with flux. When spray methods of flux application are to be used, the manufacturer's literature should be consulted regarding the recommended spray gun, nozzle, pressure, and solvent to use.

The application of flux with a sponge is perhaps the most effective and least messy method. To apply the flux, the board is firmly pressed against a sponge that has been saturated with liquid flux.

When hand soldering, the proper amount of flux can best be applied with the use of *flux-core wire solder.* This form of solder contains a core of solid rosin flux in a single or multiple core. There is no significant advantage in using multiple-core solder since it is essentially the volume ratio (amount of flux to solder) that determines optimum soldering conditions. Core sizes are available that provide a ratio of rosin flux per unit volume of solder of 0.6 to 4.4%. These ratios can be obtained for any size of wire solder. Indications are that 60/40 rosin flux core solder with a diameter of 0.040 inch and 3.6% flux is ideally suited for hand soldering pc boards and other electronic precision work.

## 17.3 THE SOLDERING IRON

The soldering iron, shown in Fig. 17.2, consists of four basic parts: (1) *tip,* (2) *heating element,* (3) *handle,* and (4) *power cord.* Some soldering irons are equipped with either a cork finger grip or a heat deflector. These are designed to improve thermal insulation in the handle, which tends to become hot after prolonged use. In addition, the cork finger grip allows for more comfortable use of the iron.

**Figure 17.2** Temperature-controlled soldering iron. *Courtesy of Hexacon Electric Co.*

When power is applied to the soldering iron, the tip is heated by direct thermal contact with the heating element into which it is set. The solder is liquefied when it comes in contact with a tip that has reached its operating temperature.

A soldering iron is selected for a specific application by considering the following factors: (1) size and style of the tip, (2) tip material, (3) required tip temperature, and (4) tip-temperature recovery time. Where precise control of tip temperature is required, *temperature-controlled* soldering irons having tips with built-in temperature sensors are available. This type of iron is shown in Fig. 17.2. Temperature-controlled units allow for tip temperature settings from 500 to 800°F (260 to 427°C). They maintain the set idling temperature to within seconds after each soldering operation. This type of temperature control ensures consistent soldering temperature conditions.

Selecting a tip style is somewhat a matter of personal preference. However, the shape of the particular tip used must provide the largest contact area to the specific connection for maximum heat transfer while minimizing the possibility of heat damage to surrounding leads or components. Several widely used tip styles are shown in Fig. 17.3. Each of the available tip configurations is designed for a specific soldering application. *Chisel*-style tips are commonly used for hand wiring and general repair work. The large flats of these tips allow large areas to be heated rapidly. The *turned chisel* tips lend themselves very well to soldering in confined areas, such as hand soldering components to double-sided pc boards. *Conical* tips are preferable for soldering high-density wiring, eyelets, and small heat-sensitive parts.

Soldering iron tips are manufactured chiefly from copper and are available both in plated and unplated finishes. Typical platings for tips are *iron, gold,* and *silver.* These platings protect the copper tip from corrosion and pitting, which become pronounced over extensive periods of use. The plated tips should never be filed or cleaned with harsh abrasives that would remove the plating and expose the copper. The tips should instead be periodically cleaned by first dipping them cold into liquid rosin flux and then bringing them

Chisel (Long Taper)

Turned Chisel

Conical Sharp

**Figure 17.3** Widely used solder tip configurations.

up to operating temperature to loosen any surface oxidation or contaminants (burned solder and flux residues) that may be present. Solder is then applied to the hot tip. Contaminated areas are easily detected by observing those areas that resist wetting by the solder. Sufficient solder has been applied when it begins to puddle. The tip is then wiped on a moist fine-pore cellulose sponge to remove the contaminants.

Tips should periodically be removed from the soldering iron because oxide will build up on the shank or threaded portion that fits into the heating element. If this oxide formation is not occasionally removed, the tip will seize in the element, making it difficult to remove and possibly damaging the element.

The required tip temperature is based on its application. For general-purpose soldering, such as to terminal strips and solder lugs, a tip temperature of 600 to 900°F (316 to 482°C) is sufficient. Printed circuit soldering requires fast heat transfer to the foil, yet excessive heat that could upset the foil bond must be avoided. For soldering to conductor patterns having widths of $\frac{1}{32}$-inch or larger and for terminal pads having $\frac{3}{32}$-inch diameters or larger, irons with tip temperatures of 800 to 850°F (427 to 454°C) are recommended. When soldering to more delicate pc boards or extremely fine wire (28 gauge or smaller), tip temperatures of 550 to 700°F (288 to 371°C) are suggested. However, when fine-component wire is to be soldered, especially in the case of heat-sensitive devices, a tip temperature of at least 900°F (482°C) is recommended (if lead heat sinks are not used) so that rapid heat transfer occurs with a minimum amount of tip contact time.

Irons are usually rated in *watts,* and are typically available from 20 to 60 watts for electronic applications. It is difficult to compare exact tip temperature with wattage rating since the length and the diameter of the tip largely influence tip temperature. For example, a 25-watt iron with a $\frac{3}{16}$-inch tip diameter and $\frac{3}{4}$-inch tip length will produce a temperature of approximately 680°F (360°C). Another 25-watt iron with a tip having the same $\frac{3}{16}$-inch diameter but a $1\frac{1}{2}$-inch length will reach a temperature of approximately 605°F (318°C). As a rule, tip temperature decreases linearly, approximately 10°F for every 0.10-inch increase in tip length. Tip temperature is inversely related to tip diameter. However, this relationship is not linear, as is the case of length versus temperature. Therefore, empirical data, such as those provided in Table 17.1, must be consulted. This table shows comparative values of tip temperatures to specific wattage ratings for three common tip diameters. The values given in this table are for iron-clad chisel-style tips having a length of $\frac{3}{4}$ inch.

A more general relationship of wattage to tip temperature is given in Table 17.2. This table provides a comparison of the common wattage ratings to a range of tip temperature that can be expected from any style or size tip. In general, approximately a 50°F change in tip temperature can be expected between the successive wattage ratings listed.

There is a useful technique for determining the approximate tip temperature of iron-clad tips by visual inspection. When the iron is heated to its *idling* (maximum operating) temperature, the flats of the tip are alternately wiped several times across the surface of a

**TABLE 17.1**

Comparison of Wattage Rating and Tip Temperatures for Various Sizes of Solder Tips

| Diameter, D (in.) | Length, L (in.) | 20 W | 25 W | 30 W | 35 W | 40 W | 50 W | 60 W |
|---|---|---|---|---|---|---|---|---|
| $\frac{3}{64}$ | $\frac{3}{4}$ | 675°F | 725°F | 775°F | 825°F | 875°F | 925°F | 975°F |
| $\frac{1}{8}$ | $\frac{3}{4}$ | 640°F | 690°F | 750°F | 800°F | 860°F | 910°F | 960°F |
| $\frac{3}{16}$ | $\frac{3}{4}$ | 630°F | 680°F | 730°F | 790°F | 850°F | 900°F | 950°F |

**TABLE 17.2**

Soldering Iron Wattage Rating Versus Range
of Tip Temperatures

| Wattage Rating | Range of Tip Temperatures (°F) |
|---|---|
| 20* | 550–750 |
| 25* | 600–800 |
| 30* | 650–850 |
| 35* | 700–900 |
| 40 | 750–950 |
| 50 | 800–1000 |
| 60 | 850–1150 |

*Commonly selected wattage ratings for electronic applications.

moist sponge. The color of the tip is observed 2 to 3 seconds after wiping. There is a relationship between the color obtained and the approximate tip temperature. This relationship is given in Table 17.3. Tip temperature can be periodically monitored using this technique.

*Recovery time* is the rate at which a tip will return to its idling temperature after transferring heat (tip cooling) to the work during soldering. In mass production applications wherein a rapid succession of soldered connections are to be made, fast recovery time is absolutely essential. It is not of major concern in prototype work, however, since the recovery rate is normally much greater than the rate at which soldered connections are made.

The following example illustrates how to evaluate the many available irons and tips to select the most appropriate one for a specific application. An iron will be selected for soldering a *double-sided pc board having typical conductor widths of* $\frac{1}{32}$ inch with terminal pads of approximately $\frac{1}{16}$ *inch in diameter*. Ideal tip temperature for pc board soldering is 800 to 850°F (427 to 454°C). As can be seen in Tables 17.1 and 17.2, a wattage rating of 35 watts will provide this necessary temperature with a $\frac{3}{64}$- by $\frac{3}{4}$-inch tip. The style of tip should permit ready access between components. A *conical chisel* design is selected since this configuration provides the desired contact area for component lead soldering to terminal pads. Finally, the iron-clad finish is chosen for durability. The tip selected for this example is quite versatile because of its thin chisel shape, which works well with both double- and single-sided boards.

In general, the soldering iron that produces the best performance over an extended period of time is one that has high heat conductivity and low heat loss when in contact with connections to bring them up to soldering temperature. It should also have a cool, lightweight handle, a no-burn power cord with strain relief and a simple method for removing the tip and disassembling the iron to replace the heating element.

**TABLE 17.3**

Reference for Iron-Clad Tip Temperature
by Visual Inspection

| Tip Temperature (°F) | Tip Color (2 to 3 seconds after sponge wipe) |
|---|---|
| 700 | Silver |
| 800 | Gold |
| 850 | Gold with streaks of blue at tip |
| 900 | Blue to purple |
| 1000 | Ash black |

## 17.4 THERMAL PROFILE OF A SOLDERING IRON

Modern personal computer technology has simplified the ability of obtaining an accurate thermal profile of a typical soldering iron. Analog Devices, Inc. has introduced a new family of signal-conditioning modules that can connect a type J thermocouple temperature sensor directly into the serial port of a personal computer. This hardware, used in conjunction with a software package such as LABTECH ACQUIRE,* forms a complete data-acquisition system that can be used to accurately measure the thermal characteristics of a soldering iron.

The basic parts of this temperature measuring system are shown in Fig. 17.4a. The soldering iron tip is first prepared by drilling a hole equal in diameter to the width of the thermocouple. Heat is then applied and the drilled hole is tinned to provide good thermal contact with the thermocouple. The white and red leads of the type J thermocouple are connected directly to pins 3 and 2, respectively, of a 6B11 signal conditioning module. This module converts the analog input (thermocouple voltage) into a serial digital output that is accepted by the RS-232-C port of an IBM personal computer. Power for all signal conditioning is supplied from a +5-volt power supply module. The serial computer interface is made with a separate cable between the signal conditioning module and the computer.

The LABTECH ACQUIRE software package is menu driven and runs on computers compatible with the IBM PC, XT, and AT and provides the interface between the user and the instrumentation system. The software package can be installed onto a hard disc system by first creating a directory in DOS (Disc Operating System) in which ACQUIRE will run. This is done by typing from the C prompt

```
C:\DOS>MD\ACQUIRE<CR>
```

where C:> or C:\DOS> stands for the C prompt, MD for *Make Directory,* and <CR> for carriage return. Next, change the directory to ACQUIRE by typing

```
C:\DOS>CD ACQUIRE <CR>
```

where CD indicates *Change Directory.* A subdirectory within ACQUIRE is then made to save the setups. This is done by typing

```
C:\ACQUIRE>MD SETUP <CR>
```

To install the software, the ACQUIRE floppy disk is placed into drive A and the following is typed:

```
C:\ACQUIRE>COPY A*.*<CR>
```

When all files on the floppy disk are copied from drive A to drive C, the disk is removed and stored. To complete the software installation, ACQUIRE must know about the hardware configuration. From the C:\ACQUIRE prompt, type

```
C:\ACQUIRE>AQ<CR>
```

which initiates ACQUIRE and the main menu appears as

```
CONFIGURE   GO   ANALYZE   SAVE   RECALL   DELETE   OPTIONS   QUIT

set up data acquisition
```

*LABTECH ACQUIRE is a registered trademark of Laboratory Technologies, Inc.

Computer Cable
Interface

dc Power
Supply

Signal Conditioning
Module

Thermocouple

(a)

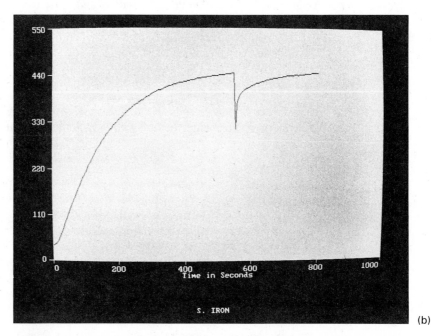

(b)

**Figure 17.4** (a) A single-channel, stand-alone, data acquisition system used to accept soldering iron tip temperature and input this information into a personal computer; (b) the thermal characteristics of a standard soldering iron.

The LABTECH ACQUIRE software is now loaded and ready for the hardware configuration. The thermal profile of a soldering iron will be used as an illustration. For this task, the only main menu options that will be used are CONFIGURE, GO, and QUIT. The ACQUIRE package has more versatility with additional menu options, such as ANALYZE, SAVE, RECALL, and DELETE. These are not discussed here.

To select from the main menu, press the left or right cursor key ($\rightleftarrows$) until OPTIONS is highlighted in reverse video. By depressing the carriage return <CR> key, the OPTIONS MENU will be displayed as follows:

```
Current Value CGA (4 Colors)

                                                              Options Menu

Display Type                                          EGA (8 Colors)
Start on Key Press                                          No
Hold Display on Screen                                     Yes
Hardware Interface Device                             Demo Board
Analyze Drive, Path Name, File Name                   C:\Lotus\123
Header Records in File?                                    Yes
```

*Display Type* will be highlighted in reverse video. ACQUIRE inspects the system and displays the type of graphics adapter card present within the system.

The up and down (↑↓) cursor keys are used to move the options menu to highlight *Start On Key Press*. The *No* is changed to *Yes* by first depressing the F1 key. A pop-up menu with the *Yes No* options will appear. The up or down cursor key is used to highlight *Yes* and this option is selected by depressing the carriage return <CD> key. A *Yes* in this field will make the system wait for any key, other than ESC (escape), to be depressed before data acquisition begins. In this way, the beginning of the system receiving data can be controlled by the operator. A *No* in this field will result in data to be taken the moment that the GO. EXE program is loaded.

The cursor keys are used to highlight *Hardware Interface Device*. This entry is used to change the designation for the hardware connected to the system. *Demo Board* and *6B* are included in the pop-up menu since there is only one 6B11 module attached. Highlighting 6B and depressing <CR> will select this option.

To return to the main menu, the escape (ESC) key is depressed. The left or right cursor (⇄) key is used to highlight CONFIGURE. This option is selected by depressing the carriage return key. The *Setup* menu will be displayed as follows:

```
Current Setting 1

                                                              Setup Menu

Number of Analog Channels [1..4]                           1
Number of Digital Channels                                 0
Time Stamp Date?                                          Yes
Sampling Rate (Hz)                                      10.000
Run Duration                                           30.000
Starting Method                                      Immediate
Trigger Channel                                            1
Trigger Threshold                                       0.000
Trigger Polarity                                         High

File Name                                             TUTOR.PRN
Number of Windows                                          1
Width of Windows in Seconds                            30.000
Window Color                                            Black
```

| Channel Number | 1 |
|---|---|

```
Channel Name                                        Ch.1
Display in Window Number [1 . . 1]                    1.
Scale Factor                                        1.000
Offset                                              0.000
Minimum Display Value                               0.000
Maximum Display Value                              10.000
Trace Color                                         RED.
```

The up and down cursor keys are used to move this setup menu down until *Sampling Rate (Hz)* is highlighted. The number on the menu is changed to 1 (one sample per second) by typing 1 <CR>. Next, *Run Duration* is highlighted and the setup number is changed to 800 <CR>. The system is now ready to take data once every second for 800 seconds, or about $13\frac{1}{3}$ minutes.

Dropping to *File Name,* IRON <CR> is entered. For *Window Width in Seconds,* the horizontal axis of the graph is matched to the run duration by typing 800 <CR>. At *Window Color,* the F1 key is depressed and a pop-up window with eight colors will appear on the screen. The up and down cursor keys are used to highlight white and are selected by depressing <CR>.

Finally, *Maximum Display Value* is highlighted and the setup value is changed by typing 550 <CR>. The vertical axis is now limited to 550°C.

The configuration of the display is complete. To return to the main menu, ESC is depressed. To initiate GO (i.e., load the GO.EXC program), <CR> is depressed when GO is highlighted in reverse video. Any key can then be depressed at the same time that the soldering iron is plugged in. As shown in Fig. 17.4a, data on tip temperature (°C) as a function of time (seconds) begin to be collected.

The thermal characteristic curve of the soldering iron is shown in Fig. 17.4b. This curve shows that the soldering iron reached a maximum tip temperature of 440°F in ~500 seconds. At ~500 seconds, the iron was wiped on a dampened sponge to simulate an actual soldering operation. Tip temperature dropped quickly to about 300°F, and recovery time to restore it to 440°F was about 200 seconds.

To exit ACQUIRE, the ESC key is depressed to recall the main menu. QUIT is highlighted and the carriage return <CR> is keyed. A system similar to the one just described greatly simplifies the task of data collection and display.

## 17.5 HAND SOLDERING PRINTED CIRCUIT BOARDS

Of prime importance in soldering is that the surfaces to be joined are clean and void of oxides. For pc board soldering, this is especially true, for a poorly cleaned terminal area would prolong heating time excessively when soldering, which could upset the foil bond. Not only should the terminals be cleaned, but it is imperative that liquid flux be applied to all joints prior to soldering even if terminal areas have been tinned. The flux may be applied to the pc board by any of the methods discussed in Sec. 17.2. However, wiping is preferable in prototype work when a solder mask is to be employed. This method allows the flux to be selectively applied only to those terminal areas that are to be soldered. The pc board is first secured in a *circuit board holder* and tilted to a horizontal position. Liquid flux is then applied with a fine artist's brush or toothpick onto the areas to be soldered. This arrangement is shown in Fig. 17.5. When all the surfaces to be soldered have been coated with flux, soldering may begin.

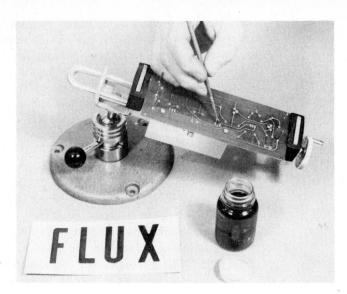

**Figure 17.5** Application of liquid flux to selected areas.

To solder leads to pc boards, it is important that heat and solder be applied quickly and accurately. A small amount of solder is initially applied to the flat of the soldering iron tip (solder bridge) to promote rapid heat transfer to the joint (Fig. 17.6a). The entire flat of the tip surface is then placed in contact with the terminal area and lead simultaneously. The more surface area of the tip contacting the terminal pad and lead, the more efficient the transfer of heat (Fig. 17.6b). The joint is allowed to heat to the melting point of

(a)          (b)

(c)          (d)

**Figure 17.6** Printed circuit board soldering: (a) forming a solder bridge; (b) heating solder pad and lead; (c) solder application; (d) correctly soldered terminal.

solder, usually taking from 1 to 2 seconds for most connections. Solder is then applied to the terminal pad and lead simultaneously and at a point on the opposite side of the lead from where the tip is contacting. The correct orientation of tip and solder is shown in Fig. 17.6c. As soon as the solder begins to flow, no more should be applied. Since solder follows heat flow, it will wet the entire connection and flow over the lead, forming a smooth contour of solder around the lead and terminal pad. The soldering iron is then removed and the joint allowed to cool. The joint must never be moved until the solder has completely solidified because any movement of the lead as it is cooling will prevent positive alloying from taking place. *It is important to apply only a minimum amount of solder sufficient to ensure proper alloying.* A properly formed solder joint is smooth and shiny in appearance. All surfaces must be completely wetted and the contour of the lead on the terminal pad clearly visible. A correctly soldered lead to a printed circuit terminal pad is shown in Fig. 17.6d. A precaution should be mentioned here. When soldering the individual terminal pads, it is important to keep track of those leads that must be heat sinked. Lead heat sinking is discussed in Sec. 16.4.

Excess solder and contaminated flux residue on a tip must be removed before another solder joint is formed. Cleaning is accomplished quickly and easily by wiping the tip on a moist sponge. Soldering should never be attempted with a tip covered with excess solder or contaminants if quality results are to be expected.

Swaged terminals are soldered to pc boards in a similar manner to lead soldering, with one important exception. Since the terminal pad area for swaged terminals is larger than those used for lead connections, it is necessary to apply heat for a longer period of time. The flat area of the soldering tip is placed in contact with the swaged portion of the terminal and solder is applied to the terminal pad. When heated sufficiently the solder will completely wet the terminal pad and flow uniformly around the swaged terminal in the area where it contacts the copper foil.

Since many pc boards do not employ solder masks, and if conductor pattern plating is not economically feasible, the technican may wish to tin the entire conductor area of the board to retard oxidation. This can be done during the soldering operation. Flux is first applied to the entire board surface. A small amount of solder is applied to the flat of the tip and the tip is slowly drawn along the conductor path. A thin layer of solder will be deposited on the surface of the copper. Several passes may be necessary to completely tin a conductor path. This will prevent excessive solder from forming in addition to minimizing foil bond damage. For this technique, a soldering iron with a tip temperature of between 500 and 600°F (260 and 316°C) should be used. The application of solder via the soldering tip is restricted to tinning conductor patterns. This technique is never employed for making soldered connections.

Properly soldered connections are uniform in appearance and their quality may be judged by visual inspection. Each solder joint may be inspected immediately upon soldering or after completion of the entire board. To aid in this inspection, a 5 × to 10 × magnifying glass may be used. Following is a list of deficiencies and their causes common to improperly soldered connections. These are shown in Fig. 17.7.

1. *Solder peaking.* Characterized by a sharp point of solder protruding from a connection. Peaking is caused by the rapid removal of heat before the entire joint has had an opportunity to completely reach its soldering temperature. The solder follows the hot tip as it is removed from the connection, resulting in this peaking condition. Reheating the joint will correct this deficiency.

2. *Incomplete wetting.* Occurs when portions of the soldered connections have not been alloyed with solder and are completely visible. This may be the result of both insufficient heat and solder. It may be also be the result of contaminants on the soldering tip or terminal pad. Reheating the joint and applying additional solder will correct this fault if the condition is not caused by contaminants. In that case, de-

(a)

(b)

(c)

(d)

**Figure 17.7** Typical inferior solder connections: (a) solder peaking; (b) incomplete wetting; (c) excessive solder; (d) cold solder joint.

soldering and cleaning are necessary before a new joint is attempted. The terminal pad can be cleaned with a sharpened pencil-style typewriter eraser. (De-soldering techniques are discussed in Sec. 17.6.)

3. *Excessive solder*. Evident when the lead contour is not plainly visible. This is extremely undesirable since it prevents complete inspection of the alloying action of the solder, thus obscuring potential troubles. This condition can be rectified by removing some of the solder with the use of desoldering aids.

4. *Cold solder joint*. An inferior connection easily detected by its dull-gray, grainy appearance, or as a cluster of solder that has not properly wetted all the surfaces. This is the result of applying insufficient heat, which prevents solder from alloying with the metal parts. Cold solder joints can generally be corrected by reheating the connections. If, however, the puddling of solder is caused by a connection that has not been thoroughly cleaned, it must be de-soldered and cleaned before it can be properly soldered.

Flux residue remaining on the connection after soldering is undesirable. This residue presents a poor board appearance and contaminants suspended in this residue could cause troublesome electrical leakage paths. If more active fluxes must be used, their removal is doubly important to prevent corrosion. Flux removal can be accomplished by *hand brushing, dipping,* or with *ultrasonic equipment.*

Brushing with a stiff-bristle brush dipped in a flux solvent is often acceptable, but may not remove flux completely from inaccessible cavities characteristic of pc boards in which swaged eyelets and terminals are used. Dipping and mildly agitating the board in a solvent bath is preferable since it will improve solvent action in some areas that are not possible to reach by hand brushing (Fig. 17.8). A more efficient means of flux residue removal, especially when activated fluxes are used, is to use an ultrasonic tank containing the flux solvent. Several minutes in this system ensures the most thorough cleaning.

For most electronic soldering applications, rosin flux residue is the type most generally encountered. Isopropyl alcohol is an excellent solvent for this residue. As in the case of most solvents, certain hazards present themselves. Alcohol is flammable and should be used in a well-ventilated area and kept away from the skin. Finally, the remaining solvents are quickly evaporated by the use of an oil- and moisture-free air hose.

## 17.6 DE-SOLDERING PRINTED CIRCUIT BOARDS

When a lead is soldered to a terminal pad, it is difficult to remove without damaging the component or the terminal area. Several commercially available de-soldering aids, however, simplify this task. These aids are the *solder wick, de-soldering bulb, solder sucker, de-soldering tips,* and *extraction tools.*

A solder wick is made of finely woven strands of copper wire. This flattened wick is placed over the terminal area and lead to be de-soldered. This arrangement is shown in Fig. 17.9a. The soldering iron tip is placed in contact with the solder wick and pressed down against the connection. As the heat from the iron is transferred to the wick and connection, the solder will melt and flow in the direction of the heat transfer. The solder is thus trapped by the solder wick as it flows up through the weave. The result is shown in

**Figure 17.8** Mild agitation for flux removal is accomplished by use of a rocker table.

(a)

(b)

(c)

(d)

**Figure 17.9** De-soldering techniques: (a) position of solder wick and soldering tip; (b) de-soldered terminal pad using solder wick; (c) de-soldering bulb, *courtesy of Ungar, Division of Eldon Industries, Inc.;* (d) solder sucker.

Fig. 17.9b. The solder can be completely removed from the joint with this method. The used portion of the wick is discarded. Solder wick is available in rolls for many de-soldering applications. With the solder removed from the connection, the lead can be bent away from the terminal pad and the component easily withdrawn.

The de-soldering bulb and solder sucker can also be used for removing excess solder or for de-soldering component leads. Both tools have hollow Teflon tips with high heat resistance and will not scratch or mar delicate pc board conductors. The de-soldering bulb, shown in Fig. 17.9c, is employed by depressing the bulb and then placing the hollow tip alongside the soldering iron tip on the joint. As the solder begins to melt, the bulb pressure is released. The liquid solder is drawn up into the bulb by the suction. Both tools are then removed from the connection for inspection. If solder remains, the process must be repeated. The solder sucker shown in Fig. 17.9d is employed in basically the same manner except that the suction is produced by a spring-loaded piston. The piston handle is first pushed downward. The handle is then rotated to engage the release pin. As the solder begins to melt, the pin is disengaged and the solder is drawn up through the hollow tip as the piston snaps upward inside the tubular handle. Both tools are easily disassembled to remove the accumulated solder. These de-soldering tools do not generally remove sufficient solder on the first attempt to allow components to be removed, thus rendering the solder wick the preferable method.

For removing ICs, the previously discussed de-soldering aids would have to be used on each lead individually, making it a time-consuming process. De-soldering *tips* specifically designed to simultaneously heat all the leads for removing devices quickly and easily are available for this purpose. De-soldering iron tips for TO-5-style cases and dual-in-line configuration ICs are shown in Fig. 17.10a.

The de-soldering tip is placed in contact with all the device leads simultaneously. As the leads are de-soldered from the foil side of the board, the device is removed from the component side with an extraction tool such as that shown in Fig. 17.10b. With this tool, the device is lifted away as the solder is melted. The remaining solder present on the terminal pad may obstruct the access holes and can be removed with any of the de-soldering aids discussed. The recommended type of extraction tool is one whose metal clamps grip the leads, thus providing some degree of heat sinking. It must be mentioned that lead heat sinking is just as important when de-soldering as it is during soldering to avoid damage.

## 17.7 WAVE SOLDERING

A much more elaborate system of production line pc board soldering is the process of *wave soldering*. This method is widely used in the industry because of its large soldering capacity, its ability to accurately regulate time and temperature exposure of the pc board, the scrubbing action of the solder wave, which aids in complete wetting and soldering of each joint area, and the consistently high-quality results obtained.

A typical wave-soldering system consists of three main sections: the *fluxer, preheater,* and *solder wave.* An integral part of the system is a conveyor, which is used to accurately control the movement of the pc board over the fluxer, the preheater, and the solder wave. Precleaning and postcleaning cycles of the pc board are not a part of the system and are performed separately.

A typical bench-type wave-soldering machine is shown in Fig. 17.11. The conveyor speed is adjustable from 1 to 15 feet per minute at angles of between 5 and 9 degrees with the surface of the bench. The board is introduced into the system by attaching it to the fingers of the conveyor. The circuit side is first run across a foam fluxer. At this station, a head of fine bubbles rise from the fluxer. These bubbles are typically generated by pass-

(a)            (b)

**Figure 17.10** Device de-soldering equipment: (a) de-soldering tips for TO-5 and DIP-style cases; (b) de-soldering DIP-style devices. *Courtesy of Ungar, Division of Eldon Industries, Inc.*

**Figure 17.11** Bench-type wave-soldering machine. *Courtesy of the John Treiber Company.*

ing air through a submerged porous ceramic cylinder that is impervious to the flux. A chimney and brushes aid in uniform flux distribution. The bursting flux bubbles rise by capillary action and penetrate into plated-through holes. Rosin-based flux, such as Kenco No. 465 Resin Flux, having 25% or higher rosin content, is generally preferred in the wave-soldering process. This flux not only promotes uniform solder wetting and formation of a strong intermetallic bond with the copper, it also reduces the surface tension of the liquid solder. Surface tension reduction is important in preventing solder bridges and icicles (solder peaks) from forming.

After the board is fluxed, the conveyor continues its travel over a preheater station. The purpose of this station is threefold. First, preheating evaporates excess flux solvent on the board to prevent splattering and entrapped gases, which could cause pinholes in connections when the board is soldered. Second, it conditions the board and assembly against thermal shock. Third, preheating the board helps to overcome the heat-sink effects on large components caused by the solder wave. Preheaters must be capable of providing a range of heat, as measured on the component side of the board, from 100 to 275°F (38 to 135°C). Overheating the flux must be avoided because it degrades its effectiveness by reducing its mobility.

With the board preheated, it is next conveyed over the solder wave, which quickly produces large numbers of high-quality and well-contoured solder connections simultaneously. These connections are both electrically and mechanically sound. The wave is formed by pumping molten eutectic solder (63/37) vertically upward through a solder nozzle having a large plenum chamber that rests in a solder pot. A high-capacity centrifugal pump propels the molten solder through the nozzle to form a standing wave. The most commonly used wave shape is bidirectional, that is, solder flows in two directions.

Bridges and icicles will usually result if the solder system has too narrow a solder wave and the boards are run on a horizontal conveyor (see Fig. 17.12a). A web of solder tends to extend beyond the wave in the direction of travel, which creates these problems. If the solder wave is narrow, inclining the conveyor to an angle of 5 to 9 degrees will increase solder *peel back* and therefore reduce the solder web and significantly reduce solder defects (see Fig. 17.12b). For best results, however, a solder system having a deep, wide wave (3 to 4 inches along the path of travel at 1 or more inches above the nozzle edges) and an inclined conveyor will virtually eliminate the formation of a solder web which causes bridging and icicles. This is shown in Fig. 17.12c. Inclining the conveyor causes the board to be introduced onto the solder wave at a high solder-velocity point. This takes advantage of the natural scrubbing action of the wave at this point in its velocity profile. A wider wave used together with an inclined conveyor also allows the soldered conductor pattern to exit the wave at a practically zero velocity point which not only increases solder *peel back* but reduces heat transfer to the components being soldered.

To obtain the smoothest solder wave and minimum solder deposits having a bright,

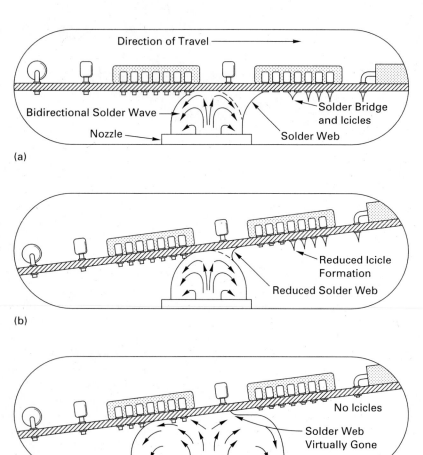

(a)

(b)

(c)

**Figure 17.12** Use of wide solder wave with oil and an inclined conveyor eliminates solder bridges and icicles: (a) narrow solder wave with horizontal conveyor; (b) narrow solder wave with inclined conveyor; (c) wide solder wave with inclined conveyor.

shiny appearance, the formation of dross (oxides of tin and lead) must be kept to an absolute minimum. One method of reducing the dross formation is to use an oil (soldering fluid) as a dross blanket in the solder pot. As solder is pumped, the oil forms a thin film on the surface of the wave. Another effective method is to intermix oil, such as Hollis No. 225 Soldering Fluid, by precision metering with the wave at the impeller of the pump. This causes the oil to disperse uniformly within the solder wave. In both methods, the oil reacts with the dross to form harmless tin and lead soaps without degrading the solder. In addition, the surface tension of the solder is reduced which allows excess solder to drain more easily from the board. The soldered boards also become coated with the oil, which intermixes with the rosin flux and protects the soldered connections from oxidation. This mixture of oil and rosin is easier to remove than rosin alone in the cleaning process, assuring bright soldered joints.

Although wave soldering is an automated process, a degree of skill and knowledge of soldering is necessary to properly operate the system. This knowledge includes an understanding of the proper conveyor speed, type of flux, preheater temperature, width of the solder wave, solder temperature, the use of an oil blanket or oil intermix, and the established quality standards. When the pc boards have been assembled, soldered, and cleaned, they may be tested for operational performance and modifications can be made if necessary.

## 17.8 THERMAL PROFILE OF THE WAVE SOLDERING PROCESS

One essential function of a process engineer may be to evaluate and modify a wave soldering system in order to quantify both the magnitude and duration of temperatures that contact a pc board as it passes through the various operations. By tightly controlling the thermal profile of this process, high-quality results, with minimum rejects, will be obtained. The data acquisition system described in Sec. 17.4 is ideally suited to evaluate the thermal characteristics of a pc board as it passes through the wave soldering system.

Typically, during the wave soldering process, a pc board will be subjected to a temperature variation of from about 25°C (room temperature) to as much as 250°C. For this reason, a type J thermocouple is selected and constructed by twisting and welding together iron and a constantan (40% nickel and 60% copper alloy) wire. This thermocouple is capable of accurate, repeatable measurements of temperatures from −184 to +760°C.

The thermocouple is first inserted into a small via hole of the board. (Refer to Chapter 8 for a description of a via hole.) The thermocouple must enter from the component side and should extend into the hole until it just begins to exit the circuit side. This arrangement will allow the temperature changes to be monitored and will simulate those changes to which a component lead would be subjected. The thermocouple extension leads pass above the component side of the board and to the input of the 6B11 signal conditioning module discussed in Sec. 17.4.

A thermal profile of a conventional wave soldering process obtained from the data acquisition system is shown in Fig. 17.13. Test panels are used to properly establish process parameters as follows: During the preheat or flux activation phase, a thermal rise of approximately 2°C/second is obtained by adjusting the conveyor speed within the typical range of 6 to 8 feet per minute. Optimum results are achieved when the circuit side surface temperature is about 183°C, the melting point of eutectic solder, as the board exits the preheater. As the board moves over the solder wave, additional heat is transferred to the board, which raises the circuit side surface temperature still higher. The board dwells in the molten solder wave for 3 to 5 seconds, causing the surface temperature to reach a maximum temperature of about 250°C. Control of the time that the board remains in the wave is a function of the profile of the wave and the depth to which the board extends into the solder. Refer to Fig. 17.13. Test panels can be run and rerun to adjust the controls for

[Conveyer Speed – 6 to 8 Feet per Minute]

Critical Temperatures

**Figure 17.13** Thermal profile of a conventional wave soldering process.

optimum results. Finally, a maximum cool-down cycle of −5°C/second is recommended to minimize thermal shock and maximize joint strength.

The data acquisition system just described can be configured quickly with a personal computer and a menu-driven software package such as ACQUIRE, which is discussed in Sec. 17.4.

## EXERCISES

### A. Questions

17.1    What are the two metals used to make solder, and what is the percentage of each in solder used for general-purpose electronic work?

17.2    Explain the term *eutectic solder*. What is its melting point temperature?

17.3    What type of flux is used for electronic work?

17.4    What is the purpose of flux?

17.5    What is the recommended method of cleaning a plated soldering iron tip?

17.6    What is the approximate temperature of a $\frac{3}{4}$-inch-long tip having a diameter of $\frac{3}{64}$ inch when used with a 25-watt heating element.

17.7    Explain the following terms: *cold solder, solder peak,* and *wetting.*

17.8    What is used to remove rosin flux residue from electrical connections?

17.9    What benefits can be derived from the generation of a thermal profile for a wave soldering process?

17.10    What is solder wick used for?

17.11    What is the major cause of solder bridge and icicle formations when wave soldering pc boards? How is this eliminated?

## B. True or False

Circle T if the statement is true, or F if any part of the statement is false.

| | | | |
|---|---|---|---|
| 17.1 | Solder is an alloy of tin and lead. | T | F |
| 17.2 | For most hand-soldering applications, 60/40 eutectic solder is used. | T | F |
| 17.3 | Solder melts at a lower temperature than either pure tin or pure lead. | T | F |
| 17.4 | Rosin flux with activating agents is noncorrosive at room temperature. | T | F |
| 17.5 | Conical soldering iron tips are used when large areas must be rapidly heated. | T | F |
| 17.6 | When hand soldering, the solder should be placed between the iron's tip and the lead to ensure proper wetting. | T | F |
| 17.7 | A cold solder joint can be corrected by simply reheating the connection without the addition of more solder. | T | F |
| 17.8 | Oil is used to reduce the formation of dross in a wave-soldering system. | T | F |

## C. Multiple Choice

Circle the correct answer for each statement.

17.1    Solder used in the electronics industry for component assembly consists chiefly of tin and (*copper, lead*).

17.2    Chloride-based fluxes are not used in the electronics industry because they are (*expensive, corrosive*).

17.3    Rosin-based fluxes are most active at (*room, high*) temperature.

17.4    For a constant soldering iron wattage rating, the tip temperature (*increases, decreases*) as the tip length increases.

17.5    A good solder joint results if a (*liberal, minimum*) amount of solder is applied to the connection.

17.6    After a solder joint has been completed, flux residue is removed with the use of a solvent such as (*alcohol, hydrochloric acid*).

17.7    The most common method used by industry for soldering large quantities of pc boards is (*hand, wave*) soldering.

## D. Matching Columns

Match each item in column A to the most appropriate item in column B.

| COLUMN A | COLUMN B |
|---|---|
| 1.  Dross | a.  Solder defect |
| 2.  Eutectic | b.  Return to idle temperature |
| 3.  Recovery time | c.  Flux solvent |
| 4.  Flux | d.  Rosin |
| 5.  Rosin joint | e.  Oxidizing scum |
| 6.  Solder wick | f.  63/37 ratio |
| 7.  Alcohol | g.  De-soldering tool |

## E. Problems

17.1 Lay out, shear, and bend an aluminum tray for holding a soldering iron tip cleaning sponge. The tray is to have inside dimensions of $\frac{3}{4}$ inch by $3\frac{1}{2}$ inches by $3\frac{1}{2}$ inches. Tabs are not required on the inside corners. These corners are sealed in Problem 25.2 to retain excess moisture. Break all sharp edges with the appropriate file.

17.2 Solder the assembled pc board for the seven-fixed-interval UJT timer designed and fabricated in Problems 6.2, 10.2, and 16.3. Chassis mounting and interconnection wiring are done in Problems 25.4 and 26.5, respectively.

17.3 Solder the pc board for the series feedback regulator designed and constructed in Problems 6.4, 10.4, and 16.4.

17.4 Solder the pc board for the tachometer circuit designed and constructed in Problems 6.5, 10.5, and 16.5. Circuit board and meter installation are accomplished in Problem 25.5. Wiring is done in Problem 26.6.

# Surface Mount Technology

*Upon completion of this chapter on surface mount technology, the student should be able to*

- Determine land sizes and footprints for many surface-mounted components and devices.
- Properly identify the side of a pc board on which surface-mounted and conventional through-hole components can be placed when both technologies are used on a common board.
- Select, test, and properly apply solder paste to surface-mounted components.
- Select, test, and properly apply adhesives to surface-mounted components.
- Be familiar with the wide variety of soldering methods used in surface mount technology.
- Hand solder surface-mounted components.
- De-solder surface-mounted components.

## 18.0 INTRODUCTION

Surface mounting stems from the technologies of hybrids where surface-mounted devices are soldered inside a ceramic or plastic package having through-hole leads. For these, it appears that the conventional methods of design and manufacturing have, to a large extent, reached the limit regarding factors of cost, reliability, size, and weight. With the emergence of surface-mounted technology (SMT), the generation of lighter, smaller, and more dense pc boards at lower costs and higher reliability has been realized.

Surface-mounted technology involves the bonding of components and devices directly to the surface of the board and connected directly to the conductor pattern, thus eliminating the requirement of drilling holes in the board.

The sizes of surface-mounted components (SMC) and devices (SMD) are very much smaller than their equivalents made for conventional through-hole installation. Because SMT is relatively new, all equivalencies are not available. In addition, there is a great deal of nonuniformity of standardization among manufacturers of SMT components and device packages. As a result, both conventional and SMT are required to be used on most pc board designs. With the improvement in the technology, more and more board area will be devoted exclusively to surface mounting.

***Caution:*** *Prior to performing any of the operations described in this chapter, approved safety glasses must be worn.* Also, the section on safety outlined at the beginning of the book should be reviewed for more specific information.

Surface-mounted conductor pattern layout does not lend itself to the relative ease of layout experienced with conventional through-hole design. This is largely due to the lack of standardization of SMT package designs and the wide variation in component tolerances in the industry. In addition, land pattern design is process-dependent, which means that the same design can experience different manufacturing defects depending on the process variables. (*Land pattern* is the term used in SMT to define the terminal pads to which the device leads are soldered.) This is brought about by the fact that each manufacturer has to establish specific land pattern designs that will function with their materials and processes. For this reason the land patterns presented in this section and the techniques for their design serve only as guidelines.

Common land patterns, also referred to as *footprints,* to be discussed include *ceramic rectangular resistors and capacitors* (small outline RC and CC); *tantalum capacitors* (small outline molded plastic and welded-stub); *tubular passive components* (metal electrode leadless face or MELFs); *transistor chip patterns* (small outline transistors or SOTs); *small outline integrated circuits* (SOICs); *small outline J lead devices* (SOJ memory and SRAM packages); and *chip carriers* (plastic leaded chip carrier or PLCC, and leadless ceramic chip carrier or LCCC).

The land pattern design parameters for ceramic rectangular resistors and capacitors are shown in Fig. 18.1. The dimensions of component length ($L$), width ($W$) and height ($H$) and the termination length of the metal bands (MB) are measured first. To determine the width of the land ($A$) and its length ($B$), the following relationships are used:

$$For\ capacitors:\ A_c = W + 0.01\ inch$$
$$B_c = H + MB$$
$$For\ resistors:\ A_r = W + 0.01\ inch$$
$$B_r = H + MB + 0.01\ inch$$

Resistors are much thinner than capacitors and require a longer land to prevent misalignment due to their lower mass. The space between the lands ($G$) is the same for both components and is determined as follows:

$$G = L - 2MB - 0.02\ inch$$

The overall length of the footprint would thus be $G + 2B$. Two examples of land patterns for ceramic resistors and capacitors are shown in Table 18.1.

The procedure to determine the land patterns required for molded plastic and welded-stub tantalum capacitors is similar to that for ceramic resistors and capacitors, as shown in Fig. 18.2. Once the dimensional information is obtained for each component type, the following are used to determine the footprint dimensions and spacings:

$$For\ molded\ plastic:\ A_{mp} = w + 0.010\ inch$$
$$B_{mp} = h + MT$$
$$G = l - 2MT - 0.020\ inch$$
$$For\ welded\ stub:\ A_{ws} = W + 0.010\ inch$$
$$B_{ws} = H + MT$$
$$G = L - 2MT - 0.020\ inch$$

**Figure 18.1** Land pattern design parameters for resistors and capacitors.

**TABLE 18.1**

Resistor and Capacitor Land Pattern Dimensions

| Component | A | B | G | G + 2B |
|-----------|-----|-----|-----|--------|
| R1206* | 0.065 | 0.060 | 0.070 | 0.190 |
| C1206† | 0.065 | 0.060 | 0.070 | 0.230 |

*$L = 0.130$, $W = 0.064$, $H = 0.030$, $MB = 0.020$

†$L = 0.130$, $W = 0.064$, $H = 0.030$, $MB = 0.020$

Dimensions are in decimals of an inch.

Two examples of tantalum capacitor footprints are shown in Table 18.2.

Surface-mounted tubular passive components (MELFs) are resistors, ceramic and tantalum capacitors, and diodes. These packages have metal end caps that are soldered to the lands. Because these packages are round and there is a tendency for them to roll out of position, a slot must be provided in the lands if they are to be soldered using the vapor reflow method. If the component is to be wave soldered, an adhesive is used which eliminates the need for the slots.

The measurements used for the design of MELF patterns are shown in Fig. 18.3. The following are used to determine the footprints for this package style:

$$A_{MELF} = \text{diameter } (D) + 0.010 \text{ inch}$$
$$B_{MELF} = \text{diameter } (D) + MC$$
$$G = L - 2MC - 0.020 \text{ inch}$$

If a notch is required, notch depth $= B/2$ and notch width $= 0.010$ inch. An example of MELF land pattern dimensions is given in Table 18.3.

Some of the most commonly used small outline transistor (SOT) chip patterns are the SOT23, SOT89, and the SOT143 packages. These footprints are shown in Fig. 18.4. When designing the land patterns for SOTs, the distance between land centers is generally made equal to the on-center dimensions of the device leads. At the same time, the land should be made 0.015 inch longer than the lead on each of its ends. Many footprints are available from graphic suppliers, such as Bishop Graphics, that will accommodate a variety of designs.

(a) Molded plastic

(b) Welded stub

**Figure 18.2** Land pattern design parameters for tantalum capacitors.

**TABLE 18.2**

Tantalum Capacitor Footprint Dimensions

| Component | A | B | G | G + 2B |
|-----------|-----|-----|-----|--------|
| *Molded plastic package B\** | *0.090* | *0.060* | *0.050* | *0.170* |
| *Welded stub package B†* | *0.060* | *0.075* | *0.085* | *0.235* |

*L = 0.130, W = 0.080, H = 0.030, MT = 0.030, H = 0.075, W = 0.110.

†L = 0.160, W = 0.059, H = 0.045, MT = 0.030.

Dimensions are in decimals of an inch.

The criteria for selecting commercially available footprints for small outline integrated circuits (SOICs), or R-packs, are much less complicated than for those designs discussed previously. The land width and length should be 0.030 and 0.075 inch, respectively. The on-center spacing of the lands is typically 0.050 inch. This will result in about a 0.015-inch space between adjacent lands. To determine the space (*S*) between rows of lands, the relationship *S* = body width − 0.01 inch is used.

The footprint of a SOIC, together with typical dimensions of various other packages with different lead counts, is shown in Table 18.4. The footprint design for plastic leaded chip carriers (PLCCs) and leadless ceramic chip carriers (LCCCs) is shown in Fig. 18.5. Note that the PLCC chip has leads that turn *under* the package. This formation is referred to as *J leads* as opposed to those typical of SOT packages, which are called *gull wings*. The footprint is determined by the following:

$$\text{on-center lead spacing} = 0.050 \text{ inch}$$
$$\text{land width} = 0.025 \text{ inch}$$
$$\text{land length} = 0.075 \text{ inch}$$
$$\text{dimension } A \text{ and } B = X + 0.030 \text{ inch (for PLCC)}$$
$$\text{dimension } A \text{ and } B = X + 0.070 \text{ inch (for LCCC)}$$

Note that for square chip carriers, dimension *A* = dimension *B*. This is not true of rectangular packages. These dimensions for a variety of square and rectangular chip carrier footprints are given in Table 18.5 (see pg. 345).

Surface-mounted components are soldered to the pc board by either vapor phase, solder wave, or infrared processes that are discussed in Sec. 18.5. Variations in land pattern dimensions may be required for the same component for a different soldering process. One common problem encountered is the varied degree of solderability at termination ends. This can cause misalignment of components with fingerprints. Also, components such as resistors and capacitors, instead of remaining flat, may end up in a standing position, called *tombstoning*. If these components are positioned too close to each other, a large component may shadow a smaller one, causing insufficient solder or solder bridges to result. For these reasons, it is essential that all land pattern designs first be evaluated through the processing of test boards.

**Figure 18.3**  Land pattern design parameters for MELFs.

**TABLE 18.3**

Sample of MELF Land Pattern Dimensions

| Component | A | B | G | G + 2B |
|-----------|-----|-----|-----|--------|
| MLL41* | 0.109 | 0.114 | 0.140 | 0.368 |

*L = 0.200, diameter = 0.099, MC = 0.015.

Dimensions are in decimals of an inch.

**Figure 18.4**  Small outline transistor patterns.

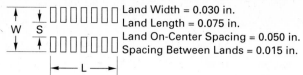

Land Width = 0.030 in.
Land Length = 0.075 in.
Land On-Center Spacing = 0.050 in.
Spacing Between Lands = 0.015 in.

**TABLE 18.4**

SOIC and R-Pack Footprint Dimensions

| Package | L | W | S |
|---------|------|------|------|
| S08 | 0.175 | 0.290 | 0.140 |
| S014 | 0.325 | 0.290 | 0.140 |
| S016 | 0.375 | 0.290 | 0.140 |
| S0L24 | 0.575 | 0.450 | 0.300 |
| R-pack 14 | 0.325 | 0.350 | 0.200 |
| R-pack 16 | 0.375 | 0.350 | 0.200 |

Dimensions are in decimals of an inch.

Some recommended spacing considerations for SO, SOJ, R-packs, and PLCC packages are shown in Fig. 18.6. These spacings also take into account automatic assembly and testing requirements.

The following societies may be consulted for additional design information: The *Institute of Interconnecting and Packaging Electronic Circuits (IPC), Electronics Industries Association (EIA),* and the *Society of Manufacturing Engineers (SME).*

## 18.2 SURFACE MOUNT ASSEMBLY

The surface mount packages discussed in the preceding section are shown in Fig. 18.7. Figures 18.7a through d show leadless resistor and capacitor packages with metal end caps or terminations, and Figs. 18.7e, f, g, and h illustrate SOT and SOIC packages with gull and modified gull wing leads. A leadless chip carrier with recessed edge conductors is shown in Fig. 18.7i, and Fig. 18.7j is a plastic chip carrier package having J leads.

As mentioned previously, a wide variety of components are not available for surface-mounting applications, resulting in pc board designs incorporating this technology together with conventional through-hole insertion. An exception to this is in the design of memory boards, where all required packages are available for surface mount.

As a result of the requirement for a combination of both technologies, SMT assemblies have been divided into three categories which describe their application. These are shown in Fig. 18.8 (see pg. 347). Type I assembly contains only surface-mounted compo-

Dimensions in Decimals of an Inch

Dimension A = X + 0.030 and B = Y + 0.030 for PLCC
Dimension A = X + 0.070 and B = Y + 0.070 for LCCC

**Figure 18.5**   PLCC and LCCC footprint dimensions.

**TABLE 18.5**

Overall Dimensions of PLCC and LCCC Land Patterns

| Package Style | Pin Count | Dimension A | Dimension B |
|---|---|---|---|
| Square | 16 | 0.370 | 0.370 |
| Square | 24 | 0.470 | 0.470 |
| Square | 44 | 0.720 | 0.720 |
| Square | 68 | 1.020 | 1.020 |
| Square | 124 | 1.720 | 1.720 |
| Rectangular | 22 | 0.350 | 0.550 |
| Rectangular | 32 | 0.520 | 0.620 |

Dimensions are in inches.

nents on a single-sided, double-sided, or multilayer pc board. Where components are placed on the bottom side, the tack of the solder paste applied to the lands prior to assembly is sufficient to hold them in place during board handling and baking prior to reflow soldering. If the reflow soldering is done by vapor-phase or infrared processes, the solder paste will continue to hold them in position. If wave soldering is used, all components must be held in place with an adhesive on any side that will be passed over the wave. Because many devices are heat sensitive, one side of the board may be vapor-phase reflowed while the less temperature-sensitive components on the other side of the board are wave soldered. This procedure is more common with type II and III assemblies that combine surface mount with through-hole designs. Type II assemblies place SMCs on one or both sides of the board with through-hole insertion components on the top only. Type III as-

**Figure 18.6** Minimum land-to-land or land-to-pad clearances.

Scale

0   0.100"   0.200"   0.300"   0.400"   0.500"

(a) Ceramic resistor or capacitor — Leadless Termination

(b) Tantalum capacitor

(c) Tantalum capacitor/ welded stub

(d) MELF

(e) SOT 23

(f) SOT 143

(g) SOT 89

(h) SOIC — Gull Wing

(i) LCCC — Recessed Edge Connector

(j) PLCC — J Lead

**Figure 18.7**  Surface mount component and device packages.

sembly places through-hole insertion components on the top side with SMCs on the bottom side.

Due to their size and delicate mechanical features, SMCs are mounted manually with the aid of a vacuum pipette or, for volume production, by automatic component placement. Prior to placement, solder paste and adhesive are applied as required. These processes are discussed in detail in Secs. 18.3 and 18.4, respectively. The two primary functions of automatic equipment for SMT assembly are picking up the components and placing them onto the correct lands. For this reason, this type of equipment is generally termed "pick and place." Assembly equipment having single or multiple pick and place

PLCC      SOIC      PLCC

Plated-Through Via Hole (PWR)  Ceramic Resistors  SOICs
(a) Type I exclusively surface-mounted components and devices on top or on top and bottom

DIP  SOIC      PLCC

Ceramic Capacitors  Ceramic Resistors  SOIC  Via Hole (Gnd)
(b) Type II insertion components on top with surface-mounted components on one or both sides

Four-Layer MLB      DIPs

Ground Plane    Ceramic Capacitors    Voltage Plane
(c) Type III insertion components on top with surface-mounted components on the bottom

**Figure 18.8** Type I, II, and III SMT assemblies.

vacuum heads are available. Components are placed sequentially with X, Y, and Z movements of the vacuum head. Automatic machines are usually fed with components that are bulk packaged on reels or in magazines called DIP sticks, which are plastic tubes that channel parts into the machine.

When SMCs are manually installed, they should be handled by their bodies, not by their terminations or leads. In addition, care should be exercised so as to avoid smearing solder paste or adhesive onto the packages that would interfere with solder joint formation and result in short circuits.

In any process of assembly used, the component terminations or leads must register squarely on lands. Even though the surface tension of the melting solder paste tends to pull misregistered components into place, this should not be counted on. If misregistration is not detected on SMCs that have been secured with adhesive, correction is difficult and time-consuming.

## 18.3 SOLDER PASTE

To solder SMCs to pc boards, solder paste or solder cream is first applied to the footprints by automatic dispensing with a syringe, screening, or stenciling, after which the components are assembled. Solder paste is the standard term, although rosin-based materials are referred to as creams. Solder paste is a homogeneous combination of minute prealloyed solder particles that are suspended in a flux system. The solder particles are spherical as a result of blowing or splashing melted solder onto a spinning wheel in a controlled atmosphere. Their round shape has less tendency to clog screens as well as having minimum surface oxidation.

The flux system is made up of a flux base and a gelling or suspension agent that are blended together. The formulation of this flux system is a well-guarded secret in industry. The proprietary ingredients involved give the paste body, the ability to withstand soldering temperature and time without breaking down or losing efficiency (thermal resistance), chemical strength (effectiveness), and a consistency that makes it suitable for application by screening, stenciling, or syringe dispensing.

Factors that govern the selection of a solder paste for a specific application are *particle size, flux activators, metal content,* and *rheological properties* such as viscosity, tackiness, slump, and working life. Each of these factors will be described in detail.

If the screening process of flux application is to be used, the particle size should be small enough to pass through a 200-mesh (200 holes per square inch) screen. Larger particle sizes may be used if stencils or syringes are the method of application.

The flux promotes wetting action between component leads and the surface mount lands. Solder paste is available with three different types of flux: rosin flux (R), mildly activated rosin (RMA), and fully activated rosin (RA). The R flux contains no activators and may not have an aggressive enough cleaning action for acceptable solderability. The RMA fluxes are the most generally used for surface mount applications. The RA fluxes are used only when a more potent cleaner is required. They should be avoided whenever possible, however, since the characteristics of surface-mounting techniques prevent thorough flux cleaning under and around components.

The metal content of the solder paste determines the size of the solder fillets that will form about the component leads. Solder pastes have different melting temperatures. For the soldering of plastic surface-mounted packages, a lower-melting-point paste should be used compared to that used for ceramic packages. If higher shear and tensile strength are required at elevated temperatures, tin–antimony alloy solders are used. Tin–lead alloy solders are not compatible with gold-plated terminations. For this application, indium–lead alloys are recommended. Compatibility with silver-plated leads requires a tin–lead solder containing 2% silver. The most widely used solder pastes for mounting plastic components to FR-4-type pc boards are those having either 63% tin and 37% lead (eutectic) with a melting point of 360°F (183°C) or 62% tin, 36% lead, and 2% silver, which has a melting point of 355°F (179°C).

The rheological properties of the solder paste are controlled by the gelling or thickening agents and secondary solvents. The solvent dissolves the flux and controls the tackiness of the paste as a result of its evaporation rate. The degree of tackiness is an important factor in the selection of paste since it holds the SMCs in place during placement, baking, and handling before reflow soldering.

The viscosity of solder paste changes when agitated by either syringe dispensing, screening, or stenciling. This is due to the fact that these pastes are *thixotropic* fluids; that is, they become more fluid when worked and more viscous when allowed to rest. The recommended viscosity of solder pastes as related to their method of application is as follows:

*Syringe dispensing:* 300,000 to 400,000 centipoise (cP)
*Screening:* 450,000 to 550,000 cP
*Stenciling:* 550,000 to 750,000 cP

These viscosity measurements may be taken with a Brookfield viscometer using a TF spindle at 5 revolutions per minute.

The term *slump* refers to the ability of the paste to spread, yet stay within the bounds of footprints after it has been deposited. The degree of slump is determined mainly by the percentage of metal in the paste. Sample test boards with land patterns are used to evaluate how far the paste spreads after it is applied. This is to ensure that it does not spread onto adjacent lands before or after component assembly.

The working life of solder paste is defined as the length of time from when it is removed from the container and goes through all of the fabrication processes—its application, SMC placement, and baking—to the solder reflow process. The degree of tackiness determines if the working life of the paste has elapsed.

When pc boards with SMT components and devices are soldered by the vapor phase or the infrared reflow process, solder paste often throws small spherical particles of solder referred to as *solder balls*. These can measure 2 to 5 mils in diameter. Since they usually end up on the insulated surface around footprints, they are easily dislodged and create a potential short-circuit problem.

Solder balls form either as small particles that float away with the melting flux before the larger mass of paste melts or as heavily oxidized particles that cannot be removed by the flux and activators. These particles are pushed aside as the molten solder melts and flows about the connection.

To test for paste acceptability in terms of solder ball formation, several 6- to 10-mil-thick paste patterns are screened onto a suitable sample of nonmetallic material, such as FR-4 pc board or ceramic—or frosted glass. The test panel is then put through a solder reflow process and the paste examined. It is acceptable if no solder balls are generated and the solder deposits are large shiny spheres having smooth surfaces. A few particles having diameters of 5 mils or less are acceptable. Anything larger, or a halo or clusters of particles are unacceptable. Some of the methods used to eliminate the formation of solder balls are better control of the amount of paste applied and using fresh paste with the most active flux allowable.

The application of the solder paste to the lands by any method must be checked for coverage and for registration. A number of acceptable and unacceptable configurations of paste deposits on land patterns are shown in Fig. 18.9. Figures 18.9a through d are acceptable, with part (a) preferred. Those labeled as unacceptable (Figs. 18.9e through h) would result in short circuits or unreliable solder joints. Soldering of pc boards is discussed in Sec. 18.5.

| | | |
|---|---|---|
| (a) Preferred | | Optimum Shape of Paste Deposit with Proper Alignment |
| (b) Acceptable | | Excess Solder Slump but Less Than 2 Times Land Area and No Bridging |
| (c) Acceptable | | Deficiency of Paste but 75% of Land Covered |
| (d) Acceptable | | Misalignment but 75% of Land Covered |
| (e) Unacceptable | | Smeared Print Causes Solder Bridges and Solder Balls |
| (f) Unacceptable | | Skipped Print, Insufficient Paste Causes Incomplete Solder Joints |
| (g) Unacceptable | | Ragged Edges Cause Nonuniform Solder Joints |
| (h) Unacceptable | | Misalignment with Less Than 75% of Land Covered |

**Figure 18.9**  Acceptable and unacceptable solder paste deposits.

## 18.4 ADHESIVES

The two most common types of adhesives used for SMCs are *epoxies* and *acrylics*. Unlike thermoplastic adhesives, which harden and soften with temperature cycles, these adhesives, once hardened, will not soften or lose their bond strength with elevated temperatures. Because epoxies and acrylics are classified as thermosetting adhesives, they will withstand the heat of the solder wave and will retain the components tightly to the pc board as it passes over the wave.

To be most effective, the adhesive selected must be properly applied and cured. In addition, the adhesive must have specific precure, cure, and postcure properties that are compatible with SMT. The adhesive must also have sufficient *green strength,* which is similar to the tackiness of solder paste, to hold the component in place before curing. It should also have the proper viscosity to fill the gap between the component and the board without spreading onto the footprints.

For most applications, one-part adhesives are preferred over the two-part (mixing) type, which requires more time to prepare and has a short working life since it begins to harden when the two parts are mixed. Even though one-part adhesives have a shorter shelf life, refrigeration extends this life.

An adhesive that cures in the shortest time at lower temperatures is preferred, provided that it possesses sufficient bond strength. Greater bond strength results from elevated temperatures with the same curing time. However, the temperature must not be so high as to adversely affect heat-sensitive components or deform the board.

The adhesive should have minimum shrinkage properties when cured to prevent excessive stress on component bodies. Curing time should not be accelerated, since voids may occur which would decrease the effectiveness of the adhesive. If outgassing occurs during the cure process, flux may become entrapped from the solder paste, which will result in difficult cleaning problems.

The curing of adhesives is accomplished in infrared (IR) ovens for epoxies and with a combination of UV light and heat for acrylics. The thermal-cured epoxies are popular since the IR oven can also be used for reflow soldering without requiring the expense for a UV light system. Acrylic adhesives have a faster curing time, however, and are preferred for high-volume production.

Acrylic adhesive must extend beyond the component so that the UV light will strike it and begin the chain reaction of curing. If this polymerization does not occur, heating alone will allow uncured adhesive under the component, resulting in voids and entrapped flux during the soldering process. Ultraviolet thermal systems use a 2 kilowatt lamp set at a distance of about 4 inches from the board. The adhesive is exposed for about 15 seconds, and the full cure will result in less than 2 minutes in an IR oven.

Epoxies have a curing time of about 5 minutes when the temperature is gradually increased to about 305°F (150°C). The board is then allowed to cool slowly to room temperature for another 10 minutes. A typical thermal profile curve of epoxy adhesive cure in an IR oven is shown in Fig. 18.10. (The procedure for obtaining a thermal profile is discussed in Chapter 17.)

After the reflow soldering process has been completed, the adhesive no longer serves a useful function. Its postcure properties are essential, however, in that it must not absorb moisture and must remain nonconductive and noncorrosive.

The application of adhesives is accomplished by three common methods: screening, pin transfer, and syringing. Syringing is generally preferred because in many instances, there are components on the board before the adhesive is to be applied. In these situations, screen and pin transfer applications cannot be used. Syringing forces the adhesive through

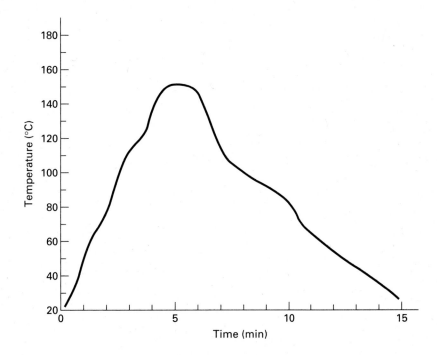

**Figure 18.10** Thermal profile of epoxy cure in an IR oven.

an orifice by electric, hydraulic, or pneumatic pressure. This system, although not as fast as other methods of application, controls the rate of adhesive flow very accurately to ensure uniformity of deposit.

The pin transfer method employs a grid of pins that are dipped simultaneously into the adhesive. This array of pins is then automatically positioned and lowered to the board. The adhesive has a greater affinity for the nonmetallic substrate than for the metal pins. Gravity causes a uniform amount of adhesive to be deposited from the pins.

Screening is done using the same techniques as the silk-screen process described in Chapter 10. The application of surface mount adhesives, however, requires a stainless steel screen and the appropriate mesh number. As in the use of any adhesive, the manufacture's instructions should be carefully observed.

The correct application of adhesive for large and small components is shown in Fig. 18.11 together with the correct dot (adhesive deposit) sizes for various land sizes. The deposit of adhesives for a large component after the application of solder paste is shown in Fig. 18.11a. Note that there are two adhesive dots on the substrate between the lands and a smaller dot on the bottom of the component. This will ensure that the space under the component is filled with adhesive. However, excess adhesive should be avoided since it will interfere with proper solder joint formation. Approximate adhesive dot sizes used for typical lands are shown in Figs. 18.11b, c, and d.

(a) Multiple adhesive dots for large SMCs

(b)

(c)

(d)

**Figure 18.11**   Adhesive application for large and small components.

## 18.5 SOLDERING FOR SMT

The soldering process in SMT is termed *reflow soldering*. The joining of metal surfaces of the lands with the terminations of the components and devices is completed by the mass heating of the predeposited solder paste. When the paste melts, solder fillets are formed between the lands and the terminations. Automated volume production methods include *wave soldering, vapor-phase soldering,* and *infrared reflow*. Laser soldering is another automated process which is faster than hand soldering but not as fast as the other production processes. Because laser soldering is a relatively new process, it is more expensive and the technical problems involved with solder defects and component damage have not all been resolved. For these reasons, laser soldering is confined to specialized applications and is viewed as a complement to rather than a replacement of the other soldering processes.

Wave soldering for pc boards has been discussed in Sec. 17.7. Some modifications of the process are required for boards that incorporate both conventional and surface-mounted components and devices. These modifications require special features to accommodate higher board densities as well as components being mounted on the bottom side of the board. These features, shown in Fig. 18.12, are primarily *air knives,* and a *dual solder* wave.

The first air knife overcomes the problem of not being able to smooth the applied flux with a brush roller, due to the surface-mounted components on the bottom side of the board. The air knife effectively removes excess flux and smooths the flux coating over lands and component terminations.

The dual soldering wave is a required feature because of the components mounted to the bottom of the board. The first wave is turbulent, which ensures sufficient distribution of solder across the board as it passes. The velocity of the solder allows it to penetrate between tightly spaced components and leads. The solder jet is pointed in the direction of the conveyer travel, which results in no solder skips. However, the applied solder is very uneven and excessive, causing bridges and icicles. The second wave is the conventional type used for through-hole components. As the board passes over this smooth wave, excess solder is removed along with any bridges and icicles.

Immediately after the board leaves the second solder wave, it passes over a hot air knife. A jet of hot air, whose temperature is above the melting temperature of solder, is directed at the soldered joints. A properly wetted solder connection will be unaffected by the jet of air since the surface tension of the solder exceeds the force of the air. Any solder that is not bonded, such as bridges, icicles, and excess amounts that were not removed by the second solder wave, is removed by the hot air knife. To prevent the blowing of solder up through holes in the board and, at the same time, remove just the right amount of solder, the air knife's spacing from the wave, and its angle to the board, in addition to the air temperature, volume, and velocity, must be accurately controlled. The hot air process also

**Figure 18.12** Air knives and dual solder wave reduce solder defects of surface-mounted components.

acts as a method of quality control since the superheated air stress-tests each solder fillet and uncovers any land or pad that is not properly soldered.

To maintain and monitor precise temperature control of each phase of the wave soldering process, a temperature profile of the board is required. This is done by placing one or more thermocouples at strategic locations on the board and recording the temperature variations as it passes through the preheat and wave soldering stages. The recommended temperature parameters for preheat rate, temperature differential between preheat and solder wave, time of the board in the wave, and cool-down rate are 2°C/second maximum, 100°C maximum, 3 to 4 seconds, and −5°C/second, respectively. The procedure for obtaining a temperature profile is discussed in Chapter 17. A temperature profile for a pc board as it is passed through a single solder wave is shown in Fig. 17.13.

Vapor-phase and infrared soldering processes are much different from wave soldering, where the source of solder and most of the required heat are provided by the wave. This significant difference for both processes will be described in detail.

For type I assemblies, vapor-phase or infrared soldering processes are used exclusively, while types II and III assemblies require the use of wave soldering as well. The source of heat for vapor phase comes from the condensation of vapors from a boiling liquid onto the metal terminations, solder paste, and lands. Since heat is also transferred by convection, the work is heated uniformly. Also, the board is protected from heat damage because its temperature cannot exceed that of the boiling point of the liquid. This automatic vapor-phase soldering system makes use of either *batch* or *in-line* equipment. These are shown in Fig. 18.13. In the batch system shown in Fig. 18.13a, the work is lowered through a secondary vapor layer, usually Freon TE, into the primary vapor, which is an inert perfluorocarbon such as FC-5311 from the 3M Corporation. The board remains in the primary vapor zone, whose temperature is approximately 419°F (215°C), for 20 to 30 seconds. The board is then raised into the secondary vapor zone and allowed to drain off the primary fluid (30 seconds to 1 minute). The temperature of this zone is about 225°F (107°C). Because of the secondary vapor zone, the cooling coils at the top of the chamber, and the high density of both vapors, the loss of expensive perfluorocarbon is kept to a minimum. The high density also acts to reject air and moisture from entering the process.

The in-line system, shown in Fig. 18.13b, does not have a secondary vapor zone. The board is passed through a saturated vapor zone in 20 to 30 seconds. There are vapor condensing surfaces at the top of the chamber as well as at the input and output locations. With an optional vapor recovery system added to the machine, the vapor loss can approach that of the batch system.

The infrared reflow soldering process passes the boards along a conveyer between top- and bottom-mounted IR quartz iodide lamps having ceramic reflectors for higher wavelength or IR panels called area source emitters. The basic arrangement of this system is shown in Fig. 18.14. Both natural and forced-air convection are used on these systems. Test panels should be run through the process to ensure that the emitted wavelength is correct, so that nonuniform heating of the board will be avoided. Also, the optimum air convection needs to be determined to cure the solder paste properly and prevent the formation of solder balls or splatter. Infrared soldering is the generally preferred process because it is more controllable for higher yields and lower operating costs.

Some automated soldering systems provide a forced cooling stage as the final process in the soldering operation. This has been shown to improve the strength of solder joints.

For small soldering tasks, hand soldering is commonly employed. It is also used as a means of reworking or correcting soldering defects resulting from an automated process. The power of the soldering iron used on SMCs, such as ceramic resistors, capacitors, and diodes, should not exceed 22 watts. These components could easily be damaged if subjected to extreme heat. A pointed soldering tip is preferred and its temperature should not exceed 700°F (370°C). Excessive pressure against connections should be

(a) Batch type

(b) In-Line style

**Figure 18.13** Batch and in-line vapor-phase soldering systems.

avoided to prevent delamination of lands. In addition, the tip should be in contact with the connection for only 1 to 3 seconds, depending on the type of component.

Because of their small size and light weight, along with the fact that they have no leads inserted through drilled holes, SMCs tend to move off of the land when hand soldering. This problem can be overcome by placing a droplet of adhesive onto the substrate between lands. Using solder that has a diameter of 0.031 inch is ideal for transferring the adhesive to the board. Any cyanoacrylate adhesive will prove effective. These adhesives are

Figure 18.14　In-line infrared reflow soldering system with forced cooling.

more commonly known by their trade names, such as Crazy Glue, Instant Glue, or Super Glue. It is important that the deposit of glue be accurately placed and that the amount be controlled so that it does not spread onto the lands when the component is pressed into place. This would inhibit the formation of a proper solder connection. Since this adhesive is fast acting and very strong, care must also be taken to ensure that the component is properly placed. It would be virtually impossible to move it after the adhesive has set. Once the adhesive has cured, soldering of the connection may begin. Solder having a maximum diameter of 0.031 inch is shaped to a fine point for this delicate work. The time to solder each connection should be no more than 2 to 3 seconds for chip resistors and 1 to 2 seconds for delicate IC leads.

Because of the small size of the parts and the requirement for precision craftsmanship, a magnifying lamp, such as that shown in Fig. 18.15, is essential. Flow charts for type I, II, and III assembly and processes are shown in Fig. 18.16.

## 18.6　DE-SOLDERING FOR SMT

Seldom is the yield 100% where there are no components damaged or solder joint defects after the assembly and soldering process. For SMT, rework and repair are less damaging to the components and the pc board than for conventional through-hole boards. Also, less operator skill is required for working with surface-mounted boards.

Figure 18.15　Soldering and rework of small surface mount technology devices are best accomplished under a magnifying glass.

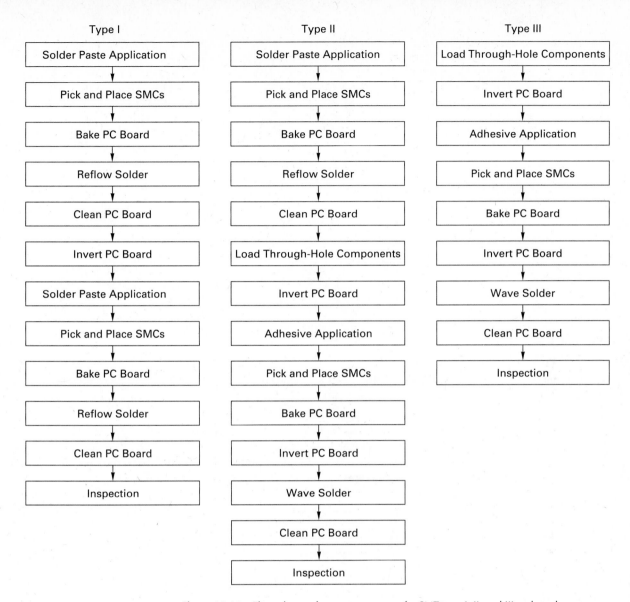

| Type I | Type II | Type III |
|---|---|---|
| Solder Paste Application | Solder Paste Application | Load Through-Hole Components |
| Pick and Place SMCs | Pick and Place SMCs | Invert PC Board |
| Bake PC Board | Bake PC Board | Adhesive Application |
| Reflow Solder | Reflow Solder | Pick and Place SMCs |
| Clean PC Board | Clean PC Board | Bake PC Board |
| Invert PC Board | Load Through-Hole Components | Invert PC Board |
| Solder Paste Application | Invert PC Board | Wave Solder |
| Pick and Place SMCs | Adhesive Application | Clean PC Board |
| Bake PC Board | Pick and Place SMCs | Inspection |
| Reflow Solder | Bake PC Board | |
| Clean PC Board | Invert PC Board | |
| Inspection | Wave Solder | |
| | Clean PC Board | |
| | Inspection | |

**Figure 18.16**  Flow charts of process sequence for SMT type I, II, and III pc boards.

De-soldering equipment for SMT falls into two general categories: *soldering iron conductive tip configurations* and *hot air systems.* Conductive tips are much less expensive than hot air systems.

In using conductive tips to remove SMCs, all of the solder joints must be melted simultaneously with uniform heat. The component is then removed: first, by a slight twisting and then, by lifting off while the solder is still molten. Twisting is important to prevent damage to the leads, especially if the component is held with adhesive. As the tips heat the solder joints, the adhesive is raised to its *glass transition* temperature ($T_g$). This is the temperature at which the adhesive changes from its hardened state to a viscous condition. The twisting motion shears the adhesive. If no adhesive had been used, the twisting is still necessary, since it keeps the component flush to the surface of the board as the leads are separated from the molten solder and prevents deforming them.

Some of the common tip styles used for de-soldering SMCs are shown in Fig. 18.17. Figure 18.17a shows a tip style designed for leadless chip packages such as ceramic resistors, capacitors, and MELFs. They are also available in widths that will de-

**Figure 18.17**  De-soldering tips.

solder SOT packages. The tip style shown in Fig. 18.17b is used on SOIC packages, and the arrangement in Fig. 18.17c is designed for de-soldering flatpacks and LCCC and PLCC packages. The tip style selected must have the correct dimensions to span all of the leads of the specific package size.

When using conductive tips for de-soldering, certain precautions must be taken to avoid damage. A liquid RMA flux should first be applied to the solder joints and then the tips placed against the connections with minimum pressure. The length of time for the solder to become molten should not be more than 2 seconds. The tweezer action of the tips, along with their being wetted with molten solder, will easily allow the component to be removed from the board. It should then be placed onto a damp sponge and the lands cleaned with a suitable solvent such as isopropyl alcohol. The tips should also be wiped on the sponge to remove excess solder and flux residue.

Hot air systems are available for removal and repair of both passive and active de-

**Figure 18.18**  Hot air de-soldering systems.

vices. As a result of the special designs of these systems, hot air is directed onto the solder joints through a nozzle. This prevents the bodies of the components from overheating. The temperature reached during the de-soldering process will be less than that involved in either the vapor-phase or IR reflow operations. The hot air temperatures reach as high as 400°F (205°C). Two hot air system nozzle designs are shown in Fig. 18.18. The open wall nozzle shown in Fig. 18.18a is useful for de-soldering gull wing style terminations. When fitted with a precision insert or pilot that fits over the case of the device and a vacuum tube with a suction cup, the system can also be used to remount devices. To remount, a microscope with controls is required to result in perfect lead-to-land registration. Where there is concern for overheating the solder connections of adjacent components that will not be de-soldered, the vented nozzle design, shown in Fig. 18.18b, is employed.

Many systems also have the feature of heating the bottom surface of a board in preparation for de-soldering components on the top surface. The purpose of this is to reduce thermal shock to the board. The temperature used, however, should not be so excessive as to adversely affect bottom surface solder joints or components. Heating the board for 30 to 40 minutes at a temperature of 200°F (94°C) has been found to eliminate thermal shock problems while not causing any damage to joints or components.

After assembling components and hardware to the printed circuit boards, the next phase of a project's development is to construct the sheet metal chassis elements. In the next unit, Unit 5, information will be given on sheet metal fabrication techniques. Topics include metal shearing, layout, drilling, filing, bending, and finishing.

## EXERCISES

### A. Questions

18.1      What do the terms *SOIC, MELF,* and *PLCC* represent?

18.2      Describe the soldering defect called *tombstoning*. How is it avoided?

18.3      How are conventional and SMT components oriented on type II assemblies?

18.4      What factors govern the selection of solder paste for a specific application? Briefly discuss each factor.

18.5      What are *rheological properties?*

18.6      What are some of the methods used to prevent the formation of solder balls during reflow soldering?

18.7      What is the difference between *pot life* and *shelf life?*

18.8      What is the advantage to the use of epoxy and acrylic adhesives when compared to the thermoplastic type?

18.9      Why does a wave soldering system for type II assemblies use a dual solder wave?

18.10      Why is it recommended to heat the bottom surface of pc boards prior to de-soldering SMCs?

### B. True or False

Circle T if the statement is correct, or F if any part of the statement is false.

18.1      Resistors are much thinner than capacitors and require a longer land to prevent misalignment.          **T    F**

18.2  Lands for MELFs are partially slotted to prevent these devices from rolling out of position when adhesive is not used.      **T**    **F**

18:3  Type III assemblies have SMCs exclusively on one or both sides of the board.      **T**    **F**

18.4  Adhesives must be used to hold components and devices to the bottom side of the board for any soldering process used.      **T**    **F**

18.5  RMA flux is the most active of the three types used in solder paste.      **T**    **F**

18.6  Solder paste to be deposited by stenciling requires a higher viscosity than that applied by screening.      **T**    **F**

18.7  When solder paste is applied, a minimum of 75% of the land must be covered.      **T**    **F**

18.8  Epoxies and acrylics are classified as thermoplastic adhesives.      **T**    **F**

18.9  Acrylic adhesives require both UV light and heat for curing.      **T**    **F**

18.10  The maximum soldering iron tip temperature for use on surface-mounted components is 750°F.      **T**    **F**

## C. Multiple Choice

Circle the correct answer for each statement.

18.1  Manually assembled SMCs should be handled by their (*bodies, terminations*).

18.2  Plastic SMCs require solder paste having (*lower, higher*) melting temperatures.

18.3  Epoxy and acrylics are classified as (*thermoplastic, thermosetting*) adhesives.

18.4  Acrylic adhesives can be cured in about (*2 hours, 2 minutes*).

18.5  Solder joints should not remain in a solder wave for more than (*4, 10*) seconds.

## D. Matching Columns

Match each item in column A with the most appropriate item in column B.

| COLUMN A | COLUMN B |
|---|---|
| 1. Air knife | a. Footprint |
| 2. Land pattern | b. Syringe |
| 3. MELFs | c. Flux |
| 4. Acrylic | d. Rheological property |
| 5. Type I assemblies | e. Round bodies |
| 6. RMA | f. SMCs only |
| 7. Slump | g. Adhesive |
| 8. Solder paste | h. Wave solder |

# Unit

# FIVE

# Chassis Fabrication and Metal Finishing Techniques

# Shearing

*Upon completion of this chapter on sheet metal shearing, the student should be able to*

- Properly and safely use the foot-operated squaring shear, the hand-operated squaring shear, and the hand notcher to accurately shear sheet metal.
- Make the necessary adjustments in these metal cutting shears for obtaining precise cuts and notches.

## 19.0 INTRODUCTION

Industrial descriptions of machine tools include sheet metal *shears* in the category of *metal-forming machines*. The action of a shear is primarily to cut pieces of sheet and bar stock (or structural shapes) to desired dimensions. The construction of electronic equipment begins with metal-cutting operations. Selecting and properly using the most appropriate tools for metal cutting is essential for quality workmanship. Bulk sheet stock must be cut into blanks of the desired overall dimensions for the chassis (or enclosure elements) before layout work begins.

Two basic types of shears are typically used in electronic fabrication. These are the *blade shear* and the *notcher*. The vertical drive for shear machines is either manual or power-operated. Some shear machines are also equipped for *reciprocating* vertical motion. Shears are capable of producing a neat, straight cut without edge imperfections.

This chapter considers in detail the alignment and practical operating information for the foot-operated and hand-operated *squaring shear* and the *hand notcher*.

***Caution:*** *Prior to performing any of the operations described in this chapter, approved safety glasses must be worn.* Also, the section on safety outlined at the beginning of the book should be reviewed for more specific information.

## 19.1 FOOT-OPERATED SQUARING SHEAR

The foot-operated squaring shear, shown in Fig. 19.1, is used for (1) *cutting* large pieces of sheet metal into *blanks,* (2) *trimming* stock to obtain a finish cut or reference edge, and (3) *squaring*. The basic parts of this shear, shown in Fig. 19.2, are two *cutting blades, bed, foot treadle,* and four *alignment gauges*.

Accurate and professional chassis layout begins with blank stock that is not only cut to the correct overall dimensions, but also is cut squarely. Before the 90-degree alignment of the shear blades with the side gauges is tested, the following inspections of the shear should be made: (1) uniform hold-down bar pressure, (2) proper shear blade clearance,

**Figure 19.1** Foot-operated squaring shear. Proper positioning of hands and feet is illustrated together with the correct implementation of the back stop.

and (3) even tension when the foot treadle is depressed. Each of these tests is described below.

A test piece of scrap metal of uniform thickness is used to test for uniform hold-down bar pressure. With the test piece under the hold-down bar and the foot treadle fully depressed, proper adjustment is made via two adjusting screws located on the front side of the hold-down bar. The screws are set so that the work is held firmly, with care taken so that excessive pressure does not deform the metal.

Proper adjustment of shear blade clearance between the upper and lower blades is a uniform gap, typically from three to five thousandths of an inch. Adjustment screws for varying clearance to this distance are located on the front end of the shear frame. To ensure a clean cut, there must be uniform translation of motion between the foot treadle and

**Figure 19.2** Parts identification of foot-operated shear and correct metal positioning for shearing (top view).

the upper cutting blade. This tension adjustment is made by turning the two *turnbuckles* that link the *foot treadle* and the *crosshead.*

When the foregoing tests and adjustments are completed, the 90-degree alignment of the blades and the side gauges may be tested. To determine this alignment, a *carpenter's square* is placed on the shear bed, with the longer leg flush with the lower cutting blade. The shorter leg of the square is placed against one of the side gauges. For square cuts, the side gauges must be perpendicular to the cutting edge and flush against the square. Any necessary correction can be made by loosening the two *securing bolts* holding the side gauge in place and positioning the gauge so that it rests flush against the leg of the square. When the securing bolts are tightened, the shear is properly aligned for square cuts. *Both* the gauges on the shear bed should be checked for alignment before starting any cutting operations. Once the side gauges are properly adjusted, stock placed against them for cutting will have a right angle between the edge of the stock against the gauge and the cut edge. Although either side gauge may be used, the gauge on that side of the shear where the cutting action *first* occurs is used when cutting small pieces of stock.

Many shears are equipped with *back gauges* and a *stop* for *back shearing,* as shown in Fig. 19.2. Although front shearing is more accurate when using the side gauges, the back gauges are convenient when a large quantity of stock is to be cut to the same dimensions. The back gauges are calibrated in sixteenths of an inch measured from the rear edge of the lower blade. The gauges also serve as the supporting arms of the back stop. To ensure that the back gauge pointers are properly set, a long piece of stock of a known uniform width is inserted between the back stop and the rear edge of the lower blade. The pointers should be readjusted if they do not read the correct dimensions.

The desired depth of cut is easily set by moving the stop along the back gauges to the required dimension using the clamping screws to secure the back stop in place. Metal to be cut is placed on the shear bed and moved under the upper blade until it contacts the back stop. With the metal held firmly in position, pressure is applied to the foot treadle. The hold-down bar will contact the metal first, securing it firmly against the shear bed. Additional pressure applied to the treadle engages and depresses the upper blade until the metal is cut. Removing the pressure from the treadle releases the blade and then the hold-down bar, in sequence. To aid in shearing large pieces of stock, *extension arms* are available. These arms bolt to the front edge of the shear bed and support the material.

Careful attention to all the adjustments mentioned in this section results in a neatly cut edge free of waves and burrs. The *cutting capacity* of the shears should *never be exceeded.* The cutting capacity of the type of shear shown in Fig. 19.1 is approximately 16 gauge for mild steel and 12 gauge for the softer, nonferrous metals (aluminum, brass, and copper), with a maximum cutting length of 36 inches.

Several safety precautions must be observed when using a foot-operated squaring shear. Of course, fingers must be kept clear of the hold-down bar and cutting blades at all times. In addition, the operator's feet must be positioned so that no injury occurs during depression of the foot treadle.

## 19.2 HAND-OPERATED SQUARING SHEAR

For work on smaller pieces of stock, the *hand shear* is usually preferable since it is more accurate and easier to operate than the foot shear. The hand shear consists of a *table,* a *side gauge, back gauges,* and two *cutting blades.* A *hand lever* causes two *eccentric cams* to move the *crosshead.* This imparts the vertical motion to the *upper blade.*

Use of the hand shear is basically the same as that of the foot shear. Hand shear tables vary in size from 6 inches to 36 inches and the shearing capacities are similar to those of foot shears. Sheet metal is cut by holding the stock firmly against the side gauge or back gauge stop with one hand and pulling the shear handle forward with the other. This operation is shown in Fig. 19.3.

**Figure 19.3**   Using the hand-operated shear to trim chassis blanks. Metal to be cut is positioned against the right-side gauge and held firmly with the left hand. The micrometer back gauge is also shown.

The hand shear should also be checked for correct adjustment prior to use. The adjustment procedures are basically the same as those for the foot shear (Sec. 19.1). If the shear table is not wide enough to accommodate a carpenter's square, a piece of sheet metal that has been previously squared serves the purpose.

The *side gauge* on the shear table is calibrated from the *cutting edge* of the *lower* shear blade. This gauge should be checked for accuracy before any cutting operations by placing a piece of square stock of known dimension on the shear table with one edge flush with the lower blade and the other against the side gauge. The known length of stock should exactly measure the dimension shown on the gauge. If adjustment is necessary, the securing screws on the gauge are loosened and the gauge is moved until the correct calibration mark exactly aligns itself with the edge of the stock. With the side gauge properly adjusted, measurements can be taken directly from drawings and stock can be cut without the need for layout marks scribed into the work. The degree of accuracy of any cut is directly related to the accuracy to which the side gauge is adjusted. The side gauge is normally calibrated in sixteenths of an inch. If the drawing requires a closer tolerance than permitted by this calibration, lines scribed into the stock may be used to determine the correct cutting position. For an even greater degree of dimensional accuracy, *micrometer back gauges* are available with precise settings as low as 0.001 inch.

## 19.3  HAND NOTCHER

In the construction of electronic equipment, it is often necessary to shear *notches* of various sizes and shapes into a piece of stock. The *hand notcher* is a useful tool for this purpose. The basic parts of the notcher are the *operating handle, table, dies,* and *blades.* The operating handle is attached to an *eccentric cam.* A *V-ram* and a *vertical block,* to which the blades are attached, move vertically as the hand lever is operated.

The notcher is used for shearing short lengths of stock, cutting 90-degree notches, and making various angular trim cuts. The two cutting blades are mounted at right angles to each other and positioned over a pair of dies that are secured to the table. The edges of

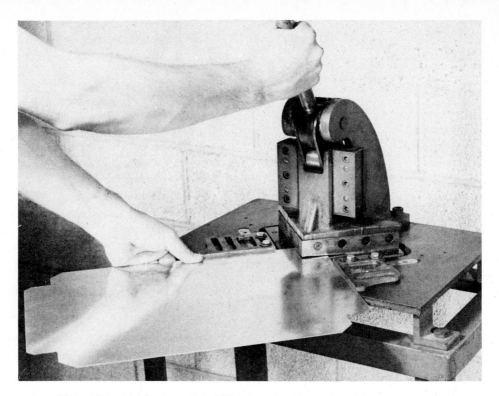

**Figure 19.4** Metal to be cut is held firmly against the notcher table and stops with one hand, while the handle is operated with the other. The stops assist in positioning metal quickly for duplicating work.

the blades and dies are at 45-degree angles to the sides of the table. Figure 19.4 shows a hand notcher and illustrates its use.

For notches having angles *less than* 90 degrees, a *bevel square* or *protractor head* provides a means of transferring drawing dimensions directly to the work. The bevel square or protractor head is set to the desired angle of cut. The base of the square or head is then placed against the bevel blade or rule. While maintaining the desired angle, the work is moved along the table scale for the required depth of cut. For notches having angles *greater than* 90 degrees, it is necessary to perform *two* cuts.

When quantity duplication of work is performed, it is tedious to use the table scales alone or in combination with a square to align the stock. Stops are available, similar to those shown in Fig. 19.4, which are easily bolted to the notcher table over both scales. Each stop is slotted for ease of adjustment.

Extreme care must be taken to keep fingers away from notcher blades. Unlike the hand- and foot-operated shears, which provide some degree of protection to the operator, the notcher is *not* equipped with a hold-down bar. The correct cutting position of the notcher blades and dies must be continually maintained; otherwise, both will be damaged. The maximum capacity of the hand notcher is approximately the same as for the foot shear shown in Fig. 19.1 (16 gauge for mild steel and 12 gauge for the softer metals), with a cutting length of 6 inches.

## 19.4 HYDRAULIC AND MECHANICAL POWER-DRIVEN SHEARS

Hydraulic or mechanical power-driven shears have a wider range of metal capacities than those of the manually operated type. Power shear capacities range from $\frac{1}{8}$ inch by 24 inches to 1 inch by 30 feet mild steel with cutting pressures up to 8000 tons. Power shears for even thicker and wider stock are also available for special applications.

Hydraulic shears offer some advantage over the mechanically powered motor-driven models. The hydraulic drive provides both variable speed and variable blade pressure. This allows wider adjustments to obtain optimum cutting action for any specific type of stock material.

Power shears have distinct advantages over manually operated models if quantity cutting is a major concern. These shears may be set for single or multiple shearing operations, thereby increasing cutting rates and lowering costs.

## EXERCISES

### A. Questions

19.1   What are the two basic types of shears used in sheet metal fabrication for electronic packaging?

19.2   What are the four basic parts of a foot-operated squaring shear?

19.3   Answer Question 19.2 for a hand-operated squaring shear.

19.4   What is the maximum thickness, in decimal form, of aluminum and of mild steel that can be cut without exceeding the capacity of the foot- or hand-operated shears?

19.5   Explain how a properly adjusted side gauge on a hand shear can save considerable time when shearing stock to drawing specifications.

19.6   How does a hand notcher differ from a hand-operated shear?

19.7   What is used with a hand-operated shear when a greater degree of dimensional accuracy is required?

19.8   What are typical metal thicknesses that can be cut with hydraulic and mechanical power shears?

19.9   Why is it important that the hold-down bar pressure of a foot-operated shear be properly adjusted? What could happen if this pressure is not sufficient? What could happen if the pressure is excessive?

19.10  What safety precautions need to be emphasized when using the foot-operated shear?

### B. True or False

Circle T if the statement is true, or F if any part of the statement is false.

19.1   Cutting 12-gauge mild steel on the squaring shear will not exceed its capacity.                                                                    T   F

19.2   Micrometer back gauges make the hand-operated shear more accurate.                                                                             T   F

19.3   Micrometer back gauges are available down to precise settings of 0.025 inch.                                                                      T   F

19.4   The hold-down bar on a hand notcher must be uniformly pressing down on the metal before it is sheared.                                            T   F

19.5   It is possible to cut a 125-degree notch with the hand notcher.         T   F

19.6   The maximum width of stock that can be cut on a foot-operated shear is 4 feet.                                                                    T   F

19.7   The maximum length of cut using a hand notcher is 6 inches.            T   F

## C. Multiple Choice

Circle the correct answer for each statement.

19.1    Cutting large pieces of sheet metal into blanks requires the use of a (*notcher, squaring shear*).

19.2    The cutting capacity of a foot-operated squaring shear is given in units of (*metal length, metal thickness*).

19.3    The side gauges of a squaring shear are (*parallel, perpendicular*) to the cutting edge.

19.4    Extension arms on a squaring shear are bolted to the (*front, rear*) edge.

19.5    The hand notcher requires a minimum of (*two, three, four*) cuts to produce a sheared angle of greater than 90 degrees.

## D. Matching Columns

Match each item in column A to the most appropriate item in column B.

|   COLUMN A | COLUMN B |
|---|---|
| 1.  Cutting capacity | a.  Foot treadle |
| 2.  Cutting pressure | b.  Clamp |
| 3.  Side gauges | c.  90-degree cuts |
| 4.  Hand notcher | d.  8000 tons |
| 5.  Squaring shear | e.  90 degrees to cutting edge |
| 6.  Cutting length | f.  16 gauge |
| 7.  Hold-down bar | g.  36 inches |

## E. Problems

19.1    Test the 90-degree alignment of the side gauges to the cutting blades, side gauge 1-inch graduation distance from cutting blades, and parallel alignment of back stop to cutting blades for identical back gauge settings on a foot-operated shear.

19.2    Check the accuracy of the 45-degree angle between the cutting blades and front edge of a sheet metal notcher. Also inspect the alignment of the scale zero graduation marks and the leading ends of the cutting blades.

19.3    Fabricate a piece of 16-gauge aluminum into the configuration shown in Fig. 19.5 for a transistor heat sink. The heat-sink layout will be completed in Problem 20.2, drilling and deburring in Problem 21.2, and bending in Problem 23.2.

**Figure 19.5**

# Chassis Layout

## LEARNING OBJECTIVES

*Upon completion of this chapter on laying out sheet metal chassis elements, the student should be able to*

- Properly use layout dye for marking sheet metal.
- Properly use the layout table for sheet metal layout.
- Use common layout tools, such as scribes, scales, hermaphrodite calipers, combination squares, punches, and dividers to produce good-quality sheet metal layouts.
- Lay out an uneven dimension into a number of equal parts.

## 20.0 INTRODUCTION

In Chapter 19, stock was sheared into *blanks* and all the information relative to hole size, hole position, bend lines, and so on, resulting from the planning stage will now be transferred from the chassis layout drawings to the work. Techniques and necessary tools for prototype development are considered in this chapter and attention is given to methods for producing accurate layout. Selecting and using the proper tools is stressed to guide the technician in developing greater skills to minimize the inherent errors associated with this type of work.

Specific information is provided in this chapter on *layout dye, layout tables,* and procedures using the *scribe* and *scale.* Also included are discussions on the proper use of the *hermaphrodite caliper, combination square, punches,* and *dividers.*

## 20.1 LAYOUT DYE

Layout dye is a convenient aid in sheet metal layout work. Commercial dyes are generally *blue* or *yellow*, the selection depending on the application. Blue dye is preferable for chassis layouts since it provides more contrast with the light metal surface. Layout dyes are fast-drying and are easily removed. For this reason, care must be exercised to prevent oil or solvents from contacting the surface during the layout operations.

The surface to be dyed must be free from dirt and oil in order for the dye to adhere. Any common detergent will serve as a cleaning agent. A light coat of dye can be easily applied with a small nylon brush or cotton swab. The dye should be applied to the side of the metal that will *not* be visible after the metal is bent to the desired form. After approximately 10 minutes drying time, the surface is ready for working. Light pressure applied with a pointed object on the dye results in a smooth, clean line where the dye is removed. The resulting contrast between the dark color of the dye and the exposed underlying metal facilitates rapid and accurate location and identification of layout marks for punching, drilling, cutting, and bending operations. If the applied pressure is excessive, the surface will be gouged. Enough pressure must be applied, however, to ensure the removal of a suitable amount of dye to achieve the desired contrast. Removal of the dye, after the chassis fabrication operations have been completed, requires the application of a suitable solvent, such as wood alcohol, and light rubbing with a cloth or cotton swab.

## 20.2 LAYOUT TABLE

For precision chassis layout work, a *layout table* will aid in producing good-quality results. Layout tables are commonly made of cast iron and are generally referred to as *bench plates* or *surface plates*. These plates are machined on the top and sides with the edges perpendicular to the top surface and at right angles to each other. This level surface can serve as an excellent means of testing for flatness. Any rocking motion of a surface placed on the bench plate indicates a variance from true flatness.

Figure 20.1 shows a typical layout table. In scribing layout lines on a chassis blank, the following procedure is followed. The blank is placed in one of the corners of the layout table with two of its adjacent edges exactly flush with two adjacent edges of the table. *Parallel clamps* firmly secure the metal to the table. Excessive clamp pressure should be avoided to prevent damage to the work. Usually, finger-tight pressure is sufficient. It is often necessary to reposition the clamps when they obstruct the layout operations. With the metal thus properly positioned, the edges of the layout table serve as a convenient reference for measuring, tool positioning, and alignment.

## 20.3 SCRIBE AND SCALE

The *scribe* serves as the instrument to mark construction lines on the metal surface. The scribe, as shown in Figure 20.1, is a sharp-pointed implement that makes a fine groove in the dye covering the metal. The handle is knurled to reduce the possibility of slipping. The tip has a taper of approximately 15 degrees, resulting in an extremely fine point, to facilitate a high degree of accuracy in scribing lines. Many scribes have *tungsten carbide*

**Figure 20.1** Transferring dimensional information to the sheet metal with scales and scribe.

tips and seldom require sharpening. Tips made from hardened tool steel need to be sharpened occasionally. Care must be exercised in this sharpening process to maintain the original taper over the entire length of the tip to avoid *rounding*. Rounding prevents the tip from getting as close as possible to the scale edge when laying out measurements.

Dimensional information from the chassis layout drawing is transferred to the blank stock with the aid of a scale. The end of the scale should never be used as a measuring reference unless it is established that the dimension from each end to the first major graduation is accurate. It is common practice to use one of the major numbered graduations as the reference. This procedure is shown in Fig. 20.1. A small *light* scratch is made on the metal with the point of the scribe to the desired measurement on the scale. To position a hole, *two* perpendicular dimensions transferred from the drawing must be constructed on the work. The point of intersection of these two lines represents the center of the hole.

To minimize errors, all measurements should be made from a reference edge of the metal. Measuring from a previously marked point could compound errors. It may be necessary, then, to add or subtract measurements from the drawings before transferring them to the metal.

When scribing lines such as bend lines along the entire length of the metal, two *location marks,* spaced as far apart as possible, are first made. The edge of the scale is then placed along these marks. The scribe, with the point along the scale edge and tilted for close contact with this edge, is drawn *lightly* across the metal. The resultant line drawn on the dye should completely include or cover the location marks to ensure accuracy. If a correction has to be made, a new line must be drawn. To avoid later confusion brought about by a double line, the incorrect line should be blotted out with dye.

## 20.4 HERMAPHRODITE CALIPERS

A simple way of producing scribed lines parallel to the edge of the metal blank is to use the *hermaphrodite caliper*. This instrument is shown in Fig. 20.2a. It consists of an *adjusting knob* and *two legs,* one having a *scribe point* and the other a *hooked end* for alignment along the metal edge. The legs are joined at the top with a *spring crown* that provides tension. To set the desired width of scribe, a scale is first placed on the layout table. The scribe tip of the caliper is then set on a convenient reference graduation. The adjusting knob is rotated until the face of the hooked end exactly aligns with the desired measurement of the scale. A scribed line is made on the dye by holding the hooked end perpendicular to the edge of the blank and resting it on the surface of the layout table.

Care must be taken to maintain the correct position when using the caliper and also to apply the proper amount of pressure. The caliper should be tilted at approximately 45 degrees toward the operator (Fig. 20.2b). While maintaining this 45-degree angle, a constant downward pressure is applied with the index finger to the scribe leg as the caliper is drawn across the work from the farthest end. Although the hermaphrodite caliper legs are rigidly positioned, lateral pressure can cause the tip distance to change as the legs tend to flex. If an error in marking is made, it can be blotted out with dye and another line drawn. An effort should be made to draw the scribe tip across the work only once. Otherwise, a double line could result, inviting error. Since alignment and pressure must be maintained simultaneously for optimum results, experience will provide the necessary "feel" for this instrument.

## 20.5 THE COMBINATION SQUARE

The combination square, shown in Fig. 20.3, is a useful and versatile tool in layout work. It consists of four parts: the *grooved blade,* a *squaring* and *miter head,* a *bevel protractor head,* and a *centering head.* The blade is a steel rule normally graduated on all four edges,

(a)

(b)

**Figure 20.2** Hermaphrodite calipers; (a) nomenclature; (b) correct pressure and position are important.

set between blade and head and another line is scribed perpendicular to the first. The intersection of the scribed lines locates the center of the hole.

To scribe 45-degree angle lines, the 45-degree angle edge of the miter head is used. With the inside face of the head against the reference edge, the blade makes a 45-degree angle with respect to that edge (Fig. 20.5). Lines may be scribed at this angle using either edge of the blade.

For angles other than 45 or 90 degrees, the bevel protractor head is used. This head is adjustable through 180 degrees, calibrated from zero degrees center scale to 90 degrees at either scale end (Fig. 20.6). With the edges of the blade parallel to the face of the protractor head when placed against the reference edge of the blank, the scale reading is zero degrees and is aligned with the index mark on the frame of the head. As the blade is rotated, the angle made may be read directly as it aligns with the index mark. Before using the combination square with any of the heads, care must be taken to assure that the blade is secured to the head by tightening the thumbscrew. Otherwise, angular and scalar errors may occur.

## 20.6 PUNCHES

To facilitate the drilling process, as discussed in the next chapter, a conical indentation must be made in the metal at the intersection of the location marks of all hole centers. These indentations are made with the use of a *hammer* and *punch.*

Punches are classified as *prick* and *center,* depending on the angles of the point. Both punches have hardened steel points and knurled handles for ease in gripping. The prick punch has a point ground to 30 degrees and serves to introduce a small, round indentation primarily for location purposes. The center punch has a point ground to 90 degrees. Its function is to provide a larger-diameter indentation in the metal at the proper angle to accept the narrowest part of the drill bit web for alignment or centering. This indentation also prevents the drill bit from "walking" from the center position of the hole during the drilling operation.

With the punch held perpendicular to the work and the point at the center of the intersecting lines, it is *lightly* tapped with a hammer. An 8-ounce ball peen hammer is sufficient for use in punching. It is important to place the work on a hard, flat surface for complete support during the punching operation. Use of a heavy hammer or excessive striking force will damage the work as well as the punch. It is advisable to test for the correct striking force on a sample piece of stock, applying only one strike with the hammer. Any error in the location of the indentation will result in a corresponding error in the location of the hole. To correct an off-center error in punching, the tip of the center punch is placed in the indentation at an angle such that when struck the hammer will tend to drive the point *toward* the intended intersection.

*Automatic punches* are available with prick or center punch tips (Fig. 20.7). These punches eliminate the need for a hammer because they contain an adjustable spring-loaded driving mechanism in the handle that forces the tip into the work when pressure is applied to the handle. The amount of striking force required for the proper indentation is determined by the type and hardness of the metal used. The pressure is adjusted by turning the upper portion of the punch clockwise to compress the spring for increased pres-

**Figure 20.7**   Automatic punch.

sure or counter-clockwise to expand the spring for reduced pressure. To use the automatic punch, it is positioned perpendicular to the work surface with the tip exactly at the point of intersection of the location marks. A steady downward pressure is applied to the handle until the spring is released, driving the punch tip into the work. As with most hand tools, practice on sample stock with the automatic punch will help in selecting the optimum setting.

## 20.7 DIVIDERS

Dividers serve several purposes in layout work. In addition to locating centers of evenly spaced holes and dividing lengths into equally divided parts, they also are used to scribe circles about hole location marks to ensure accuracy of drilling or chassis-punching operations. The common divider, as shown in Fig. 20.8, consists of *two legs* with scribe points, *spring crown,* and *adjusting knob.* The adjusting knob is used to set the distance between the divider scribe points.

To set the divider for scribing a circle, one tip is placed at a convenient graduation mark on a scale, such as the 1-inch mark. The adjusting knob is then rotated until the other tip is on the graduation mark corresponding to the radius of the circle. To position the divider for use, one tip is placed in the indentation previously made in the punching operation with the other tip contacting the work surface. While maintaining the first divider tip firmly in place, the other is swung around the pivot point inscribing an outline of the circumference of the hole in the dye. The divider should be positioned so that the spring crown is perpendicular to the work surface as the divider is rotated. To divide an uneven dimension into several equal parts, the divider is used with the technique illustrated in the solution to the following example.

*Example.* It is necessary to position five holes along a front panel, the total length of which is 14 inches. The spacing between each hole and the distance from each end of the panel must be equal.

Spring Crown

Adjusting Knob

Legs

Scribe Points

**Figure 20.8** Dividers.

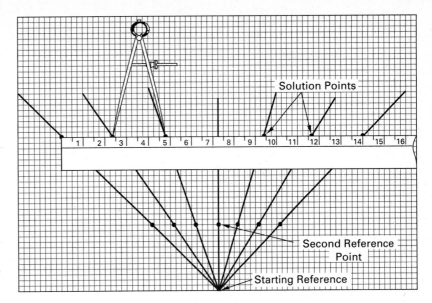

**Figure 20.9** Construction technique for equal spacing of location marks.

*Solution.* A sheet of quadruled paper is placed lengthwise and a point is marked on a grid intercept close to the bottom and halfway between either side (Fig. 20.9). (Although 17- by 22-inch paper with four divisions to the inch is used for this solution, the size and divisions per inch are not important as long as linear graph paper is used.) At a point approximately 3 inches directly above the first reference point, another point is marked. Several points are then marked at 1-inch intervals both to the left and to the right of the second point on the same horizontal grid line. For this problem, seven points, including the center point, were made to account for the six spaces necessary. Straight lines are now drawn through each of the seven points and intersecting at the first point at the lower edge of the paper. It can now be seen that each horizontal grid line above the first reference point represents a division of six equal parts, progressively increasing with each vertical increment. To determine the correct distance between hole centers for this problem, the edge of a scale is placed over the construction line on the extreme left. With the scale aligned with the horizontal grid lines and maintaining the reference point on the scale edge, the scale is moved up until the 14-inch mark on the scale intercepts the construction line on the extreme right. (Refer to Fig. 20.9.) Dividers can now be set to one of the equal horizontal dimensions on the edge of the scale and this dimension can be transferred to the metal. A scribed center line on the work is necessary to locate the height of the position of the holes. The dimension set on the divider can now be transferred along this center line. One scribe tip is brought gently to the edge of the panel where the center line runs off the surface. Care must be exercised that this tip does not drop below the face of the panel; otherwise, an error in layout will result. With the first scribe tip held carefully in this position, the other scribe tip is brought down on the center line and a small arc is drawn on the dye. The same scribe tip is maintained at the intersection of that arc and the center line and the divider is gently rotated until the other scribe tip is brought along the center line to cut another arc. By such successive 180-degree rotations of the divider, the required number of hole centers will be accurately located.

In the following chapter we provide detailed information concerning the proper selection and use of drills, bits, and chassis punches.

## EXERCISES

### A. Questions

20.1    How should a scale be positioned on the sheet metal surface when transferring dimensions using a scribe?

20.2    How does the use of a hermaphrodite caliper reduce layout time?

20.3    What care must be taken when using a hermaphrodite caliper to avoid the introduction of layout errors?

20.4    When a hermaphrodite caliper is not available, what other method is used to rapidly locate and scribe lines parallel to a reference edge?

20.5    What tool is used to locate angles other than 45 or 90 degrees on sheet metal?

20.6    Why is a center punch used to make an indentation at the intersection of layout lines at the center of all holes to be drilled?

20.7    How are errors in center-punched locations corrected?

20.8    List three purposes that dividers serve in sheet metal layout.

20.9    Why is layout dye used in sheet metal layout work?

### B. True or False

Circle T if the statement is true, or F if any part of the statement is false.

20.1    Layout dye should be removed from the metal surface before any bending is done.                                                             **T    F**

20.2    Sheet metal layout is done from a top view.                            **T    F**

20.3    Hermaphrodite calipers should be held perpendicular to the metal surface when scribing layout marks.                                       **T    F**

20.4    A prick punch makes a 90-degree indentation for drill bit alignment.    **T    F**

20.5    Layout tables are commonly made of hardened wood.                       **T    F**

### C. Multiple Choice

Circle the correct answer for each statement.

20.1    The preferred color of layout dye used on aluminum is (*yellow, blue*).

20.2    Layout lines made with a scribe require (*light, heavy*) pressure.

20.3    The quickest way to scribe layout lines parallel to the edge of sheet metal is with the use of (*a combination square and scribe, a hermaphrodite caliper*).

20.4    The layout of angles other than 45 or 90 degrees with a combination square requires the use of a (*bevel protractor, miter*) head.

20.5    Automatic punches (*require, do not require*) use with a ball peen hammer.

### D. Matching Columns

Match each item in column A to the most appropriate item in column B.

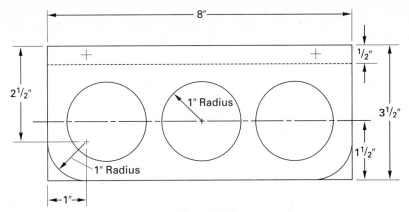

**Figure 20.10**

| COLUMN A | COLUMN B |
|---|---|
| 1. Hermaphrodite calipers | a. 22.5-degree layout angle |
| 2. Dividers | b. Center |
| 3. Bevel protractor head | c. Tungsten carbide bit |
| 4. Miter head | d. Scribing circles |
| 5. Punch | e. Combination square |
| 6. Layout dye | f. Parallel layout lines |
| 7. Scribe | g. Yellow |

**E. Problems**

20.1 Using the layout techniques discussed in this chapter and quadruled graph paper, locate six evenly spaced hole locations across 11 inches.

20.2 Lay out the bend lines on the stock cut for the heat sink in Prob(a) 9.3. The bend lines are positioned from *A* to *B*, *B* to *C*, *C* to *D*, and *D* to *A*. This type of heat sink is suitable for device case styles such as DO-4, DO-5, TO-3, TO-61,

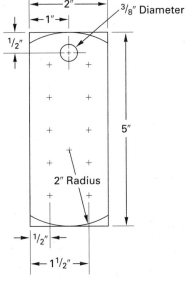

**Figure 20.11**

and TO-68. Determine the hole spacing required for the TO-3 style and locate and center punch this hole location pattern on the center section of the heat sink. Drilling and deburring are performed in Problem 21.2 and bending in Problem 23.2.

20.3 Shear a $3\frac{1}{2}$- by 8-inch piece of 16-gauge aluminum to lay out an oil pressure, battery, and temperature gauge panel shown in Fig. 20.10. The gauge hole locations are to be evenly spaced across the panel. Drilling and deburring are completed in Problem 21.1, hole cutting and bandsaw operations are performed in Problem 22.3 and bending is done in Problem 23.1.

20.4 Lay out a 2- by 5-inch piece of 14-gauge steel as shown in Fig. 20.11 for a drill gauge. Using the layout techniques discussed in this chapter, center-punch five pairs of evenly spaced hole locations along the length of the gauge in addition to the $\frac{3}{8}$-inch hole position. Drilling and deburring of the gauge are done in Problem 21.3, and radius cutting, punching, and filing are performed in Problem 22.2.

# Drilling, Reaming, and Punching

21

## LEARNING OBJECTIVES

*Upon completion of this chapter on the drilling, reaming, and punching of sheet metal, the student should be able to*

- Know the characteristics of twist drills.
- Understand how to use a drill gauge and a micrometer to determine the size of a drill bit.
- Use the drill press to perform drilling operations.
- Use the depth stop to limit the depth of a drilled hole.
- Change the spindle speed of a drill press.
- Use the proper lubricant for a specific drilling task.
- Properly use deburring tools.
- Properly use a hand expansion reamer.
- Properly use a hand-operated sheet metal punch.
- Understand the use of draw and turret punches.

## 21.0 INTRODUCTION

In preceding chapters we have discussed the detailed techniques for preparing sheet metal for chassis construction. All the layout work, including the accurate positioning of all required holes, has been located on the metal blanks. Cutting operations can now begin. This chapter presents the basic cutting operation techniques of *drilling, reaming,* and *punching.*

In this chapter drill bit design is discussed, including *spiral flutes* for twist drills, *straight* and *taper* shanks, *standard drill sizes,* and *drill gauges.* Drill point configurations are also presented. To implement the drill bit, the *drill press,* together with its associated hardware and general operating instructions, is considered with respect to obtaining accurate hole size for common sheet metal applications. In addition, the proper lubrication for drilling specific metals is tabulated.

Deburring tools are also discussed, in addition to the techniques of reaming, including the use, advantages, and disadvantages of the *hand-expansion reamer.* Practical operational information on punches, including the *hand, draw,* and *turret* punches is presented. Techniques for using *square* and *rectangular* punches are discussed and punching and drilling are compared.

***Caution:*** *Prior to performing any of the operations described in this chapter, approved safety glasses and ear protectors must be worn.* Also, the section on safety outlined at the beginning of the book should be reviewed for more specific information.

*Drill bits* are manufactured of *carbon* or *tungsten–molybdenum* steel rods. The carbon steel bits are made for drilling into soft metals such as aluminum, copper, brass, or mild-tempered steel. These bits become dull after a relatively short period of use. Tungsten–molybdenum bits, commonly referred to as *high-speed* (HS) bits, are used for drilling either soft metals or high-tempered steel, stainless steel, and other hard metals. High-speed bits will produce more satisfactory results than carbon bits when used on fiberglass, Bakelite, or other plastic materials. Some plastics, especially glass-cloth laminates, tend to dull carbon bits very quickly.

A drill bit is manufactured by milling two *spiral grooves* called *flutes* into the rod stock. The flutes are positioned directly opposite each other, beginning at the *point* and extending along the entire *body length* of the bit. The area between the flutes is called the *web*. Figure 21.1 identifies the basic parts of the twist drill.

Along the outer surface of the *land* generated when the flutes are milled, a *margin* is ground to provide clearance on the cutting edge and also to ream and maintain drill alignment. When the point is ground to the correct angle, a *chisel edge* is formed at dead center in addition to two *cutting lips* at the beginning of each flute.

Twist drills are available in *short lengths* for center-hole or starting purposes, *jobber lengths* for common drilling applications, and *extended lengths* for deep-hole drilling. Twist drills are also available in a variety of shank configurations, the most common of which are the *straight* and *tapered* shanks. These shanks are extensively used in sheet metal and panel stock drilling applications in electronic packaging. Smaller-diameter drills are made with a straight shank since they are accommodated by the conventional drill press or powered hand-drill chucks. Although more expensive for the smaller sizes (less than $\frac{1}{2}$ inch), the taper shank has several advantages. It is more convenient to insert the bit by pressing it into a *tapered collet* on a drill press or lathe. In addition, bits can be removed and replaced much more quickly from a collet than can straight-shank bits used with a key-type chuck. The prices of comparable size drills of each type in the larger sizes are equivalent.

Three size systems are used for drill bits: *fractional, numerical,* and *letter*. The three lengths described previously with straight or tapered shanks are available in each system. *Fractional* bits range from $\frac{1}{64}$ (0.0156) inch to $\frac{1}{2}$ (0.500) inch in diameter in increments of $\frac{1}{64}$ inch. A complete set of drills in this system would therefore include 32 bits. The *numerical* system consists of 80 bits ranging from number *80,* whose decimal equivalent size is 0.0135 inch in diameter, to a number *1* bit, 0.228 inch in diameter. *Letter* designation of bits starts with *A* size (0.234 inch in diameter) through *Z* size (0.413 inch in diameter). For a composite tabulation of the standard drill size systems, showing the fractional, number, and letter designations along with their decimal equivalents, refer to Appendix II.

The larger drill bits have their size stamped on the shank in accordance with the particular system used. For the smaller bits, *drill gauges,* such as the one shown in Fig. 21.2a, must be used to determine the size because the stamping would be too small and

**Figure 21.1**   Twist-drill nomenclature.

**Figure 21.2** Tools for drill size measurement. *Courtesy of General Tool Company and the L.S. Starrett Company, Athol, MA.*

therefore difficult to read. Each hole in the drill gauge is designated by the size bit that will fit snugly into it. Drill sizes are checked by passing the flute portion of the bit into the gauge hole. The shank end is never used for measuring, since any burr or wear incurred in its use will result in an incorrect size determination. Drill gauges are also available for the larger bit sizes to determine diameters when the stamping on the shank has worn from extensive use. Drill bit diameters may also be measured with a micrometer. It is important, however, that this measurement be taken across the margins. This procedure is shown in Fig. 21.2b.

## 21.2 DRILL BIT POINTS

The points of twist drills have several features that must be examined in order to completely understand the cutting phenomenon. Figure 21.3a identifies the drill point nomenclature. When drilling, the bit is introduced into a center-punched hole, and the chisel edge at dead center of the point is the first cutting edge to come in contact with the work. As the rotating bit is forced against the work, the cutting lips begin to chip metal away. These chips are channeled from the work through the flutes. The flutes also allow cutting lubricants to be applied directly to the work area at the chisel edge. Once the drill bit has penetrated deeply enough into the work so that the entire length of the cutting lips are making contact, the sharp leading edge of the margin will begin to ream the inside of the hole to the final dimension. Notice in Fig. 21.3b that the body clearance is achieved by the margin being *slightly raised* above the surface of the land. Lip and body clearance are essential so that the bit will not jam or bind in the work during the drilling operation. Once the point has passed through the work, the bit should be slightly raised and lowered before removing it so that the hole formed will be completely reamed.

The degree of drill point angle depends on the type of material being worked. In general, the harder the material, the flatter the point. For general-purpose applications, such as drilling aluminum and soft steels, a drill point angle of 118 degrees, a lip clearance of 12 to 15 degrees, and a standard chisel edge angle of 120 to 135 degrees is recom-

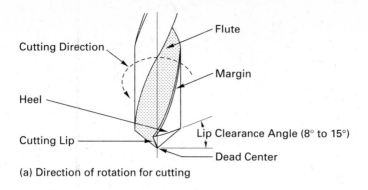

(a) Direction of rotation for cutting

(b) Cutting edges and clearances

**Figure 21.3**   Drill point nomenclature.

mended. When drilling materials such as Bakelite or epoxy-glass (used extensively in printed circuit work), a steep drill point angle of 90 degrees and a lip clearance of 12 degrees is recommended.

The point configurations described above result in reasonably round holes, especially in the smaller sizes. Many times, however, chassis construction requires the drilling of larger holes (greater than $\frac{3}{8}$ inch). Because aluminum is a soft metal, conventional drills tend to cut a hole that is out of round, especially in the larger sizes. The *sheet metal bit* (also referred to as a *brad point bit*), as shown in Fig. 21.4, reduces this problem with its *double-spurred* cutting lips. The point or chisel edge at the center of the web is used to align the bit in the center-punched hole. As the bit is forced against the work, the sharpened spurs on each side of the lands cut into the material in a circular fashion. Continued pressure on the bit causes the spurs to completely cut through the metal, leaving a slug of the diameter of the hole. The margin then reams the hole to the desired dimension.

Since common sizes of bits used in printed circuit work are in the order of No. 55 to No. 67, difficulty may be encountered when attempting to rigidly secure and align these small shanks into a conventional chuck. The enlarged shank on the printed circuit drill bit overcomes this problem. However, the shorter-length bit increases the possibility of breaking. This problem is overcome by using straight-shank bits. Extreme care must be exercised when using small-diameter drills because of their delicate structure.

## 21.3  DRILL PRESS

One of the most useful power tools employed in electronic fabrication is the *drill press*. A bench-type drill press is shown in Fig. 21.5. Mounted on the vertical *column* is a horizontal *base table* to support and secure the press to the workbench. The base table may also

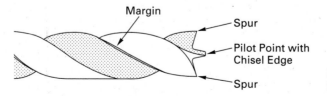

**Figure 21.4**   Sheet metal drill point configuration.

Figure 21.5   Bench-type drill press.

be used as a supporting surface when unusually long pieces are to be drilled. Uppermost on the column is the *head*. Between the base table and the head, a movable *worktable* is positioned on the column and held in place by a *lever-operated clamp*. The work to be drilled is first positioned on the worktable, which is moved vertically along the column by first loosening the lever clamp. When the table is conveniently positioned and the hole in the center of the table is aligned with the *spindle* axis, the clamp is tightened, rigidly securing the table.

Most standard drill press heads are equipped with a three-jaw *Jacobs chuck,* accommodating straight shank drill bits up to $\frac{1}{2}$ inch. The chuck jaws are opened by a counterclockwise rotation of the *chuck key* inserted into the chuck. When the shank of the drill bit is inserted between the jaws of the chuck, clockwise rotation of the key will securely clamp the jaws onto the drill.

Chucks for both straight and tapered shanks are secured to the lower threaded end of the spindle. The drilling pressure is applied to the spindle by pulling the *feed handle* toward the operator. This action provides the *downward* motion of the normally stationary spindle and revolving chuck toward the work. The feed handle is coupled to the spindle with a *spindle return spring* that causes the spindle to return to its normal position after the downward force is removed. This upward motion should be guided with the feed handle to prevent needless jarring of the spindle and head assemblies. When the spindle does not automatically return to its normal position, *spring return tension adjustment* is necessary. The *locking screw* or *pin* on the left side of the head is loosened and the spring return tension knob is turned counterclockwise to increase tension. Since the knob is under constant spring tension, this adjustment should never be made when the drill press is on or when a drill is in the chuck. Should the knob be inadvertently released with the locking screw loose, the spindle will plunge downward (if the feed handle is not held), resulting in damage to the work, the drill, and possibly to the operator.

When a specified depth of hole is to be drilled into the work rather than a complete hole, a *depth gauge* is provided along the right side of the head. Figure 21.6 shows how the depth gauge limits penetration of the drill below the work. The depth gauge has graduations in $\frac{1}{16}$-inch divisions and is threaded to provide for two *locking thumb nuts*. As the feed handle is turned for downward motion, the depth gauge moves through the stop, which is a machined hole in the head casting. The gauge moves with the spindle. With the thumb nuts positioned on the gauge, downward motion stops when they contact the *depth stop*. This prevents any further downward motion beyond the desired position. This position is set by feeding the spindle downward with the drill in the chuck to the desired drill point depth and then tightening the locking nuts until they contact the depth stop. Two locking nuts ensure that the vibration during drilling operations will not change the adjustment.

The spindle is rotated by a belt-driven pulley assembly powered by a motor mounted to the rear of the head. The motor base is secured to two steel rods that pass through machined holes to the rear of the head casting. The position of the motor base can be adjusted by loosening the *clamping bolts* that hold the rods in place. The belt tension between the *cone pulleys* on the motor and upper end of the spindle should not be excessive in order to avoid needless spindle and motor bearing wear. An acceptable means of testing belt tension is to depress either side of the belt at its midpoint. Proper tension exists if approximately $\frac{1}{2}$ inch of inward deflection is attained. More or less than this optimum deflection indicates a need for adjustment. Belt tension is increased when the motor is moved away from the head and is decreased when brought toward the head.

The spindle speed can be changed by moving the belt directly opposite to the motor and spindle pulleys. The belt must always operate in a horizontal position. When the belt position on the cone pulleys is to be changed, the motor should be brought toward the head to provide sufficient slack to avoid stretching or damaging the belt. When the largest motor pulley and smallest spindle pulley are used, the fastest spindle speed is obtained,

**Figure 21.6** Depth gauge set for drilling sheet metal to eliminate unnecessary spindle travel.

which is approximately 3600 revolutions per minute (rpm). The slowest spindle speed of approximately 400 rpm is obtained using the smallest motor pulley and largest spindle pulley. These combinations are achieved by the motor pulleys and the spindle pulleys being inverted with respect to each other.

## 21.4 DRILLING OPERATION

All drilling operations begin by aligning and securing the stock to the worktable. *C clamps* provide good stability for drilling metal used in chassis fabrication. This arrangement is shown in Fig. 21.7. The center hole in the worktable should always be aligned with the spindle axis. In addition, when drilling sheet metal, a piece of wood of uniform thickness is used between the worktable and the stock. This allows the drill bit to completely pass through the work without contacting the worktable and provides complete support so that burring is minimized as the drill point breaks through the work (see Sec. 21.5). When drilling small parts or components, a *drill-press vise* provides the necessary support. For holes less than $\frac{1}{4}$ inch, holding the vise manually against the table is usually sufficient. However, for larger holes, the vise should be mechanically secured. Drill press vises are available with a slotted side or end sections that can be bolted to the table through these slots. For a further measure of security against work rotation, a *safety bar* is employed. This bar threads into the front of the vise and must rest against the *left side* of the column.

The correct drill bit size, drill point angle, and lip clearance must be compatible with the material to be drilled. Table 21.1 shows the type of drills with cutting speed, drill point angles, lip clearance angles, and proper lubricants to use for various materials. The proper drill is secured in the drill press chuck with the drill point serving as reference to align the work. Without securing the work to the table, the spindle is lowered so that the drill point nearly touches the work surface. With the drill point held stationary in this position, the work is adjusted so that the center-punched hole is aligned with the dead center of the drill bit. The power is then turned on and the drill point is allowed to gently touch the center hole and remove a small amount of material. The spindle is then returned to its normal position and the power is turned off. Without moving the work, the test cut is inspected for uniformity about the center lines of the hole. If the alignment is proper, the work is carefully secured to the table. When working with soft metal and twist drills, it is advisable to use a small bit (No. 28 to No. 32) to drill a *pilot hole* through the material to aid in alignment. Also, to ensure as round a hole as possible, a drill bit several hundredths

**Figure 21.7** Clamping techniques for drilling.

*21.4 Drilling Operation*

**387**

**TABLE 21.1**
Drilling Recommendations for Various Materials

| Materials | Cutting Speed for High-Speed Drills* (ft/min) | Drill Point Angle (deg) | Lip Clearance Angle (deg) | Lubricants |
|---|---|---|---|---|
| Stainless steel | 40 | 136–150 | 8–12 | Sulfur and lard oil |
| Carbon tool steel | 60 | 136–150 | 8–12 | Sulfur-based oil, mineral lard oil |
| Epoxy glass† | 80 | 90–110 | 12–15 | Dry |
| Cold-rolled steel | 90 | 120–130 | 12–15 | Sulfur-based oil, mineral lard oil |
| Bakelite | 100 | 60–90 | 12–15 | Dry |
| Cast iron | 120 | 90–118 | 12–15 | Dry or air jet |
| Copper | 200 | 90–110 | 12–15 | Kerosene or dry |
| Hard brass | 200 | 110–125 | 12–15 | Dry, kerosene, or lard oil |
| Bronze | 200 | 110–125 | 12–15 | Dry or mineral oil |
| Aluminum | 250 | 90–118 | 12–15 | Kerosene, lard oil, or turpentine |

*Drilling speeds for carbon steel drills are approximately one-half of those for high-speed drills.

†Use carbide tip drill for long drill tool life.

of an inch smaller than the final dimension should be used before the final hole is drilled. This progressive *undercutting* will allow the lips to cut rather than the chisel edge and thus avoid the possibility of the drill's "walking." Of course, the use of a sheet metal drill will eliminate the need for undercutting. However, the use of a pilot hole is recommended for either type of drill.

From Table 21.1 the average cutting speed is selected for the given material. (Notice that the maximum cutting speeds for HS drills are approximately twice those for carbon–steel drills.) This information is given in units of *feet per minute* (fpm). These speeds should be considered *maximum* and should not be exceeded in order to minimize wear and not overheat the point. It is therefore important to adjust the spindle speed to the closest available speed without exceeding the maximum as prescribed in Table 21.1. This is done by the proper choice of cone pulleys on the motor and spindle. To convert average cutting speed from feet per minute to rpm, the following relationship is used:

$$\text{rpm} = \frac{12 \times \text{cutting speed (fpm)}}{\pi \times \text{diameter of the drill (inches)}} \approx \frac{4 \times \text{cutting speed}}{\text{diameter}}$$

Continuous lubrication is necessary during operation to minimize friction between the point and the work, thus reducing wear. Table 21.1 lists the recommended cutting compounds and lubricants for several common materials. Lubrication should *not* be used sparingly if the life of the drill is to be prolonged.

The following example illustrates the use of the information in Table 21.1.

*Example.* It is necessary to drill a $\frac{3}{8}$-inch hole using twist drills through a piece of 16-gauge aluminum. The drills selected are as follows:

*Pilot size:* No. 31 (0.120 inch)

*Undercut size:* P (0.323 inch)

*Final:* $\frac{3}{8}$ (0.375 inch)

From Table 21.1, the necessary information is obtained as follows:

| Drill Bit Size | HS Cutting Speed (ft/min) | Drill Point Angle (deg) | Lip Clearance (deg) | Lubrication | RPM | Closest RPM on Drill Press |
|---|---|---|---|---|---|---|
| No. 31 | 250 | 118 | 12–15 | Kerosene or turpentine | *7959 | 3600 |
| P | 250 | 118 | 12–15 | Kerosene or turpentine | 2956 | 2800 |
| $\frac{3}{8}$ in. | 250 | 118 | 12–15 | Kerosene or turpentine | 2546 | 2400 |

*Notice that although this calculated rpm is 7959 and that the highest available speed on the conventional drill press is 3600 rpm, this is acceptable because the maximum recommended cutting speed will not be exceeded.

## 21.5 DEBURRING

The fabrication of a chassis requires numerous *through* holes in which the drill body passes completely through the stock. In soft metals, such as aluminum, copper, and brass, the result is a rough edge generated on the reverse side of the stock. This *burr* is the result of the flex in the remaining thin metal as the trailing edges or *heels* of the drill push this material aside along the circumference of the hole instead of cutting it away as the chisel edge breaks through the work. The amount of burr can be minimized by using a wooden block as a support. In addition, by lowering the spindle more slowly, the remaining metal will tend to flex *less* and allow the drill to chip more of the remaining material away. This reduces the size of the resulting burr.

Since burrs can prevent flush mounting of components, in addition to marring rubber or plastic hardware, they must be removed. Two common hand deburring tools are the *countersink tool,* shown in Fig. 21.8a, and the *deburring tool,* shown in Fig. 21.8b. The standard $\frac{1}{2}$-inch countersink is attached to a handle by a press fit for positive gripping action and control, and is constructed with three to five flutes. These flutes do not spiral as on a twist drill. The cutting lips run parallel to the flutes and converge at dead center of the point. The cutting lips have a clearance angle of approximately 15 degrees.

The deburring tool, shown in Fig. 21.8b, has three interchangeable *cutting heads* that pivot when fully seated in the handle. These heads are designed to deburr and form *chamferred edges* on holes in sheet metal or tubing and edges on channel stock.

The countersink is commonly employed for holes less than $\frac{1}{4}$ inch in diameter. This tool requires only moderate hand pressure to impart a circular rotation of the cutting lips against the burr. The result is a slight chamfer about the circumference of the hole. It is important to hold the countersink as perpendicular to the work as possible to ensure uniform cutting.

When harder materials need deburring, hand pressure may not be sufficient. In this application, a countersink can be used in a drill press chuck to obtain maximum pressure. The countersink should be rotated at the slowest possible spindle speed to avoid chatter, which will roughen the surface of the chamferred portion. Before the deburring operation begins on the drill press, the hole center must be aligned with the countersink point. The procedure is similar to aligning a drill point to a center-punched hole. With the drill press power turned off, the spindle is lowered until the countersink uniformly seats into the hole. The spindle is then returned to its normal position and the work is clamped securely into place. With the power on, the countersink is brought into contact with the edge of the hole to begin the deburring operation. Continual inspection is required so that only the burr is removed. Excessive spindle pressure will cause too much material to be removed from around the edge of the hole. This will result in a sharp edge around the hole on the reverse side of the work and will weaken the material around the edge of the hole because of the excessively large chamfer. Also, if the countersink is allowed to pass too far into the hole, the inside diameter will be altered.

For holes larger than $\frac{1}{4}$ inch, the deburring tool shown in Fig. 21.8b may be used.

(a)

(b)

(c)

**Figure 21.8**   Sheet metal deburring: (a) deburring small holes with a countersink; (b) deburring tool with interchangeable cutting heads; (c) deburring tool held at proper cutting angle.

The proper cutting head is inserted into the work from the *reverse side* of the burr. The cutting head has a *hook-shaped blade* that is brought down against the burr. With the handle firmly gripped and held perpendicular to the work, it is rotated in a circular motion about the circumference of the hole (Fig. 21.8c). The cutting head will pivot in the handle, allowing the blade to remove the burr. It may take several turns to completely remove the burr and produce the desired chamfer. Caution must be exercised to avoid gouging or removing excessive material about the edge of the hole.

It occasionally becomes necessary to expand the diameter of a previously drilled hole to a slightly larger size to correct an error in the selected size of a drill. Since this error is not usually detected until the assembly phase of construction, redrilling becomes cumbersome and, in some cases, impossible. For this reason the *hand expansion* reamer, shown in Fig. 21.9 is a convenient tool for enlarging holes.

Enlarging drilled holes introduces an important point in prototype chassis construction, especially if the unit being fabricated is the first of many to be produced. The undersized holes could be the result of an error on the original construction drawings. If this is the case, the drawings must be revised to show the required change in dimension of these holes so that later reference to these drawings will not result in the same error. An erroneous estimate of a hole dimension is not uncommon, but it should be corrected immediately once detected.

The hand expansion reamer consists of a *T handle* and *tapered fluted blades* running lengthwise along the body of the reamer. There are *six cutting blades* that converge at the taper point. At least six cutting blades are necessary to minimize chatter along the edge of the hole as the tool is used. Cutting is accomplished by a scraping action, and each cutting blade has a *body clearance* for relief similar to that of a countersink.

To expand a previously drilled hole, the reamer is inserted into the hole and the handle is rotated clockwise while a moderate downward pressure is applied. The result is a slightly enlarged hole tapered at the same angle as the reamer blades. When hole tolerance is critical, the tapered reamer should not be used to remove more than 0.030 inch because the resultant taper will weaken the surface of the metal about the hole.

The use of the tapered hand expansion reamer is a crude but effective method of enlarging holes. It also generates a burr, because of the downward pressure, which must be removed. Straight-sided reamers with either straight or spiral flutes are available for pre-

**Figure 21.9**  Hand expansion reamer.

cision work. These are commonly used in conjunction with a drill press or milling machine.

## 21.7 PUNCHING

Of all the methods available to the technician for fabricating holes in a chassis, the *sheet metal punch* will produce by far the neatest and most accurate results. Holes larger than $\frac{1}{2}$ to 3 inches in diameter fabricated by punching are superior to those formed by drills. The punch forms a hole that is not only more circular but produces a negligible burr. (A slight chamfer around the edges of punched holes may still be desired to break the sharp edge in preparation for finishing.)

The common *round hand-operated punch* is shown in Fig. 21.10a, consisting of a *punch* with a threaded hole, a *die* with an unthreaded hole, and a *screw*. The punch is constructed of tempered steel with the cutting edges having two sharpened *spurs* aligned 180 degrees apart. As the screw is tightened, drawing the punch and die together, the spurs and cutting edges *shear* the metal.

Punching with a hand punch begins by drilling a guide hole to accept the screw. This guide hole is drilled in the center of the hole to be formed. The screw size is either $\frac{1}{4}$, $\frac{3}{8}$, or $\frac{3}{4}$ inch in diameter, depending on the size of the punch. The guide hole thus drilled should provide a $\frac{1}{16}$-inch clearance for the screw. Table 21.2 shows the screw size and guide hole size for punching holes from $\frac{1}{2}$ to 3 inches.

The punch has two parallel flat sides at its base to secure it in a vise so that it will not turn. Once the correct guide hole has been drilled and the punch secured, the screw is passed through the closed end of the die. As shown in Fig. 21.10b, the sheet metal is positioned over the punch. The screw is then passed through the guide hole and threaded into the punch. The screw is tightened to clamp the punch and die against the stock, making sure that the die is centered. (If the die is not centered, the resulting hole will only be off-center by a maximum of $\frac{1}{32}$ inch, which, for most applications, is tolerable.) The head of the screw is square and should be tightened with a properly fitted open-end wrench. As the screw is further tightened, the wide edge of the die applies uniform pressure on the sheet metal about the circumference of the hole being punched. As the metal is forced into

(a)                                          (b)

**Figure 21.10**  Hand-operated chassis punch: (a) round chassis punch; (b) chassis punch properly assembled for cutting.

**TABLE 21.2**

Screw and Guide Hole Sizes for Chassis Punches

| Chassis Punch Size (in.) | Screw Size (in.) | Guide Hole Size (in.) |
|---|---|---|
| $\frac{1}{2}$ | | |
| $\frac{9}{16}$ | $\frac{1}{4}$ | $\frac{5}{16}$ |
| $\frac{5}{8}$ | | |
| $\frac{11}{16}$ | | |
| $\frac{3}{4}$ | | |
| $\frac{13}{16}$ | | |
| $\frac{7}{8}$ | | |
| $\frac{15}{16}$ | $\frac{3}{8}$ | $\frac{7}{16}$ |
| $1$ | | |
| $1\frac{1}{16}$ | | |
| $1\frac{1}{8}$ | | |
| $1\frac{5}{32}$ | | |
| $1\frac{3}{16}$ | | |
| $1\frac{7}{32}$ | $\frac{3}{4}$ | $\frac{13}{16}$ |
| $1\frac{1}{4}$ | | |
| $1\frac{5}{16}$ | $\frac{3}{8}$ | $\frac{7}{16}$ |
| $1\frac{3}{8}$ | | |
| $1\frac{1}{2}$ | | |
| $1\frac{5}{8}$ | | |
| $1\frac{3}{4}$ | | |
| $1\frac{7}{8}$ | | |
| $2$ | | |
| $2\frac{1}{8}$ | $\frac{3}{4}$ | $\frac{13}{16}$ |
| $2\frac{1}{4}$ | | |
| $2\frac{1}{2}$ | | |
| $2\frac{25}{32}$ | | |
| $2\frac{3}{4}$ | | |
| $3$ | | |

the punch, the spurs start the initial cut. When the punch clears through the stock, a definite "snap" will be heard and the stock will drop down about the punch. When this occurs, the screw should not be tightened further. This would cause the cutting edges to be forced against the inside of the die resulting in unnecessary damage.

In addition to the round configuration, hand punches are also available in *square, key,* and *D* styles in various common sizes. These special-application punches consist of four parts: the *punch, die, screw,* and *nut. Keys, keyways,* or *flats* are located on portions of the dies, punches, and screws for automatic alignment and to hold the parts in the correct cutting position. These punches operate on the same principle as the round hand punches. Although the use of hand punches requires drilling a guide hole and assembling the various parts, the result obtained completely justify the time and effort involved.

*Draw punches* and *turret punches* are preferable when many round holes of various sizes are to be punched. The draw punch assembly is secured by bolts to a work bench. This lever-operated tool has a punch and die, but the screw common to hand punches is replaced with a *draw bolt*. The draw bolt is fastened to an *off-center moment arm* of the lever. The alignment of the draw punch is identical to that just described for the hand punch in terms of positioning the metal between the punch and the die. Some draw punches have a punch with a threaded center hole, whereas others use an additional knurled thumb nut to secure the punch in position on the draw bolt. To punch the hole, the lever is drawn toward the operator, which forces the punch through the metal and into the die. Die and punch sets are interchangeable on the draw punch, providing a wide range of hole sizes.

The most versatile of all the punches is the *turret* punch. No guide hole is necessary and a large number of die and punches are mounted on rotating drums and are readily available to the operator. A typical lever-operated turret punch is shown in Fig. 21.11. Both manual and hydraulically powered turret punches are available. The action of each is identical, with the exception of the application of power. Before any punching can be performed, a matched die and punch set of the desired size must be properly aligned. To accomplish this, the *pin locks* of both the die and punch turrets are disengaged. The two turrets are then rotated so that the selected die and its associated punch size are aligned and positioned directly below the lever arm. The locking pins are then engaged and the punch is ready for use. To operate, the metal is inserted between the punch and die. To position the metal for proper hole location, the operating lever is brought downward until the center point on the punch fits into the center-punched hole in the metal. Further downward motion of the lever engages the *cam* and drives the punch through the sheet metal into the die. The sheet metal then drops and clears the die for immediate duplication of the operation. In addition to providing for rapid setup or size changes, no guide holes are required with the turret punch and no wrench or vise is necessary. The capacities of all the punches discussed in this chapter are from approximately 12-gauge aluminum to 16-gauge tempered steel.

**Figure 21.11**   Lever-operated turret punch.

## EXERCISES

### A. Questions

21.1     What two metals are used in the manufacture of drill bits? What is the application of each?

21.2     Describe the three systems used to designate the sizes of drill bits.

21.3     What two methods can be used to determine the size of a drill bit if it cannot be read on the shank?

21.4     List the cutting speeds and lubricants recommended for drilling (a) cold-rolled steel; (b) Bakelite; (c) copper; and (d) aluminum.

21.5     Determine the following for drilling a $\frac{1}{4}$-inch hole through a piece of 16-gauge copper: (a) cutting speed (HS); (b) drill point angle; (c) lip clearance; (d) lubrication; (e) rpm; and (f) closest rpm on the drill press for this application.

21.6     Describe how deburring sheet metal is accomplished.

21.7     What is the purpose of a hand expansion reamer?

### B. True or False

Circle T if the statement is true, or F if any part of the statement is false.

21.1     A No. 28 drill has a larger diameter than a No. 36 drill.     **T    F**

21.2     Sheet metal bits should be used when drilling holes larger than $\frac{3}{8}$ inch in soft sheet metal.     **T    F**

21.3     Drills designated as HS do not dull as quickly as carbide drills and are less brittle.     **T    F**

21.4     Work should be clamped to the drill press table to avoid injury.     **T    F**

21.5     Drilling speeds for HS bits are approximately one-half of those for carbon steel bits.     **T    F**

21.6     Aluminum can be drilled almost three times faster than cold-rolled steel.     **T    F**

21.7     A guide hole must be drilled in the center of each hole prior to using the lever-operated turret punch.     **T    F**

21.8     The hand expansion reamer is used to remove burrs in drilled holes.     **T    F**

### C. Multiple Choice

Circle the correct answer for each statement.

21.1     The spiral grooves along the length of a twist drill bit are called (*webs, flutes*).

21.2     Twist drills are available in short, jobber, and (*extended, long*) lengths.

21.3     A No. 54 drill is (*larger, smaller*) than a No. 58 drill.

21.4     The recommended lubricant for drilling epoxy glass printed circuit board material is (*kerosene, dry*).

21.5     A (*countersink, file*) is the most efficient tool by which to remove burrs in drilled holes.

21.6     The most efficient tool used to enlarge a previously drilled hole is a (*file, reamer*).

## D. Matching Columns

Match each item in column A to the most appropriate item in column B.

COLUMN A

1. Feed rate
2. HS bits
3. Lubricant
4. Jacobs chuck
5. Deburring tool
6. Chassis punch
7. Twist drill
8. Reamer

COLUMN B

a. Hook-shaped blade
b. Spurs
c. Three jaws
d. Expansion
e. Margin
f. 60 ft/min
g. Air jet
h. Tungsten–molybdenum

## E. Problems

21.1  Find the clearance hole size for a No. 8 sheet metal screw from Table 25.2 and drill and deburr the mounting holes in the apron of the gauge panel of Problem 20.3. Gauge holes and radius corners will be cut in Problem 22.3.

21.2  Determine the drill sizes necessary to mount the TO-3-style case onto the heat sink of Problem 20.2. (Device lead holes should be drilled approximately 0.040 inch oversize.) All holes must be carefully deburred because flush mounting is critical. The ends of the heat sink are bent in Problem 23.2.

21.3  Tabulate the clearance and tap drill sizes for machine screw sizes 2, 4, 6, 8, and 10. Drill pairs of clearance and tap drill size holes along the length of the gauge in Problem 20.4 for the five machine screw sizes listed. Also, punch the indicated $\frac{3}{8}$-inch hole at the end of the gauge. Deburr all holes formed. Bandsaw and filing operations are done in Problem 22.2.

# Metal Cutting

---

## LEARNING OBJECTIVES

*Upon completion of this chapter on cutting sheet metal, the student should be able to*

- Be familiar with the characteristics of bandsaw blades.
- Use the correct blade speed for different materials to be cut.
- Relate the bandsaw blade width to the minimum radius of cut.
- Replace a blade in a vertical bandsaw.
- Make the necessary adjustments to a bandsaw blade.
- Properly use a vertical bandsaw to cut sheet metal.
- Properly use a hand nibbler to make inside cuts in sheet metal.
- Know the characteristics of files and the various types available.
- Use a file cleaner.
- Properly use files for cross-filing and draw-filing techniques.

## 22.0 INTRODUCTION

With the drilling and punching operations completed, the next step involves cutting irregularly shaped holes and slots and large round holes in the chassis elements in preparation for bending the metal into the desired forms. Information about the proper selection and use of each tool is provided.

In this chapter various metal cutting blades used in conjunction with a vertical bandsaw are examined together with the proper blade tension, guide settings, and cutting speeds. Practical cutting operations are illustrated on the bandsaw. The hand nibbler used in cutting irregularly shaped holes is described and operational information for proper use of this versatile cutting tool is supplied. In addition, the various types of files used to provide finish to the edges of cut metal are described.

**Caution:** *Prior to performing any of the operations described in this chapter, approved safety glasses and ear protection must be worn.* Also, the section on safety outlined at the beginning of the book should be reviewed for more specific information.

## 22.1 METAL CUTTING BLADES

The technician often encounters the problem of cutting round, extruded, or angle stock, as well as interior holes or irregular shapes in sheet metal. The shear is not suitable for these tasks. The vertically operated bandsaw with the proper metal cutting blade is ideally suited.

The shape and thickness, as well as the type of material to be cut, determine the proper blade selection. Cutting blades have three specific characteristics: *set, tooth pitch,* and *profile. Set* is the manner in which the teeth are aligned or bent away from the blade edge to provide adequate clearance between the blade and the work while cutting. Sufficient set is essential to generate adequate *kerf* (slot) so that the blade will not bind in the work during cutting. *Tooth pitch* indicates the number of teeth per inch, and *profile* is the style or shape of each tooth.

The three most common types of set are *alternate, raker,* and *wavy.* In blades with *alternate* set, every other tooth is tilted to the same side. The alternate teeth are bent in the opposite direction. This type of set is shown in Fig. 22.1a and is used primarily for cutting the softer, nonferrous metals. In the *raker* set blade, shown in Fig. 22.1b, one tooth is tilted left, the next is tilted right, and the third tooth is straight. This arrangement is continued throughout the length of the blade. The raker set blade will remove chips more easily during cutting operations than the alternate set blade and is used for cutting hard metals such as iron and steel. *Wavy* set, as shown in Fig. 22.1c, is characterized by groups of teeth tilted first left and right and then right and left along the entire blade length. Beginning with one tooth that is not offset, the first half of the group tilts increasingly until the maximum offset is reached in one direction. The second half of the group tilts decreasingly until a tooth with no offset is reached. This arrangement of outward and inward tilt is then repeated on the opposite side of the blade edge, in sequence, throughout the length of the blade. Wavy set is applicable to cutting thin sheet metal stock and tubing.

The tooth pitch must be suited to the work to result in a clean and efficient cut and to prevent damage both to the work and to the blade. For cutting thin sheet metal or tubing, blades with 32 teeth per inch are recommended. At least two teeth must contact or engage the work surface at all times. This prevents any tooth from becoming hooked on an edge or corner of the work and being stripped from the blade. The wavy set readily provides adequate kerf in thin stock. When soft or mild materials such as brass or copper or low-carbon steel are to be cut, 14 or 24 teeth per inch are recommended. When cutting large cross sections of these materials, a coarse 14-teeth-per-inch blade is best and ensures proper chip removal. When these materials are in the form of angle iron, conduit, or pipe, a finer 24-teeth-per-inch blade should be used because of the thinner cross sections. High-carbon and high-speed steels, as well as tool steel, should be cut with an 18-teeth-per-inch blade to provide for adequate chip clearance.

The profile of the blade also affects the cutting operation. Figure 22.2 shows the three most common styles of tooth configuration: *standard straight face tooth, hooked tooth,* and *skipped tooth.*

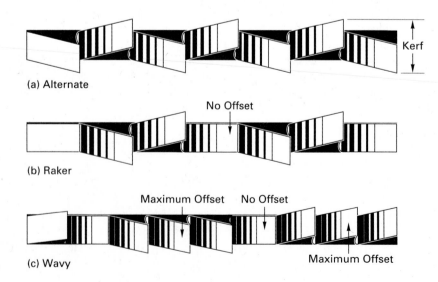

(a) Alternate

Kerf

No Offset

(b) Raker

Maximum Offset    No Offset

(c) Wavy

Maximum Offset

**Figure 22.1**  Saw blade set styles.
*Courtesy of the L.S. Starrett Company, Athol, MA.*

(a) Standard straight face

(b) Hooked

Rake

Gullet

(c) Skipped

**Figure 22.2**   Saw blade profiles.

The *standard straight face* design moves down over the work and scrapes the metal, removing the chips. This style of tooth is used for general-purpose work. Blades with *hooked teeth* are used almost exclusively for cutting large cross sections of steel, soft metals, and plastics. The slight *rake* on the teeth digs into the work instead of scraping it, thus removing metal more quickly than the standard straight-faced blade. For work on thin sheet metal, especially aluminum, the *skipped-tooth* blade is preferable. Fashioned with a large *gullet* (space) between each tooth, this style allows chips to form within the gullet and clears easier than the standard or hooked blade.

Both carbon-steel and high-speed steel blades are available. Carbon-steel blades do not maintain their sharpness as well as high-speed blades and should be used only for relatively soft materials. In addition, carbon-steel blades must be operated at slower speeds than high-speed blades to avoid impairing normal blade life. In general, hard materials should be cut slower than softer materials to prevent overheating and loss of temper. Most bandsaws are equipped with belt and pulley speed shift or gear reduction arrangements to provide a variation of blade surface speeds of approximately 50 to 4500 feet per minute. Table 22.1 lists the recommended speed for various materials.

Since the blade is often utilized to cut arcs and circles, blade width determines the minimum radius of cut. The $\frac{1}{4}$- and $\frac{1}{2}$-inch blades generally used in electronic fabrication will cut minimum radii of $\frac{5}{8}$ and $2\frac{1}{2}$ inches, respectively. The $\frac{1}{4}$-inch blade width is preferred for general-purpose work (Table 22.2).

**TABLE 22.1**
Recommended Speeds for Carbon-Steel Bandsaw Blades

| Materials | Surface Speed (ft/min) |
|---|---|
| Stainless steel | 40–80 |
| Carbon tool steel | 100–150 |
| Epoxy glass | 90–200 |
| Cold-rolled steel | 125–200 |
| Bakelite | 800–1000 |
| Cast iron | 75–150 |
| Copper | 500–1000 |
| Hard brass | 200–400 |
| Bronze | 150–300 |
| Aluminum | 3000–4500 |
| Plastics | 3000–4000 |

**TABLE 22.2**
Bandsaw Blade Cutting Radius Guide

| Bandsaw Blade Width (in.) | Minimum Radius of Cut (in.) |
|---|---|
| $1$ | $7\frac{1}{4}$ |
| $\frac{3}{4}$ | $5\frac{7}{16}$ |
| $\frac{5}{8}$ | $3\frac{3}{4}$ |
| $\frac{1}{2}$ | $2\frac{1}{2}$ |
| $\frac{3}{8}$ | $1\frac{7}{16}$ |
| $\frac{1}{4}$ | $\frac{5}{8}$ |
| $\frac{3}{16}$ | $\frac{5}{16}$ |
| $\frac{1}{8}$ | $\frac{1}{8}$ |
| $\frac{3}{32}$ | $\frac{3}{32}$ |
| $\frac{1}{16}$ | Square |

## 22.2 VERTICAL BANDSAW

A vertical bandsaw, as shown in Fig. 22.3, is a power tool found in most electronic shops. Although larger and more sophisticated models are used in production, the principles of operation and use are similar. The bandsaw consists of a *drive wheel* mounted below the *worktable* and an *idler wheel* above the table, both *V-pulley* driven. The motor is positioned within the pedestal that supports the saw. The blade is positioned around both the idler and drive wheels. The teeth are protected by the hard-rubber outer surfaces on the wheel rims that also provide friction to prevent blade slippage. To permit proper operation, the idler wheel is equipped with a *tension adjustment* and a *tracking* or *blade centering adjustment*.

To replace a worn or broken blade, the protective wheel covers are removed and the blade tension is reduced by rotating the spring-loaded tension adjustment knob counterclockwise. This will lower the idler wheel (Fig. 22.3). The used blade is removed and the new blade is passed into the *table slot*. (On some bandsaw tables, a large set screw is threaded into the table at the leading edge of the slot. This screw must be removed before a blade can be replaced.) The blade is then passed around the drive wheel and carefully over the idler wheel so that it is positioned between the *upper* and *lower blade guides*. These guides are located immediately above and below the worktable. With the blade centered on both wheels, the tension-adjustment knob is rotated clockwise, thus increasing the tension. Optimum tension is achieved when approximately $\frac{1}{4}$-inch displacement is obtained when finger pressure is applied to the unsupported portion of the blade. This displacement is typical for blades whose width is from $\frac{1}{4}$ to $\frac{3}{8}$ inch. For blades $\frac{1}{2}$ to $\frac{3}{4}$ inch wide, a displacement of $\frac{3}{8}$ inch is recommended.

It is important that work be fed into the blade from the front edge of the table when the throat of the bandsaw is to the left of the operator (Fig. 22.4). Therefore, the teeth of the blade must be facing the operator when viewed from the front edge of the bandsaw table. Also, the blade is designed so that the teeth will cut when they move in a clockwise direction. The distance in inches from the side of the cutting blade to the inside face of the column supporting the idler wheel is called the *throat clearance*. This clearance is used to designate the bandsaw sizes, which range from 16 to 60 inches. Bandsaw capacities are also given in terms of height from the table to the underside of the top blade guides when this assembly is in the uppermost position. These capacities range from $6\frac{1}{4}$ to 12 inches.

With the proper tension applied to the blade, the tracking is next inspected by manually rotating the idler wheel and observing if the blade deviates from its original center

**Figure 22.3** Blade tracking and tension adjustment controls.

Tension Adjustment Knob

Blade Guide Height Adjustment

Guide Groove

Tracking Adjustment Knob

Bevel Adjustment

**Figure 22.4** Proper positioning of hands when cutting sheet metal.

position on the wheels. Any deviation can be corrected by changing the tilt on the idler wheel by rotating the *tracking adjustment knob* clockwise to correct for a forward deviation and counterclockwise to correct for a blade that has moved back toward the rear of the wheel and table slot. Tracking is properly adjusted if the blade remains centered when manually rotated. Correct blade tracking is extremely important. Blades installed without regard for tracking could ride out from between the guides or completely off the wheels, thus damaging the work, the blade, and possibly the operator. *Note: Never test the tracking by switching on the power.*

To cut sheet metal with the bandsaw, the work is placed flat against the worktable and about $\frac{1}{4}$ inch from the blade. The *blade guard clamping screw* is loosened and the guard is lowered to within $\frac{1}{8}$ inch of the stock. The guard is secured and the power switch is turned on. After allowing sufficient time for the blade to attain its operating speed, the stock is fed into the blade with the right hand, as shown in Fig. 22.4. Notice that both hands are kept clear of the blade to prevent injury. The work should *never* be forced. If cutting requires excessive pressure, the blade should be inspected for dullness and replaced if necessary. At the end of a *through cut* (a cut that traverses the entire length of the stock), care must be taken not to close the kerf generated by the blade set. This will cause the back edge of the blade to bind and could result in damage to both the blade and the work. A through cut is made whenever possible because backing the work from the blade when the power is on could cause blade damage.

To saw a curve or an irregular shape, the layout of the cut is first scribed onto the work surface (see Sec. 20.2). Figure 22.5a is an example of a series of separate cuts necessary to remove a rectangular opening. One continuous cut is first made to remove the bulk of the rectangle. This cut is made perpendicular to the edge and close to one finished scribed line. As the first corner is approached, the smallest possible radius is cut in such a manner so that the blade is positioned close to the *inside* finish line. This process is continued to the other corner with a second radius, allowing the blade to exit along the remaining finish line. This will remove the largest possible portion of the metal from within the scribed area. To square the corners, the edge of the blade is placed along the edge of the first straight cut and the remaining curved area is fed into the blade until the 90-degree

(a) Removing bulk stock

(b) Squaring corners

**Figure 22.5**   Cutting sequence to form a rectangular opening with a bandsaw.

corner is encountered. The work is rotated and this procedure is repeated until all the metal is removed from the scribed area in that corner. These cuts are duplicated for the other corner (Fig. 22.5b).

When irregularly shaped objects are to be cut on a bandsaw, special holding and clamping devices or jigs are necessary. For example, to cut the shaft of a potentiometer or switch, a *V-block* is needed for support and as a base to guide the work. In all cases, a solid support should be provided as close as possible to the cut to minimize saw and component damage owing to unnecessary vibration.

Not only do the profile, pitch, and set of a blade affect the cutting action, but the speed at which the work is fed into the blade also influences the characteristics of the resulting cut. In general, a higher rate of penetration will result when a coarse blade and a heavy feed rate are applied. For fine finish work within close tolerances, the finest pitch allowable for the particular material should be used and the feed rate should be reduced. The appropriate feed rate is best determined not only by visual inspection of the work but also by audible strain of the motor speed. Too fast a feed rate will reduce the motor speed. At all times, attention should be directed to (1) avoiding excessive feed rates that crowd or bind the blade needlessly, and (2) being constantly alert to the inherent dangers associated with the use of this power tool.

## 22.3 HAND NIBBLER

Cutting internal openings in sheet metal is a frequent requirement in electronic fabrication. The *hand-operated nibbler* is ideally suited for cutting internal holes because of its operating simplicity.

The hand nibbler, shown in Fig. 22.6, is basically a miniature hand shear consisting of two blades. The lower blade engages the underside of the work when the cutting head is passed through and pressed against the edge of the access opening. The upper blade is actuated through a spring located in the handle. As the handle is squeezed, the upper blade is pulled down to contact the upper surface of the work. Further pressure on the handle causes the shearing action between the two blades. Each shearing stroke removes a small rectangular piece of metal approximately $\frac{1}{16}$ by $\frac{3}{8}$ inch. Thus, this tool "nibbles" a $\frac{3}{8}$-inch-wide strip through sheet metal. Removing the applied pressure from the handles automatically raises the upper blade to its original position and the metal chip freely falls away. The nibbler requires that a hole, approximately $\frac{7}{16}$ inch in diameter, be initially drilled into the area of metal to be removed to accommodate the cutting head through the stock. When the opening has been completed, the edges must be filed to remove the irregularities that result from the use of this tool. (The various types of files and their uses are discussed in Sec. 22.4.)

## 22.4 FILES

Of all the mechanical skills necessary in electronic fabrication, perhaps the one that requires the most practice for proficiency is *filing*. To gain this proficiency, a familiarity with the file characteristics is essential so that the most suitable file for the particular application and material can be selected and used properly.

Filing is primarily a finish operation and is employed to improve the appearance of a previously made cut in a workpiece. It is also used to remove irregular and rough edges or burrs present after certain metal-cutting operations. Also critical for a quality appearance and to minimize damage to components, wiring, and possible injury to the operator is the technique of *breaking* or *rounding* all sharp corners and edges. This is especially true if a workpiece is to be finished by painting. Breaking ensures a better bonding of the finish over the edges of the work and prevents chipping (see Chapter 24).

Files are generally constructed of hardened high-carbon steel and consist of six

(a)　　　　　　　　　　　　　　　　　　　　(b)

**Figure 22.6** Hand-operated nibbler: (a) a sheet metal nibbler; (b) interior stock removal simplified through the use of a nibbler.

basic parts: *handle, heel, face, edge, point,* and *tang* (Fig. 22.7). The sharp tang must be inserted into an appropriate size handle prior to its use to prevent injury to the operator. The wooden handles used for files have a metal *ferrule* around the end at which the tang is inserted. This prevents the wood from splitting when the tang is forced into the handle. The file is properly seated by tapping the end of the handle on a wooden block or bench after the tang has been firmly inserted by hand. This process is reversed to remove the file from the handle. By striking the edge of the ferrule next to the tang against the edge of a work bench, the file will loosen sufficiently to be withdrawn. Of course, this procedure can be avoided if a handle is provided for each file.

Files are classified with respect to *length, cut of teeth,* and *cross section*. The length of a file is measured from the heel to the point. For electronic fabrication uses, files vary

**Figure 22.7** File nomenclature.

(a) Single cut

(b) Double cut

**Figure 22.8** Basic cuts of files used in electronics work.

in length from 3 to 12 inches. Files have four basic cuts: *single cut, double cut, curved tooth,* and *rasp cut.* The curved tooth and rasp cuts are not normally employed in electronics work. The single-cut and double-cut files are shown in Fig. 22.8. The single-cut file is characterized by parallel rows or courses of teeth that are set between 65 and 85 degrees with respect to the axis of the file. This type of file is preferable when a smooth surface finish is desired. A double-cut file has *two* courses of teeth on the same face that are set at different angles. One course is set between 40 and 45 degrees and the second course traverses the face at between 70 and 80 degrees to the axis of the file. These angles are characteristic of files used for general-purpose applications. Double-cut files are also available with first and second courses at 30 degrees and 80 to 90 degrees, respectively, for finer work. Double-cut files are used primarily in *second-cut* work (i.e., in removing larger amounts of material prior to finishing with a single-cut file). The double-cut configuration produces a slicing effect that requires less effort than the single-cut file.

Tooth spacing determines the relative *coarseness* of a file. Files are categorized relative to coarseness as *rough, coarse, bastard, second cut, smooth,* and *dead smooth.* The tooth spacing is the largest with *rough cut* and becomes progressively smaller with the *dead smooth cut.* These spacings for the six categories are 20, 25, 30, 40, 50, and 100 or more teeth per inch, respectively. It is important to mention that each of the six grades are coarser for long files (10 to 12 inches) than for the equivalent grade of short files (6 inches). Consequently, length as well as grade of tooth coarseness determines the number of teeth per inch. When selecting a particular grade of file coarseness, the technician must consider the type of material to be worked. Steel and other hard metals are worked well with the less coarse grades, such as *second cut* and *smooth.* When soft metals such as aluminum, brass, or copper are to be worked, coarser grades, such as *coarse* or *bastard,* are more effective. These choices are dictated by a condition called *loading,* which is the accumulation of waste metal that clogs the file teeth. This condition builds up rapidly when using too fine a file on soft material. These bits of metal impede normal file operation and should be periodically removed with a *file cleaner.* This cleaner consists of a wooden handle to which are attached short, stiff wire bristles called the *card* and a regular bristle brush on the reverse side of the card (Fig. 22.9). To use the cleaner, the regular bristle brush is first employed to remove the loose waste material on the file face. If any material remains after briskly brushing parallel to the teeth, the cleaner handle is inverted and the card is rubbed over the file teeth. (To clean double-cut files, it is necessary to alternately brush the two courses of teeth.) To remove any remaining bits of material

**Figure 22.9** File cleaner.

(called *file points*) lodged between the teeth after these two operations, a pick must be used. The pick is made of a softer metal than the file so that it will not dull the file teeth. To minimize the rate of loading, *chalk* may be rubbed into the face of a file prior to its use. The chalk will not impede the filing operation.

Many types of file shapes and cross sections are available to accommodate the numerous applications of this tool. Figure 22.10 shows most of the common types frequently used. Following is a brief description of each.

**Mill File.** A single-cut rectangular file used primarily for lathe work and draw filing. It is available in round or smooth edges and tapers in thickness and width for one-third of its length. Other applications are tool sharpening and removing burrs and rough edges. This file is not recommended for filing flat work because of loading and the resulting marring of the work surface.

**Flat File.** A double-cut rectangular file used for general-purpose work in finishing flat surfaces. It has single-cut edges and tapers in thickness and width toward the heel and point from the center of the face.

**Hand File.** A double-cut rectangular file also used to finish flat surfaces. It is sometimes preferred over the flat file because of its parallel edges. One edge is *safe* (no teeth) and the other is single cut. This file is extremely useful for filing into 90-degree corners. The faces taper slightly toward the heel and point.

**Pillar File.** A double-cut rectangular file similar to the hand file and used for general-purpose work. It is slightly thicker but not so wide as the hand file.

**Warding File.** A double-cut rectangular file much thinner than those previously described and tapers to a point. It finds wide application in filing slots and notches.

**Square File.** A four-sided double-cut file that tapers toward the point and is used primarily for filing or enlarging square or rectangular holes.

**Three-Square File.** A three-sided double-cut file that tapers toward the point. It is generally used for smoothing internal angular surfaces or corners, especially those having angles less than 90 degrees.

**Barrette File.** A four-sided double-cut file, tapering to a point, having sharper edges than the three-square file. Only two faces cut; the other two faces are safe. This file is used for finishing in grooves, slots, and sharp angles or corners, especially where cutting must be done while an upper face is contacting a surface from which no further material is to be removed.

**Knife File.** A file with a safe back edge, a single-cut thin edge, and two double-cut faces. It is often used in place of the Barrette file.

**Round File.** A round double-cut file tapering toward the point and used for finishing or enlarging round holes or small radii.

**Half-Round File.** A semicircular file usually double cut on the flat face and single cut on the curved face. This unique shape makes it one of the most versatile files for general-purpose work such as finishing curved edges or large-diameter holes as well as flat surfaces.

**Crossing File.** An oval-shaped file with one face having a radius similar to the

**Figure 22.10** Standard file shapes and cross sections. *Courtesy of Nicholson File Co.*

Mill

Flat

Hand

Pillar

Warding

Square

Three-square

Barrette

Knife

Round

Half-round

Crossing

Crochet

half-round file and the other face having a larger radius. It tapers to a point in both width and thickness and is double cut on both surfaces. This file also finds application in finishing curved edges, but affords a selection of the most appropriate face for the work radius.

**Crochet File.** A flat file with rounded edges, double cut on all faces and edges. It tapers in both width and thickness and is used primarily for finishing slots, rounded corners, and filleted shoulders.

Most of the file cross sections just described are also available for extremely delicate work. *Swiss pattern files* find wide application in electronic packaging. The file selected is generally determined by the contour to be worked.

When filing, the metal to be worked should be firmly secured in a vise and protected from damage with *jaw protectors*. This is especially true when working with sheet metal. These protectors are often fashioned from scraps of soft sheet metal, such as aluminum or copper. The metal is bent at 90 degrees and inserted on each jaw to either side of the work. The 90-degree bend keeps the protectors in position when the jaws are opened to release the work. To minimize vibration of the work, which causes the file to "skip" across the surface resulting in *chatter* and an irregular surface appearance, the metal should be gripped as close to the working edge as possible.

*Cross-filing* is the technique of pushing the file, under pressure, over and along the edge of the work. The file handle is held in the right hand with the thumb on the ferrule and the file point in the left hand with the base of the thumb resting on the face. The right hand provides the pushing action while the thumb controls the direction and pressure. For best results, a smooth, even *cutting stroke* must be maintained as the file is forced *away* from the operator. Improved cutting results if the file is swung at between 30 and 45 degrees to the work and level to the filing surface (Fig. 22.11). A common mistake of the beginner is to *rock* the file on the cutting stroke, thus leading to a curvature of the surface. The file must be kept level along the entire length of cut. Cutting pressure is applied on the forward stroke *only*. No pressure is applied on the return stroke.

To obtain a fine finish such as on the edges of an exposed portion of a chassis, the method of *draw-filing* is used. The handle is gripped with the right hand and the point with the left with both hands perpendicular to the file axis and both thumbs pressing against the closest file edge. This technique is shown in Fig. 22.12. With the file axis at

**Figure 22.11** Proper file position for cross-filing.

**Figure 22.12**  Draw-filing technique.

right angles to the work surface, the file is passed *back* and *forth* with uniform pressure. Cutting is accomplished on both strokes. Single-cut files are considered superior to double-cut files for draw-filing.

Single-handed operation of a file is advisable only when using Swiss pattern files. Since these are so much smaller than the standard type, the file can be completely controlled with one hand. The knurled file handle should be held firmly with the index finger along the uppermost face or edge. The procedures for cutting are similar to those described for the standard-size files. Figure 22.13 shows a knife file being used to remove burrs along the tab edges of a chassis that resulted from the bandsaw cutting operations.

The proper care of files is just as important as their correct use. Files must not be allowed to come in contact with other files or tools in storage. This will cause tooth chipping and dulling. Also, the face or edge must never be struck on a surface to attempt to remove accumulated waste material that is loading the file. Although the proper care and

**Figure 22.13**  Swiss pattern files are used where access space is limited.

maintenance of tools is time-consuming, it is more than worthwhile in terms of longer tool life and quality workmanship.

The final process necessary for the completion of a chassis prior to applying a finish requires forming these elements into the desired configurations. Metal-forming tools and techniques are discussed in the following chapter.

## EXERCISES

### A. Questions

22.1   List and describe the three characteristics of cutting blades.

22.2   Explain the term *kerf*.

22.3   List the three most common types of saw blade sets together with their applications.

22.4   List the three saw blade profiles together with their applications.

22.5   What is the minimum radius that can be cut with a $\frac{1}{4}$-inch-wide bandsaw blade?

22.6   List the six coarseness categories of files.

22.7   List the four basic file cuts together with their applications.

22.8   How are files cleaned?

22.9   Describe the difference between cross-filing and draw-filing together with the application of each.

22.10  For what application are Swiss pattern files used?

### B. True or False

Circle T if the statement is true, or F if any part of the statement is false.

22.1   Tooth pitch describes the amount of kerf on a cutting blade.   **T   F**

22.2   The thicker the material to be cut, the fewer teeth per inch are required for proper chip removal.   **T   F**

22.3   Cold-rolled steel is cut at a speed approximately 20% of that for copper.   **T   F**

22.4   The minimum radius of cut for a $\frac{1}{8}$-inch-wide blade is $\frac{5}{16}$ inch.   **T   F**

22.5   The purpose of a hand nibbler is to form round holes in sheet metal.   **T   F**

22.6   When filing into a 90-degree corner, a file having at least one safe edge is used.   **T   F**

22.7   Cutting pressure is applied on both the forward and return strokes when cross-filing.   **T   F**

### C. Multiple Choice

Circle the correct answer for each statement.

22.1   Nonferrous metals, such as aluminum and copper, are cut primarily with a blade having (*alternate, raker*) set.

22.2   Cutting thin sheet metal is best accomplished using a blade with a (*hooked, skipped*) tooth profile.

22.3      The recommended speed of a bandsaw for cutting aluminum is (*300 to 450, 3000 to 4500*) feet per minute.

22.4      Cutting internal holes in sheet metal is best accomplished by using a (*bandsaw, hand nibbler*).

22.5      A second-cut file is (*more, less*) coarse than a bastard file.

22.6      A file having 50 teeth per inch would be classified as (*second cut, smooth*).

22.7      On a barrette file, (*two, four*) of the faces cut.

22.8      Cutting pressure is applied to both strokes when (*cross-, draw-*) filing.

## D. Matching Columns

Match each item in column A to the most appropriate item in column B.

| COLUMN A | COLUMN B |
|---|---|
| 1. Hand nibbler | a. Cleaner |
| 2. Kerf | b. 45-degree axis to work |
| 3. File | c. 90-degree axis to work |
| 4. Throat clearance | d. Second cut |
| 5. File card | e. Miniature shear |
| 6. Draw-filing | f. Tooth set |
| 7. Cross-filing | g. Blade profile |
| 8. Hooked tooth | h. Bandsaw |

## E. Problems

22.1      Determine the maximum bandsaw blade width and speed for cutting a 2-inch radius in a piece of 16-gauge aluminum.

22.2      Using a bandsaw, cut the 2-inch radii of the drill gauge laid out and drilled in Problems 20.4 and 21.3. Break all sharp edges with a file.

22.3      Using a nibbler, cut the three holes into the gauge panel laid out and drilled in Problems 20.3 and 21.1. Cut the 1-inch radii with a bandsaw, deburr the holes, and break all sharp edges with a file. Bending is performed in Problem 23.1.

22.4      List the most appropriate type, coarseness, and cut of file for each of the filing operations required in Problems 22.2 and 22.3.

# Bending 23

*Upon completion of this chapter on bending sheet metal, the student should be able to*

- Calculate bend allowance to overcome metal bending distortion.
- Make the necessary adjustments to a finger brake.
- Properly use the finger brake to bend sheet metal.
- Bend a chassis in the proper sequence to result in all bends being completed on the finger brake.

## 23.0 INTRODUCTION

After the cutting operations are completed, the next step in the fabrication of a chassis is to form the metal into the desired configuration. Industrial mass production techniques utilize hydraulic press brakes, which are very similar in mechanical action to the power shear (see Chapter 19). These presses can be set to automatically produce *all* the required bends in one step. The hydraulic presses are not considered in this chapter because the cost of tooling makes such automation prohibitive in prototype and small-volume work.

This chapter deals exclusively with the *hand-operated finger brake* (also called the *box* and *pan* brake), which is common to most sheet metal shops and allows the technician to produce a good-quality chassis. Information is provided on layout bend allowance, types of chassis configurations, bending sequence and brake nomenclature, adjustment, and operation. In addition, the critical limitations of most standard brakes are discussed.

***Caution:*** *When working with sheet metal, approved safety glasses must be worn.* Also, the section on safety outlined at the beginning of the book should be reviewed for more detailed information.

## 23.1 BEND ALLOWANCE

When sheet metal is bent, there is an inherent distortion of the metal that results in a dimensional change. To maintain a degree of precision when bending, several characteristics of the metal and the equipment must be examined. The result of these examinations will be an adjustment in layout measurements termed *bend compensation* or *bend allowance* to overcome the distortions generated by bending. Since aluminum and mild steel are the most common metals used for chassis fabrication, the factors that affect the bending of these metals with a hand-operated finger brake will be considered.

(a) Intended bend position

(b) Resulting bend characteristics

**Figure 23.1**  Dimensional changes due to metal thickness.

Bending sheet metal *stretches* the outer section of the bend and *compresses* the inner section. The effect of these modifications on the bend area is to alter the position of the metal edges with respect to the bend lines established in the layout operations (see Chapter 20). Figure 23.1 shows the deviation of an edge from the original established line after bending. Notice that in Fig. 23.1a an inside dimension of distance $x$ measured from edge $C$ is required after bending. This is impossible because the metal possesses a thickness $t$. After bending, the compression of the metal leaves an inside dimension of $x - B$ (Fig. 23.1b). Moreover, measured from edge $C$ is a new outside dimension, of $x + A$, the result of stretching. It therefore becomes necessary to determine the desired total inside and outside finished dimensions *after bending* by computing the values $A$ and $B$ and compensating for the change in measurements due to the metal-bending characteristics. If an accurate outside dimension is required, the bend line is positioned at a distance $x - A$ from the reference edge $C$. This procedure is shown in Fig. 23.2a. When bent, the resulting outside dimension will be desired dimension $x$ (Fig. 23.2b). The *stretching* of the metal induced by the bending has been compensated for. If an *inside* dimension is critical, the bend line is positioned at a distance of $x + B$ from the reference edge $C$. This compensates for the *reduction* of the inside dimensions stemming from the compression of the metal when bent. The desired dimension $x$, as shown in Figs. 23.2c and d, is accurately obtained because of this bend allowance.

The dimensions $A$ and $B$ depend on (1) the type of metal, (2) the radius of the bend, and (3) the angle of the bend. In chassis fabrication, 90-degree bends and bend radii equal to the thickness of the metal are constructed almost exclusively. Therefore, for our purposes, the values of $A$ and $B$ will be a function of metal type only.

For 16- and 18-gauge aluminum alloys of the 1100 and 3000 series, commonly used in chassis work, a value of $A$ equal to two-thirds the thickness ($t$) of the metal ($A = \frac{2}{3} t$) is a good approximation for most applications. It follows, then, that $B$ is equal to $\frac{1}{3} t$ ($B = \frac{1}{3} t$).

Steel presents a high opposition to compression variations. As a result, $A$ is considered to be the *total thickness* of the metal ($A = t$), whereas $B$ can be approximated to be zero ($B = 0$). An inside bend measurement on steel will be approximately correct if laid off directly without regard to bend allowance. To compensate for an outside measurement caused by stretching, the bend line dimensions must include the thickness of the metal. Typical sheet metal gauges, together with their associated thicknesses, are shown in Appendix I to aid in evaluating bend allowance.

Position Bend
Line Here

$t$

$x$

Edge $C$

$x - A$

(a) Bend allowance for accurate outside
dimension

$t$

$A$

$B$

Edge $C$

$x - A + A = x$

Due to Metal
Stretching

(b) Resulting bend

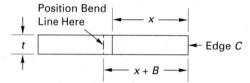

Position Bend
Line Here

$t$

$x$

Edge $C$

$x + B$

(c) Bend allowance for accurate inside
dimension

$t$

Due to Metal
Compression

$A$

$B$

$x + B - B = x$

Edge $C$

(d) Resulting bend

**Figure 23.2**  Bend allowance methods
to obtain accurate outside and inside
dimensions.

## 23.2  BENDING BRAKE

Sheet metal forming by bending is accomplished primarily in prototype work with the aid of a *hand-operated finger brake,* as shown in Fig. 23.3. The basic parts of the brake are the *table, wing, fingers, crosshead, clamping handle, angle stop, operating handles, clamping-tension adjustment screws,* and *finger-position adjustment screws.* The metal to be bent is positioned on the brake table under the fingers and secured in place by pulling the clamping handle *forward.* This engages the crosshead, to which the fingers are attached, and applies clamping pressure to the metal. Once positioned, the metal is bent by raising the operating handles attached to the wing. This causes the wing to move to the desired angle with respect to the table. The adjustable angle stops are located on either end of the brake frame. These contact the wing when it is swung up to the desired angle and impede any further motion. When the metal is bent, the wing is returned to the downward position and the work is released by again operating the clamping handle in an *upward* motion to raise the fingers and release the clamping pressure.

**Figure 23.3** Finger brake. *Courtesy of Di-Acro.*

Before any bending operations are made on the brake, *four* adjustments must be checked and corrected, if necessary, to protect the brake and the work from damage. These adjustments, which are described in the following paragraphs, are (1) *finger seating,* (2) *clamping pressure,* (3) *finger clearance (setback distance),* and (4) *wing stop angle.*

The fingers must first be checked to ensure that they are all properly seated and securely clamped to the crosshead. The seating is correct if all the fingers are parallel to one another with their leading edges in line and parallel to the edge of the wing. A *box-end wrench* should be the only type used when adjusting any of the hexagonal clamping or adjusting screws on the brake. Open-end adjustable wrenches should be avoided because they tend to slip and can damage the screw head.

After the fingers are properly positioned, the clamping pressure is then tested. This is done by first shearing a piece of stock (of the type used in forming the chassis) to a size equal to the brake's full lateral capacity and at least 2 to 3 inches wide. This sample piece is placed between the fingers and the table with at least 1 inch of the stock extending over the upper edge of the wing. The clamping handle is engaged and the pressure is checked by attempting to pull the metal from the brake. If the metal can be moved easily at either end, the clamping pressure must be increased. This is done by releasing the locknut securing the appropriate pressure adjustment screw. These screws are located on the frame at either end of the crosshead. The adjustment screw is rotated *clockwise* to lower the crosshead, thereby increasing the pressure. This test procedure of pulling the metal and adjusting the crosshead is continued until sufficient pressure is applied to the metal to just secure it to the table without damaging its surface. If the pressure appears to be excessive, the adjusting screws are turned *counterclockwise* to raise the crosshead. When the optimum pressure is achieved for the type of metal being worked, the locknuts are again secured. It is important that equal finger pressure be applied across the entire length of the metal to maintain positive gripping.

**Figure 23.4**    Finger setback clearance test.

The finger clearance distance is tested with the same sample stock used for testing the clamping pressure. The edge of this stock is placed perpendicular to the table and touching the leading edges of the fingers, as shown in Fig. 23.4. The wing is then rotated to the 90-degree bend position. The setback clearance of the fingers should be just equal to the thickness of the metal to be worked. Any distance less than this thickness will cause the fingers to dig into the metal along the bend line as the wing is engaged. This will severely weaken the work and possibly damage the fingers. A setback distance larger than the thickness of the metal will result in a "sweeping" 90-degree bend. The radius of this type of bend may be greater than normally acceptable. If adjustment is necessary, the two locknuts securing the *finger-position adjustment screws* are released. These are located at either end and to the rear of the brake frame. Clockwise rotation of the screws will drive the crosshead forward, causing the fingers to move closer to the wing. *Counterclockwise* rotation will cause the fingers to be drawn away from the wing. With the fingers uniformly positioned at the desired distance across the entire length of the table, the locknuts are tightened to secure the adjusting screws.

The adjustments for setting the bending angle stops are made via the assemblies located immediately above and to either side of the wing. A rough adjustment is first made by inserting a pin into one of four holes provided on the inside face of each assembly. These rough angles are typically 30, 45, 60, and 90 degrees. Adjusting screws secured with locknuts are also associated with these assemblies. These screws allow for fine angle adjustment among the four rough adjustments obtained with the pins. The pins extend outward from the assemblies when seated in the desired hole position and prevent any further upward movement of the wing beyond the pin's angular position with respect to the brake table. It is important to remember that because of the flexing characteristics of most sheet metal, called *"spring back,"* the metal must be bent slightly *beyond* the desired angle to compensate for this effect. The angle adjustment settings, therefore, must be *slightly greater* than the intended angle. After initial adjustments are made, a sample piece can be bent and the resulting angle tested with a protractor head and blade (refer to Chapter 20). Final angle adjustments can then be made if necessary. The four brake adjustments outlined can be made independently and do not influence one another. However, each of the adjustments affects any resulting bend.

The work must be carefully aligned and inserted into the brake for an accurate bend. To obtain a sharp bend, the scribed bend line is positioned parallel to the leading edges of the fingers and aligned so that it *bisects* the finger clearance distance. This positioning is critical to achieve the proper dimensions resulting from the bend allowance considerations previously discussed. When bending harder or thicker metals than 18-gauge mild steel, a larger radius is necessary to prevent fracture along the axis of the bend as the metal is formed. By adjusting the finger clearance distance to *twice* the thickness of the stock to be bent and bisecting this distance with the scribed bend line, a uniform sweeping bend will result.

The advantage of the finger brake, having individual fingers of varied widths, over the larger type brakes (*cornice* or *leaf type*), which operate with only the wing and one solid bending bar or leaf, will become apparent in the following section.

## 23.3 BENDING SEQUENCE

Any bending operation, irrespective of its simplicity, requires preplanning concerning the sequence and positioning of bends because of the limitations and restrictions of the brake. Random bends, without regard to sequence, may result in an impossible bending situation and waste of a time-consuming layout. Even with the simple U-shaped chassis shown in Fig. 23.5a, one critical limitation could result as a consequence of the dimensions of the work and the size of the brake. This configuration is to be bent from the scribed blank of Fig. 23.5b. The first bend can be either line *A* or line *B* and involves no problem as long as the width of the metal is less than the lateral capacity of the brake. In either case, however, the dimension *X* is critical. If the distance *X* is less than the distance measured from the table to the top of the fingers, a second 90-degree bend is impossible. Attempting to form the second bend by placing the unbent side section under the fingers will result in the previously bent section contacting the top surface of the fingers before the desired bend can be completed. A bend less than 90 degrees will result, as shown in Fig. 23.5c. To bend this type of chassis, the distance *X* must be at least equal to the height of the finger-securing bolts. Again referring to Fig. 23.5c, note that dimension *Y* is also critical if finger clamp assembly or crosshead interference is to be avoided.

An alternative method of making this second 90-degree bend is to position the chassis under the fingers with the first bent section to the *rear* of the crosshead and the second bend line aligned with the finger edges (Fig. 23.5d). This technique introduces still another limitation of the brake. That is, dimension *X* must be greater than the distance measured from the leading edges of the fingers to the rear face of the crosshead.

Even with the simple U-type configuration, the aforementioned restrictions of the brake limit the minimum size of the chassis. These restrictions should be considered during the planning stages in order to easily produce a quality chassis. When a bend cannot be made on a brake or completed to the desired angle through an oversight in the bending sequence, a great deal of effort, skill, and ingenuity with wooden blocks, clamps, and soft-faced mallets will be required to obtain an acceptable bend.

To introduce further brake limitations and bending problems, a cover, as shown in Fig. 23.6a, for the U-shaped chassis will be formed. The layout of this cover, shown in Fig. 23.6b, will require four 90-degree bends. (Two of these bends, *B* and *D*, will produce tabs.) A bending sequence that would appear to be reasonable would begin with bend *A* and continue bending clockwise or counterclockwise around the work. This sequence, however, will introduce another bending limitation. If bend *A* is formed first, the tabs associated with side *E cannot* be formed on the brake. Owing to this limitation, the following sequence must be followed to form this chassis. Bends *B* and *D* are formed first, followed by bends *A* and *C*. Figure 23.6c shows the last bend, *C*, being formed in accordance with this sequence. Since this cover has the same critical dimensions *X* and *Y* as the U-shaped chassis, similar brake limitations apply. Notice also in Fig. 23.6c that just

(a) "U" shaped chassis

(b) Scribed blank

(c) Height and depth limitations

Dimension *X* Must
Be At Least This Large

(d) Depth limitations

Dimension *X* Must
Be At Least This Distance

**Figure 23.5** Bending sequence for U-shaped configuration.

enough fingers are used to cover the bend line *A* or *C* between the previously bent tabs along lines *B* and *D*. If it is necessary to leave any length of bend line uncovered by the fingers, this distance should be limited to $\frac{1}{16}$ to $\frac{1}{8}$ inch between fingers or from the ends. This spacing will not cause any noticeable distortion along the bend.

It is always good practice to provide *relief holes* where stresses are incurred during bending, especially at corners where more than one bend is to be made. These relief holes, shown in Fig. 23.6b, prevent bulging or fracturing of the metal at these points of high stress. Notice that four relief holes were drilled at the intersections of the bend lines. The diameter of these holes should be at least equal to the stock thickness.

The bending sequence for a box-type chassis, as shown in Fig. 23.7a, will be formed from the layout shown in Fig. 23.7b. This chassis consists of a *top,* four outside *aprons,* and four *subaprons.* The bends are numerically labeled to show the proper bending sequence to produce the desired results. All four subaprons (bends 1, 2, 3, and 4) are bent first. No special order or sequence of bending is necessary for these subaprons because their relative location and short width cause no restrictions on the brake. The four inside bends (5, 6, 7, and 8) are then made to form the aprons. Notice that the first dimensional restriction of the chassis is that the subapron height must be limited to the *heel clearance* height of the fingers because the subaprons must be placed *under* the fingers

(a) Cover for "U" shaped chassis     (b) Layout for cover

$^1/_{16}$ to $^1/_8$"
Finger Side
Clearances

**Figure 23.6** Bending sequence for U-shaped cover.

(c) Final bend with finger side clearance

when bending the aprons (Fig. 23.8). A further restriction is the dimension $Y$, also shown in Fig. 23.8. This is the inside height of the finished apron as measured from the inside face of the subapron to the underside of the top. For the four aprons to be formed, the distance $Y$ must be greater than the distance measured from the leading edge to the heel of the fingers. Figure 23.8 shows the final bend being made with all the aforementioned dimensional restrictions satisfied.

## 23.4 CORNER TAB BOX CHASSIS

A box chassis with secured corners is the most widely used, since it is able to support heavy components. The U-type chassis and the unsecured box chassis are too weak to provide this support.

Figure 23.9 shows the unformed blank of a chassis element. This is a *corner tab box chassis*. The bending operations begin by forming the four outside subaprons (bends *1, 2, 3,* and *4*). As in the unsecured box chassis, these subaprons may be bent in any order. It is next necessary to make the inside bends for the tabs and aprons, beginning with bend *5*. This will include bending *both* tabs to either side of the apron associated with bend *5* as well as the apron itself. Since the wing of most brakes is a single, straight unit, the tabs cannot be bent separately. As a result, all bends along bend line *5* must be formed at the

(a) Box chassis

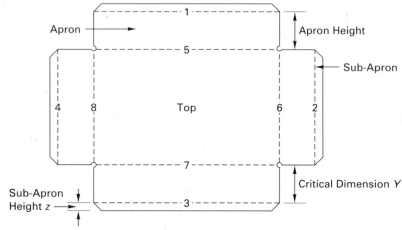

(b) Bottom view of box chassis layout

**Figure 23.7** Bending sequence for box chassis.

same time. If this bend was formed in the usual manner, bends *7* or *8* would make the edges of the corner tabs contact the adjacent end apron edges, making it impossible to complete the 90-degree bend. For this reason, some means must be devised to provide the corner tabs with sufficient clearance so that they may be bent along a line that is offset behind the apron by a distance equal to the thickness of the metal. By providing this clearance, bends *7* and *8* can now be made, and the tabs will pass *behind* the end aprons, where they may be mechanically secured. To properly form line *5* to allow for this clearance, it is first necessary to reposition the fingers with the *finger-position adjustment screws* and to use *shims* to provide the proper finger offset and bend. (A shim is nothing more than a strip of sample stock used to determine the necessary setback distances.) The finger clearance is first set to a distance of *twice* the thickness of the metal. The work is next fed into the brake from *behind* the crosshead. With bend *5* aligned with the fingers, there will be

**Figure 23.8** Finger and heel clearance restrictions.

**Figure 23.9**  Bending sequence for a corner tab box chassis with subaprons.

enough fingers to cover *just* the length of the apron without extending beyond either relief hole onto a tab. *These* fingers are then moved *forward* from the crosshead by an amount equal to the thickness of the metal by placing a shim between the crosshead and the finger supports (Fig. 23.10). The shims used in this procedure have a width of approximately $\frac{3}{4}$ inch. The shim length depends entirely on the width of the fingers that it is used with, but should be no longer than the length of the section of the bend line with which it is associated. (In this example, this length is bend line *5 within the relief holes*.) It is informative to demonstrate what would happen if the bend were performed at this point. The apron section would be bent normally, whereas the tabs would have a bend radius *twice that of the apron* and *with no setback distance*. To achieve the desired setback distance, additional

**Figure 23.10**  Finger and shim arrangement to bend the short side of a corner tab box chassis.

**Figure 23.11** Tab shims are used to obtain the offset single-thickness setback distance for tabs.

shims are placed between the tabs and the upper edge of the wing. These shims should *not* extend onto the brake table. Also, they should not extend laterally beyond a relief hole on either apron end. This technique is shown in Fig. 23.11. The bend resulting from this arrangement is shown in Fig. 23.12. Notice that the tabs are offset the exact required amount behind the apron. Bend *6* can now be formed in the same manner as bend *5*. The chassis is placed on the brake table with bend *5* (just completed) positioned to the rear of the crosshead. After bend *6* is completed, it will be necessary to remove all shims and fingers from the crosshead to retrieve the chassis.

To complete the bending of the chassis after bend *6* has been formed, the fingers are realigned and reset to a setback distance equal to the thickness of the metal. By using a total finger length that will accommodate the lengths of bends *7* and *8*, these bends can be formed. When *7* or *8* is undertaken, the subaprons previously formed by bends *1* and *3* will prevent the leading edges of the outermost fingers being placed at the extreme ends of these remaining bend lines. If bend *7* is placed under the fingers from the front of the brake, a set of fingers covering the entire length of the bend can be employed to partially set the radius. A 90-degree bend from this position cannot be formed. When the apron is bent to approximately 30 degrees, the tabs will contact the faces of the fingers. This partial bend, however, is sufficient to permit the remainder of the bend to be completed by reducing the total finger width to allow for the necessary side clearance for the subaprons. Although the subaprons now prevent the fingers from covering approximately $\frac{3}{8}$ inch at either end of the bend line, no noticeable distortion will result when the bend is completed

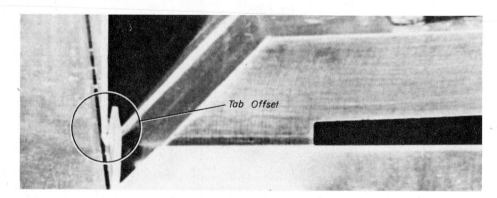

**Figure 23.12** Bend results in tab offset.

**Figure 23.13**   Chassis element bent into desired form.

because the radius was basically established by the first partial bend. Bend *8* is formed in the same manner to complete the chassis configuration.

The bending techniques discussed in this chapter are applied to form the chassis shown in Fig. 23.13. The final stage in the construction of a chassis is to apply a finish that will not only protect it but will also enhance its appearance. The various finishes and techniques of finishing are discussed in the following chapter.

## EXERCISES

### A. Questions

23.1   What is meant by *bend allowance,* and why is it an important factor in bending sheet metal?

23.2   Describe the distortions that result when bending sheet metal.

23.3   A bracket is to be bent into a U shape using 16-gauge 3000 series aluminum. Determine the inside and outside dimensions in decimals if the bend lines are 4 inches apart.

23.4   What is *setback distance?*

23.5   What is the advantage of a brake with individual fingers when compared to a cornice or leaf-type brake?

23.6   What is the purpose of *relief holes?*

### B. True or False

Circle T if the statement is true, and F if any part of the statement is false.

23.1   Bending sheet metal compresses the outer section of the bend and stretches the inner section.   **T   F**

23.2   When forming a box chassis, the subaprons are bent last.   **T   F**

23.3   Subaprons are bent first when forming a U-shaped cover.   **T   F**

23.4   The angle adjustment setting must be made slightly greater than the desired angle to compensate for the *spring-back* characteristics of sheet metal.   **T   F**

23.5   Gauge number 12 aluminum is thinner than gauge number 20.   **T   F**

## C. Multiple Choice

Circle the correct answer for each statement.

23.1     Bend allowance is required to compensate for the distortion generated during the bending operation that (*stretches, compresses*) the inner section of the bend.

23.2     On a hand-operated finger brake, the (*wing, crosshead*) holds the fingers.

23.3     Clamping pressure on a finger brake should be set using (*once, twice*) the thickness of sample stock.

23.4     Finger setback clearance for the tabs of a corner tab box chassis should be set using (*once, twice*) the thickness of sample stock.

23.5     Heel clearance limits the height of the (*apron, subapron*) of a box chassis.

23.6     The bending sequence for a standard box chassis requires that the subaprons be bent (*first, last*).

23.7     The bending sequence for a basic U-shaped chassis with tabs requires that the tabs be formed (*first, last*).

## D. Matching Columns

Match each item in column A to the most appropriate item in column B.

| COLUMN A | COLUMN B |
|---|---|
| 1. Finger brake | a. Chassis corners |
| 2. Setback distance | b. 30, 45, 60, 90 degrees |
| 3. Relief holes | c. Box or pan |
| 4. Angle stop | d. Finger relief |
| 5. Bend allowance | e. Finger clearance |
| 6. Heel clearance | f. Stretching |

## E. Problems

23.1     Bend the apron of the gauge panel laid out, drilled, and cut in Problems 20.3, 21.1, and 22.3 to 90 degrees from the front section.

23.2     Bend the four ends of the heat sink laid out and drilled in Problems 19.3 and 20.2 to 90 degrees from the center section and all in the same direction.

23.3     Considering bend allowance, design, lay out, and construct a utility chassis similar to the one shown in Fig. 23.5a, including its cover shown in Fig. 23.6a, with final outside dimensions of 3 inches by 5 inches by 7 inches.

23.4     Considering bend allowance, design, lay out, and construct a box chassis similar to the one shown in Fig. 23.9, with $\frac{3}{8}$-inch tabs and subaprons and final outside dimensions of 2 inches by 6 inches by 9 inches.

# Metal Finishing and Labeling

## LEARNING OBJECTIVES

*Upon completion of this chapter on metal finishing and labeling, the student should be able to*

- Be familiar with the types of contaminants commonly found on aluminum and steel.
- Be familiar with the technique of spray painting onto metal surfaces.
- Apply adhesive films onto chassis and panels.
- Use aluminum foil plates and dry transfers to label chassis and panels.
- Use a silk screen for labeling metal chassis.

## 24.0 INTRODUCTION

Most manufactured parts used in electronic packaging require a *finish* to enhance and protect their surfaces. This is especially true of metal elements whose corrosive characteristics require protection. A prerequisite for obtaining a sound finish is the proper removal of all surface contaminants.

The technique of spray painting the chassis is presented, followed by information on the application of literal and numerical reference markings on chassis and panels.

***Caution:*** *A chemical hazard exists in the materials described in this chapter. Approved face shields (goggles), chemical-resistant gloves, clothing, footware, and a respirator must be worn prior to performing any of these operations. Also, the section on safety outlined at the beginning of the book should be reviewed for more specific information.*

## 24.1 SURFACE PREPARATION AND FINISHING

Metal surfaces that are to be finished must be free of contaminants. The most common *soils* absorbed by aluminum and mild steel are (1) *water vapor*, (2) *fingerprint deposits*, (3) *oils*, (4) *lubricants*, (5) *paints*, and (6) *mill scale* (this term is used to describe the contaminants present on the sheet metal as a result of the manufacturing processes). If these soils are not removed, an inferior bond between the surface and the finish will result.

Cleaning is effectively accomplished with the use of *isopropyl alcohol*. Simply hand wiping or dipping the metal parts into the alcohol will remove most surface contaminants. After the metal has been allowed to dry, finishing and labeling may be applied.

Of all the materials available for finishing applications, the most economical and easiest to apply are *paints, enamels, varnishes,* and *lacquers.* They also offer the widest selection of colors and textures among all surface finishes. These finishes are made from *binders* or *vehicles* (which are oils or synthetic resins), *pigments,* and *solvents.* The composition of these finishes is as follows:

> *Paint:* vehicle and a pigment
> *Enamel:* paint and varnish
> *Varnish:* vehicle and solvent
> *Lacquer:* vehicle, solvent, and nonvolatile coloring additive

Paints and enamels are used extensively on metal surfaces common to electronic packaging. They provide a heavier and more durable coating than either lacquer or varnish and have the advantage of hiding minor surface imperfections such as scratches or nicks in metal. Paint and enamel finishes range from *lusterless* (flat) to *high gloss.* The degree of gloss is determined by the proportion of pigment to vehicle. Semigloss to flat finishes are most commonly used in electronic packaging applications. Some of the more common paint textures available are *wrinkle, crackle,* and *hammertone.* They not only hide surface imperfections but also enhance the overall appearance of the chassis. The *wrinkle*-type enamel tends to *expand* as it cures. This causes small, randomly oriented wrinkles of paint to form over the entire surface. The *crackle*-type enamel is applied over a color base. As this finish cures, it *contracts* and forms very fine random lines. The base color contrasts with the finish color producing an extremely attractive package. The *hammertone* enamel leaves a very smooth surface when it cures, yet provides a hand-hammered appearance. This finish is available in a variety of colors.

Varnishes are used when a surface is to be protected from moisture and mildew. Lacquers are often used clear or unpigmented to protect labeling from abrasion or to help protect etched or plated surfaces from staining or tarnishing. Lacquers are water resistant but will not stand up to saltwater or prolonged exposure to sunlight. These critical environments will cause the lacquer to discolor and become brittle. Plastic sprays of the acrylic or vinyl type are readily available in aerosol cans. These sprays do not possess many of the undesirable characteristics of lacquer and varnish. Plastic sprays are available either clear or in a choice of colors.

*Aerosol-pressure spray cans* are limited to painting small surface areas because of their cost. These cans are pressurized to approximately 40 pounds per square inch with a refrigerant, usually *Freon,* which acts as the propellant for atomizing the paint. Spray painting must be done in a well-ventilated spray booth that will exhaust both the solvent and excess paint particles. In addition, atomized paints are highly combustible and should be used with caution. To aid in obtaining a uniform coating, especially over irregularly shaped surfaces, the work may be placed on a turntable. This eliminates having to handle the work to paint the reverse side. For best results, the aerosol can is held perpendicular to the work surface, which is held vertically. Spraying is begun at the top of the work with the nozzle held 6 to 10 inches from the surface. The spray is swept across the work in a straight-line motion continuing beyond the edge of the work. To ensure proper coverage, each sweep of spray should slightly overlap the last pass when moving the spray gun to either the left or right. Once the surface has been completely covered, the paint must be allowed to cure. If a single coating does not provide sufficient coverage, a second coat may be applied *only after* the first coat is completely cured. For superior results, it is always good practice to apply two coats rather than one thick coat. This reduces the possibility of the paint sagging or running. For uniform paint thickness, the more difficult and unexposed areas are sprayed first and then the open surfaces, if the work is irregularly shaped. Some finishes cure at normal room temperatures but others require a drying oven. Most paints can be dried in about 20 minutes at 225 to 250°F (107 to 121°C). Manufacturers' curing specifications are available in their literature. Figure 24.1 shows a panel being spray painted using an aerosol can.

**Figure 24.1** Use of aerosol-pressure spray cans simplifies finishing small parts.

## 24.2 LABELING

The final step in completing a chassis and panel involves applying literal or numerical reference designations. The type and procedure of labeling will depend on several factors, including (1) the degree of clarity required, (2) cost, (3) time required for process, (4) durability, (5) materials and finishes involved, (6) availability of equipment and facilities, and (7) the complexity of the labeling layout pattern. The most common methods of labeling are by *decals, dry transfers,* and *silk screen.*

Pressure-sensitive decals and dry transfers, such as those shown in Fig. 24.2, provide a rapid means of labeling with crisp and sharp line characteristics. They are used extensively with the silk-screen method of labeling. Decals and transfers will adhere to practically any clean, smooth surface, such as wood, paper, cardboard, plastic, glass, or metal. To improve their durability, they should be given a protective clear coating after they have been applied.

Pressure-sensitive decals have a protective backing that must be removed before they can be applied. Because literal and numerical decals are small, it is best to use tweez-

**Figure 24.2** Typical selection of pressure-sensitive decals and dry transfers.

ers or a knife edge to remove them from the backing, which is chemically treated for easy release. This technique reduces the possibility of touching the adhesive with the fingers and upsetting the tack. Once the backing has been removed, the decal is brought into the desired position, aligned, and pressed firmly into place. Occasionally, it may be necessary to trim the clear support liner if overlapping occurs.

Dry transfers used in electronic packaging consist of drafting symbols, letters, words, numbers, and meter and dial markings printed on a transparent film with opaque black ink. This ink contains a pressure-sensitive heat-resistant adhesive. Before applying the transfer figure to a surface, the protective backing sheet is first removed. This sheet does not adhere to the film and therefore separates readily. The figure is brought into position with the inked side *toward* the work surface and is held firmly in place. The entire area around the figure is rubbed with a wooden burnisher, pencil, or ballpoint pen using overlapping strokes. Figure 24.3 shows dry transfers being applied to a chassis element. Since many figures or symbols are generally on the same sheet and closely spaced, care must be taken to avoid running over unwanted figures or portions of figures. By gently lifting one edge of the film, a check for complete transfer can be made. To assure perfectly transferred figures with sharp lines and no pinholes, the backing sheet should be placed back over the work surface and the figures reburnished. A wax residue resulting from the transfer process is generally formed on surfaces such as glass or metal. This residue can easily be removed with rubber cement thinner and a soft cloth without damaging the transfers. When dry transfers are used for direct labeling on a surface, a *plastic*-type protective film is necessary. These acrylic or vinyl films should be selected carefully because they attack certain materials. Vinyls are often preferred over acrylics because they are more resistant to flaking and chipping. These films are available in aerosol cans in both gloss and matte finishes.

The silk-screen process for labeling requires the preparation of a *clear polyester positive master* when identical duplication of work is required. Unlike the previously discussed methods of labeling, the silk-screen process is intended for mass production. (The silk-screen operation for producing marking masks as well as printed circuit conductor patterns is described in detail in Chapter 10.) Decals or dry transfers allow the positive

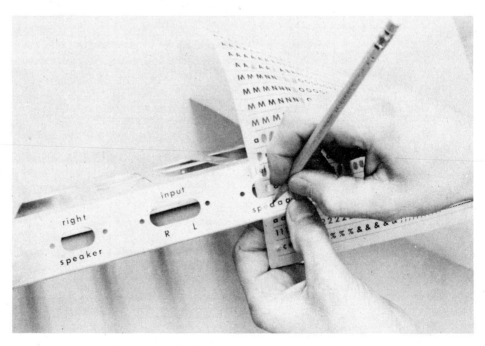

**Figure 24.3**  Application of dry transfers to metal surface.

artwork to be constructed quickly and eliminate time-consuming drafting procedures that would otherwise be necessary. A thin sheet of drafting *vellum,* which is slightly larger than the surface to be labeled, is first placed over the sheet metal layout drawing. The corners, mounting holes, and hole centers are carefully located by *pencil* on the vellum. Allowing for surface space to be taken up by mounting nuts, bezels, and control knobs, the required nomenclature and symbol locations together with the necessary spacing are *sketched* or *blocked* in. Crowding should be avoided and the spacing should be standardized to avoid confusion. Whenever possible, nomenclature should be consistently above or below components or controls. For a balanced appearance, all lettering should be centered on a vertical center line of a mounting hole. After all labeling positions have been located, the vellum is placed over a comparable size sheet of graph paper and taped into place on a drafting board. (Graph paper with 10 divisions to the inch is recommended.) A *polyester sheet* is then taped over the vellum. With the aid of the grid lines and the blocked-in nomenclature and lettering positions, all decals or transfers can be accurately positioned and applied onto the polyester using the procedure shown in Fig. 24.3.

For the silk-screen process, described in detail in Chapter 10, a *film negative* is contact-printed from the polyester master. The developed film is allowed to dry on a fine mesh silk screen attached to a frame. The screen is positioned over the metal panel to be labeled and the appropriate paint is spread across the screen with a squeegee. The paint will penetrate the screen and be deposited onto the panel only through those areas seen as

(a)

(b)

**Figure 24.4** Silk-screen method for panel labeling: (a) application of paint through the screen onto a panel; (b) screened panel.

clear. This technique is shown in Fig. 24.4a. The screen is then removed and the characters are allowed to dry. Figure 24.4b shows a panel labeled by the silk-screen method.

In this chapter only the most common types of labeling techniques have been discussed. Although many other processes are commercially available, they are not intended for prototype work. Of those labeling methods considered, decals and dry transfers are the most economical, least involved, and produce a highly acceptable appearance.

With so many variations and techniques available for finishing and labeling, the technician must consider the following factors prior to determining the most appropriate method to use: surface materials, number and complexity of markings, required sharpness of detail, equipment availability, cost, time required, and the technician's skill. An evaluation of these factors will dictate the appearance of the finished product.

The six chapters of this unit concentrated on techniques and methods for fabricating sheet metal chassis elements. Therefore, with all subassemblies for a project fabricated— printed circuit boards and sheet metal chassis elements—the technician can easily complete a project. In Unit 6 chassis hardware assembly and wiring techniques are covered in order to combine the various elements into a finished system.

## EXERCISES

### A. Questions

24.1   List the most common soils absorbed by aluminum and steel that must be removed prior to finishing operations.

24.2   Describe in detail the proper procedure for applying spray paint to a vertical surface.

24.3   What is the advantage of using a turntable when spray painting?

24.4   List the composition of (1) paint; (2) enamel; (3) varnish; and (4) lacquer.

24.5   List some of the factors that must be considered in determining the most appropriate finish and labeling to use.

# Unit

# SIX

# Chassis Hardware and Wiring

# Chassis Hardware and Assembly

## LEARNING OBJECTIVES

*Upon completion of this chapter on chassis hardware and component assembly, the student should be able to*

- Be familiar with the order in which parts and hardware are assembled.
- Know the various types of machine screws, nuts, washers, and thread-forming and thread-cutting screws.
- Use the following tools for assembling parts to a chassis:
  a. Screwdrivers
  b. Nutdrivers
  c. Wrenches
  d. Pop rivet guns
  e. Pliers
- Be familiar with the various types of special fasteners and hardware used in chassis assembly.

## 25.0 INTRODUCTION

With the completion of all subassemblies (the pc boards), the assembly of hardware and components onto the main chassis element may begin. In addition to the boards, this involves the mechanical securing of jacks, connectors, fuse holders, terminal strips, switches, potentiometers, lamps, meters, power cords, and transformers.

Before beginning to assemble parts, the overall task should be examined. If there are few components and parts to be assembled and if they are small, random mounting will pose no particular problem. If, however, the assembly involves many component parts with a wide range of sizes and weights, some order will be necessary to minimize needless effort and reduce the possibility of damaging delicate components.

Small, sturdy components such as fuse holders, terminal strips, lugs, grommets, and connectors should be mounted first. Transformers, filter chokes, and other large or bulky components should be among the last to be mounted, to avoid having to work with a heavy, bulky chassis during the initial assembly phase. Fragile components such as lamps, meters, sensitive relays, and pc boards should be the last to be mounted. As working space diminishes with the addition of components, care must be exercised to avoid damaging parts with assembly tools.

Mounting electronic components and hardware to a metal chassis may appear to be a simple task. However, efficient and quality workmanship will result only if the four major factors of packaging are considered. These are (1) mechanical security, (2) appearance, (3) accessibility, and (4) cost. Each factor is discussed in this chapter in conjunction with specific information on common and special fastening devices. Also considered are correct tool selection and assembly techniques for securing components and hardware.

*Caution: Prior to performing any of the operations described in this chapter, approved safety glasses must be worn.* Also, the section on safety outlined at the beginning of the book should be reviewed for more specific information.

## 25.1 MACHINE SCREWS

Machine screws are among the most common types of fasteners used in electronic packaging since they provide sound mechanical security and are easy to assemble and remove. Machine screws are designated by (1) *head style*, (2) *drive configuration*, (3) *diameter*, (4) *threads*, (5) *length*, (6) *material*, (7) *finish*, and (8) *fit*.

Some of the common head configurations are the *fillister, binder, pan, round, flat, oval,* and *truss* styles. These are shown in Fig. 25.1a. Although somewhat of a personal preference, the head style selected is generally determined by appearance, application, and mechanical security. Fillister, binder, and pan heads are usually preferred over round heads for appearance sake. These head styles are mechanically superior to the round head because there is more material at the driver blade pressure points. These head styles have approximately the same-size sealing plane (surface contact area) for a particular screw size, although the fillister style has the smallest head diameter. Both the flat and oval heads require that the work surfaces be countersunk prior to assembly. The flat head is used for flush mounting and has sealing plane angles of 82 or 100 degrees. Neither the flat nor the oval heads are recommended for use with thin sheet metal, wherein countersinking might weaken the assembly. The oval head is often used with a cup-style washer for securing control panels to racks. Countersinking in this case is unnecessary since the shape of the washer provides the required seating. This arrangement tends to provide both a decorative appearance as well as excellent mechanical security. The truss head style, common to sheet metal screws, has a larger sealing plane that provides greater surface contact area with the work.

The drive configuration refers to the style of head for tool access. Those most common to electronic assemblies are the (1) *slotted*, (2) *Phillips*, and (3) *hexagonal socket* or

**Figure 25.1** Machine screw styles and drives: (a) machine screw head styles; (b) head-style drive configurations; (c) drive configurations requiring special driving tools; (d) machine screw thread measurements.

*25.1 Machine Screws*

**433**

*Allen* design. These are shown in Fig. 25.1b. The slotted style is the most widely used. The Phillips head has the advantage over the slotted type in that it provides four positive-drive pressure points completely within the perimeter of the head. It is often used when the screw is visually obstructed during assembly, since it allows the screwdriver to be self-aligned. In addition, the recessed style of the Phillips head greatly reduces the possibility of the driver slipping from the screw head, which could damage the work surface. The Allen configuration allows for even more positive tool positioning and gripping and is commonly used on set screws for securing knobs to control shafts. Other less common drive configurations that may be encountered are shown in Fig. 25.1c. Each of these requires special driving tools. The various driving tools common to electronic assembly are discussed in Sec. 25.4.

The maximum thread diameter (dimension $G$ in Fig. 25.1d) varies from 0.086 to 0.216 inch for general electronic assembly. The minimum diameter, shown as dimension $T$ in Fig. 25.1d, will vary from 0.0628 to 0.1619 inch. Small sizes of machine screws are generally designated by *gauge number*. For the thread diameters referred to, the gauge numbers vary from No. *2,* having a diameter of 0.086 inch, to No. *12,* having a diameter of 0.216 inch. Larger thread diameters are given in fractions of an inch. Smaller-diameter screws, with gauge numbers of *1* or *0,* are not common in electronic assembly. A listing of some common thread gauge numbers together with their decimal equivalents is shown in Table 25.1. A more complete listing is provided in Appendix III. One way of determining the gauge number for an unknown screw size is simply to measure the thread diameter with a micrometer and compare the reading thus obtained with the corresponding gauge number of Table 25.1.

Machine screws are manufactured with *right-* and *left*-hand threads. Right-hand threads are secured by clockwise (cw) rotation and extracted by counterclockwise (ccw) rotation. Left-hand threads are rotated in the opposite directions and are used only in special cases wherein vibration or rotation of the system would tend to loosen the conventional right-hand thread.

Machine screws are also designated by the number of threads per inch. The simplest method of determining this is by placing a scale against the threads and counting the number of threads in 1 inch. For a particular screw gauge or diameter, the number of threads per inch will depend on the screw thread system used. Although there are other systems of screw designations, machine screws having either *Unified Coarse* (UNC) or *Unified Fine* (UNF) designations are predominantly used for electronic assembly. Referring to Table 25.1, a UNC screw having a gauge No. 4 diameter has 40 threads per inch, whereas a UNF screw with the same gauge number has 48 threads per inch. It is apparent that the UNC system has fewer threads per inch than the UNF system. The coarseness of the screw selected will depend on its application. Generally, UNC threads are preferable for rapid assembly since thread meshing is a less exacting process. This coarser screw is also used when threading into soft material. Since a machine screw of a given size with a coarse thread has a smaller minimum diameter than the fine thread, it will mesh deeper

**TABLE 25.1**

Machine Screw Sizes Most Frequently Used in Electronics

| Gauge Number | Decimal Equivalent (in.) | Threads per Inch | |
| --- | --- | --- | --- |
| | | UNC | UNF |
| 2 | 0.0860 | 56 | 64 |
| 4 | 0.1120 | 40 | 48 |
| 6 | 0.1380 | 32 | 40 |
| 8 | 0.1640 | 32 | 36 |
| 10 | 0.1900 | 24 | 32 |
| 12 | 0.2160 | 24 | 28 |

*Chassis Hardware and Assembly*

into the material surrounding the tapped hole. This advantage greatly reduces the possibility of thread shearing when an axial load is placed on the screw. If vibrations are a consideration, fine threads are preferable. Because of their greater minimum diameter, they can be torqued more than comparably sized coarse thread screws and will not shake loose as readily.

For all head styles except the flat and oval designs, the thread length is measured from the sealing plane to the end of the screw (Fig. 25.1d). Flat-style and oval-style screw lengths are measured from the top of the sealing plane to the end. Screw lengths between $\frac{1}{8}$ and $\frac{5}{8}$ inch are available in increments of $\frac{1}{16}$ inch. Lengths from $\frac{5}{8}$ to $1\frac{1}{4}$ inches are graduated in $\frac{1}{8}$-inch increments, whereas those from $1\frac{1}{4}$ to 3 inches are graduated in $\frac{1}{4}$-inch increments.

Screws for electronic assembly are made from *steel, brass, aluminum,* or insulating materials such as *nylon.* The finishes available include *black anodizing, cadmium, nickel, zinc,* or *chrome* platings. The finish selected will depend on the appearance and the necessary corrosion protection.

Screws are also designated with a specific relationship (fit) between mating threads of nuts or tapped holes and machine screws. There are four classes of fit between threaded parts: *Class 1* is an extremely loose fit with considerable "play" between the threaded parts; *Class 2,* sometimes designated as a "free fit," can be readily assembled with finger pressure; *Class 3* has more of a snug fit than Class 2 but requires only finger pressure for initial assembly; *Class 4* is characterized by the tightest available fit, requiring screwdrivers and wrenches for assembly of the threaded parts. Selecting the most appropriate class of fit will depend on such factors as (1) speed of assembly, (2) type and style of washer employed, and (3) the amount of looseness between threaded parts that can be tolerated. For most electronic applications, the Class 2 fit is standard.

Machine screws may be completely identified by the following abbreviated description:

$$\#6\text{–}40 \times \tfrac{3}{8} \text{ UNF—2A LH Pan Hd. Steel C.P.}$$

where

#6 is the major thread diameter (equivalent to 0.138 inch, refer to Table 25.1).

40 is the number of threads per inch.

Three-eighths is the length of thread in inches measured from the sealing plane to the screw end.

UNF is the type of thread (Unified Fine). If the type was UNC, this designation would normally be omitted and assumed to be understood.

2A—the 2 represents the class of fit between threaded parts and the A designates that the thread is external. A B suffix designates an internal-type thread for machine nuts.

LH is the left-hand thread. If a right-hand was specified, no designation is given since it is assumed to be understood.

Pan Hd. is the head style.

Steel is the material.

C.P. is the plating (cadmium-plated).

Clearance hole sizes for several machine screws are given in Appendix IV. The clearance hole size is determined by the maximum diameter of the screw thread. This type of hole is drilled when components and hardware are to be assembled with screws and nuts. The diameter of a clearance hole must be sufficiently larger than the thread diameter to allow complete unobstructed passage of the screw through the hole. The amount of

6–32 Machine Screw

$G = 0.1380''$
Total Clearance 0.0060''

$G$

Clearance Hole #27 Drill
Size (0.1440'')

**Figure 25.2** Clearance hole for a 6-32 machine screw.

clearance will depend on the materials to be fastened. However, the values given in Appendix IV are average sizes, which are satisfactory for most applications. A clearance hole together with its tolerance is shown in Fig. 25.2.

## 25.2 NUTS AND WASHERS

Nuts used with machine screws are designated by (1) *style*, (2) *chamfer*, (3) *hole diameter and threads*, (4) *fit*, (5) *material*, (6) *finish*, (7) *width*, and (8) *thickness*. Nuts common to electronic assembly are shown in Fig. 25.3a. The hexagonal style is by far the most widely used and is generally preferable to the square nut for reasons of appearance and ease of assembly. This nut may have *single chamfer, double chamfer,* or *single chamfer* and *washer face* (Fig. 25.3b). When a single chamfer nut is used, the flat side should be against the work to provide the best appearance with no exposed sharp edges. If the work surface is finished, double chamfer or washer face nuts should be used. When tightening the nut against the work surface, the chamfer or washer face will prevent any visible surface marring, such as the circular abrasions formed by the points of a single chamfer square or hexagonal nut.

The hole diameter and thread type are, naturally, consistent with the mating screw. This is also true for class of fit, material, and finish. For each size and style of nut, different widths (measured across pairs of opposite flats) and thicknesses are available. These external dimensions are shown in Fig. 25.3c. Tabulated sizes for common hexagonal nuts are given in Appendix V. Nuts with larger widths are normally used for ease of handling and rapid assembly, whereas those with smaller widths are selected if there is a possibility of interference with other hardware or if assembly clearance is limited.

Hexagonal  Jam  Knurled Thumb Nuts

Single Chamfer  Double Chamfer  Washer Face

(b)

Interference Stop Nut  Square  Cap Nut  Spring Stop Nut

(a)

$\leftarrow W \rightarrow$  $T$

(c)

**Figure 25.3** Machine screw nuts and fasteners: (a) nuts common to electronic assembly; (b) hexagonal nut face styles; (c) width and thickness measurements for hexagonal nuts.

The special-purpose nuts, also shown in Fig. 25.3a, have styles that dictate their application. A description and application of each follows:

*Jam nut:* Basically the same as a hexagonal nut, except that it is from $\frac{1}{16}$ to $\frac{3}{8}$ inch thinner, depending on the hole diameter. This nut may be used alone, but it is more often employed under a regular hexagonal nut for a more secure assembly. When this nut is tightened, its flats align with those of the hexagonal nut, providing the desired locking action.

*Interference stop nut:* Used to prevent loosening due to vibration. It has an internal unthreaded plastic or fiber collar. When the nut is tightened, the screw threads are forced into the inside surface of this tapered collar. Being elastically deformed, the collar squeezes against the screw threads, providing a secure locking action and eliminating the need for a lock washer.

*Square nut:* Often used in securing channel stock in steel cabinets or racks. The flat of the nut often fits snugly or jams against the inside member when being tightened, thus requiring merely a screwdriver for assembly.

*Knurled thumb nut:* Allows for quick finger tightening. It is commonly used on terminal posts such as those on some primary cells and telephone terminal blocks.

*Cap nut:* Used primarily for purposes of appearance. It completely covers the end of the screw and its shape eliminates sharp corners or edges. Chrome-plated cap nuts add a decorative appearance to a cabinet.

*Spring stop nut* (speed nut): Pressed onto the screw with only minimal final tightening required. It is self-retaining as well as self-locking. Because it greatly reduces assembly line operations, the speed nut is often used in radio and television receivers. It may also be used to secure smooth studs or tubing.

*Washers* are classified as either *flat* or *lock type* (Fig. 25.4). Flat washers are used in contact with a screw head or the chamfer of a nut to provide more surface area to distribute the forces imposed on the screw and nut. Lock washers are used to secure square or hexagonal nuts from loosening due to vibrations or handling. For electronic assembly, the *tooth-type* lockwasher is used extensively since it provides maximum gripping action. Normally, the use of one lock washer is sufficient. However, under severe vibrations lock washers should be placed under both the nut and screw head. The tooth-type lock washer should not be reused when removed. The gripping edges of the teeth are easily dulled and the spring action becomes fatigued. These washers have from 10 to 24 combined points of contact, depending on the number of teeth. *Internal* tooth washers are preferred over the *external* type from the standpoint of appearance since they cause no visible surface marring when assembling components such as the potentiometer shown in Fig. 25.5a. External tooth washers are generally used with flat washers for mounting components having oversized clearance holes. This arrangement is shown in Fig. 25.5b.

The *spring-type* or *split* lock washer, also shown in Fig. 25.4, uses the temper of the steel to maintain locking action. As the washer is compressed, the spring tension exerts a

**Figure 25.4** Washer styles.

Washer    External    Internal    Split

(a)

(b)

**Figure 25.5**  Washer assembly techniques: (a) washer assembly for potentiometer mounting; (b) washer assembly for oversized clearance holes.

force that prevents counterrotation. This type of washer is less effective than the tooth type because the locking action results only from spring tension. Flat steel washers are used in conjunction with tooth-type lock washers when securing soft materials, whereas resilient washers, such as lead or fiber, are used to provide a cushion effect when assembling brittle material.

The inside diameter of a washer must coincide with the clearance hole diameter of the screw with which it is associated. The outside diameter will depend on the application. Recommended washer sizes are given in Appendix VI.

## 25.3 THREAD-FORMING AND THREAD-CUTTING SCREWS

Thread-forming and thread-cutting screws are widely employed when the use of nuts or the tapping of machine threads is not practical. These screws do not require a prethreaded hole for fastening. Manufactured from hardened steel, they form or cut their own threads as they are turned and driven into a hole. They cannot be driven into a material harder than their own. These screws are available in slotted, hexagonal, and Phillips drives with head styles common to machine screws, in addition to the washer head style shown in Fig. 25.7. Three of the popular thread designs are shown in Fig. 25.6. These are the sheet metal (thread-forming), machine screw thread-forming, and self-tapping (thread-cutting) types. The sheet metal screw, shown in Fig. 25.6a, when forced into the proper-sized predrilled hole, will form the metal around the hole into a threaded shape matching its own thread. This screw has a widely spaced thread extending to a point. The machine screw thread-forming type, shown in Fig. 25.6b, has a slightly tapered end but does not extend to a point. It has standard machine screw threads, which require greater force to drive than a sheet metal screw. This screw is used in place of the sheet metal screw when the possibility of loosening from vibration exists. The self-tapping screw, shown in Fig. 25.6c, has one or more slots in the threaded end for chip relief. As this screw is driven into a hole, it cuts threads in basically the same manner as a tap. The self-tapping screw is not used on sheet metal but is intended for fastenings to castings and soft materials, such as Bakelite, copper, aluminum, and plexiglass.

(a) Sheet metal

(b) Machine

(c) Self-tapping

**Figure 25.6**  Thread-forming and thread-cutting sheet metal screw styles.

**TABLE 25.2**

Recommended Drill Sizes for Sheet Metal Screws

| Screw Size | Thread-Forming Machine and Sheet Metal Screws | | | | | | Self-Tapping (Thread-Cutting) Screws | | | | | | |
|---|---|---|---|---|---|---|---|---|---|---|---|---|---|
| | 28 | 26 | 24 | 22 | 20 | 18 | 24 | 22 | 20 | 18 | 16 | 14 | 12 |
| *Aluminum* | | | | | | | | | | | | | |
| 4 | 45 | 44 | 42 | 42 | 42 | 40 | 45 | 44 | 44 | 44 | 44 | 43 | 43 |
| 6 | 40 | 39 | 39 | 39 | 38 | 36 | 38 | 38 | 37 | 37 | 37 | 36 | 35 |
| 8 | 34 | 34 | 33 | 32 | 32 | 31 | 33 | 32 | 32 | 31 | 30 | 29 | 28 |
| 10 | 32 | 31 | 30 | 30 | 30 | 29 | 28 | 28 | 27 | 27 | 27 | 27 | 26 |
| 12 | 29 | 28 | 27 | 26 | 25 | 24 | 21 | 21 | 20 | 20 | 20 | 19 | 17 |
| *Steel* | | | | | | | | | | | | | |
| 4 | 44 | 44 | 42 | 42 | 40 | 38 | 43 | 42 | 42 | 41 | 39 | 38 | 37 |
| 6 | 39 | 39 | 39 | 38 | 36 | 35 | 36 | 36 | 35 | 34 | 32 | 31 | 30 |
| 8 | 33 | 33 | 33 | 32 | 31 | 30 | 32 | 31 | 31 | 30 | 29 | 28 | 25 |
| 10 | 31 | 30 | 30 | 30 | 29 | 25 | 27 | 27 | 26 | 24 | 24 | 22 | 20 |
| 12 | 28 | 26 | 26 | 25 | 24 | 22 | 19 | 19 | 19 | 18 | 16 | 14 | 13 |

*Metal Gauge Numbers*

These drill sizes are intended only as a guide and may vary with application.

Sizes of thread-forming and thread-cutting screws are similar to those for machine screws. To result in optimum mechanical security, the size of the predrilled hole is important. The size of this hole will depend on the *screw size, gauge of the metal,* and *type of metal.* Generally, for the same size screw, a larger hole must be provided in harder materials. For example, if a No. 6 self-tapping screw is to be used to secure sheet metal whose thickness is 0.060 inch, a No. 32 drill bit is recommended for steel, whereas a No. 36 would be necessary for aluminum. Information on hole sizes for thread-forming and thread-cutting screws is provided in Table 25.2.

Thread-forming and thread-cutting screws do not provide the mechanical security obtainable with machine screws with lock washers and nuts. However, when vibration is not a factor, they overcome accessibility problems in areas where nuts cannot be installed. Access or cover plates are an example of this application.

The proper securing of two pieces of sheet metal with a sheet metal screw is shown in Fig. 25.7. The hole drilled in the metal closest to the screw head must be a clearance hole to ensure positive metal-to-metal contact when the threading is completed and the sealing plane of the screw head is flush with the metal. For sheet metal applications, screws with larger sealing planes, such as the washer-head type shown in Fig. 25.7, should be used.

Serration to Ensure Positive Gripping

*Clearance Hole Size

+Tap Hole Size

**Figure 25.7** Recommended method for sheet metal assembly.

Washer Head Style Sheet Metal Screw

*Reference Appendix IV
+Reference Table 23.2

*25.3 Thread-Forming and Thread-Cutting Screws*

Screwdrivers are the most common tools used for securing the screws discussed in Sec. 25.3. They are available in a variety of *handles, shank lengths* and *shapes,* and *driver styles.* A popular-style screwdriver is shown in Fig. 25.8 together with the nomenclature of its parts. The handle is usually made of hardwood or plastic. In addition, a rubber grip may be provided for additional electrical protection, positive gripping, and comfort. For those screwdrivers without a rubber grip, large flutes are formed in the handle for improved gripping. The shank is manufactured from a tempered steel alloy specifically designed to withstand distortion caused by the excessive torque to which these tools are subjected. The handle and shank, however, will not withstand severe forces when the screwdriver is misused, such as for prying or bending. The shank may have either a round or square cross section. The square design is usually found on larger-size screwdrivers, where severe torque is expected. Shank lengths normally vary from 3 to 16 inches. For general electronic assembly, shank lengths up to 10 inches find wide application.

Driver styles are available for all screw-head drive configurations shown in Figs. 25.1b and c. The three most common types are the *standard, Phillips,* and *hexagonal* drivers. The standard driver has a tapered tip design that fits into the drive slot of a screw head. Common blade widths are $\frac{3}{32}, \frac{1}{8}, \frac{5}{32}, \frac{3}{16}, \frac{1}{4}$, and $\frac{3}{8}$ inch. When selecting the most suitable screwdriver to avoid damage to the screw or the work surface, the width of the blade should closely match the slot length but should not extend beyond the perimeter of the screw head.

Phillips screwdrivers are designed to fit the tapered cross-slots on screw heads made for this type of drive (Fig. 25.1b). Since more contact points are available with this design than with the slotted type, more driving force can be applied with less possibility of slippage. This style is commonly available in five number sizes: 0, 1, 2, 3, and 4. These will fit a large variety of Phillips screwhead sizes. Table 25.3 shows the screwdriver number and the screw size to which each will accommodate. Those most frequently used in electronic assembly are numbers 1, 2, and 3. This screwdriver must be held in exact alignment with the drive configuration of the screw head to be effective.

The Frearson or Pozidrive (a trade name of Phillips International) styles (see Fig. 25.1c) are similar to the Phillips drive except that they allow for a much more precise fit between the driver and the screw head. This fit enables the screw to be supported unaided on the end of the driver, which can be a time-saving advantage in assembly work in which accessibility is limited.

The *Allen* drive, as well as the special *Bristol* type (whose screw heads are shown in Fig. 25.1), also have a distinct advantage over the slotted style in that they minimize the possibility of the driver slipping out of the screw head and damaging the work surface. The Allen drive is encountered more often in electronics assembly than the Bristol style. A selection of Allen drives from $\frac{3}{64}$ to $\frac{3}{8}$ inch is suitable for most electronic applications. These sizes vary by $\frac{1}{64}$ inch for sizes from $\frac{3}{64}$ to $\frac{3}{32}$ inch, by $\frac{1}{32}$ inch for $\frac{3}{32}$ to $\frac{1}{4}$ inch, and by $\frac{1}{16}$ inch for $\frac{1}{4}$ to $\frac{3}{8}$ inch.

For extremely small set screws, the Bristol drive is often preferable to the Allen style, since its multiple spline design prevents rounding of the driving surfaces of either

**Figure 25.8** Screwdriver nomenclature. *Courtesy of Stanley Tools.*

**TABLE 25.3**

Phillips Screwdriver Selection

| Number Tip Size | Machine Screws | | Sheet Metal Screws |
| | Flat or Oval | Round | Flat, Round, Oval, or Binder |
| --- | --- | --- | --- |
| 0 | 0,1 | 0,1 | — |
| 1 | 2–4 | 2–4 | 2–4 |
| 2 | 5–9 | 5–10 | 5–10 |
| 3 | 10–16 | 12–16 | 12–14 |
| 4 | above 16 | above 16 | — |

the drive or the screw head through continued use. Bristol drives are available with four or six flutes.

For small and delicate work, *jewelers'* screwdriver sets may be used. The barrel of the handle is knurled with a swivel-top finger rest. This configuration allows the driver to be turned using the thumb and forefinger while applying pressure with the index finger (Fig. 25.9).

Screwdrivers for special applications are shown in Fig. 25.10 and are available in all the standard drive configurations. The "*stubby*" design, shown in Fig. 25.10a, is used if there is limited access to the screw head. These drivers are available in lengths from 1 to 2 inches.

*Offset* drivers, such as the one shown in Fig. 25.10b, are also used for work in re-stricted areas where even the stubby style would not fit. When the driver is of the blade type, alternate ends of the offset screwdriver must be used successively in order for the screw to be completely tightened by alternate partial turns. When using the offset Phillips or Allen drivers, it is unnecessary to use alternate ends. An improvement on the offset dri-ver is the *ratchet* design shown in Fig. 25.10c. This tool is similar to a ratchet-type socket wrench. A screw is tightened or loosened by a backward-and-forward motion of the han-dle. A two-position lever on the drive end of the handle is set to obtain either clockwise or counterclockwise ratchet driving action.

Screwdrivers having special screw-holding devices, such as that shown in Fig. 25.10d, retain the screw head firmly against the driver, allowing the full length of the shank to be used in gaining access to otherwise impossible assembly positions. The screw

**Figure 25.9** Properly held jeweler's screwdriver.

*25.4 Screwdrivers and Nutdrivers*

**441**

**Figure 25.10** Screwdrivers for special applications: (a) "stubby"; (b) standard offset driver; (c) offset ratchet; (d) driver with screw-holding mechanism. *Courtesy of Stanley Tools.*

head is grasped by a spring-loaded dual-leaf device. When this assembly is forced against the driver tip, the spring compresses and the leaves spread outward. The screw head is then inserted against the driver and the leaves cup the head under its sealing plane as the spring assembly is released and forced toward the handle.

The proper operating position of any screwdriver is always in alignment with the axis of the screw. This position minimizes the possibility of slippage damage and provides maximum surface contact area between the driver and the screw head.

*Nutdrivers* are used in a similar manner to screwdrivers but are designed to accommodate hexagonal-style machine nuts or screw heads. A set of nutdrivers is shown in Fig. 25.11a. They are available in sizes 6, 7, 8, 9, 10, 11, 12, 14, 16, and 18. The number represents the size in 32nds of an inch. For example, a number 6 will fit a $\frac{6}{32}$ $\left(\frac{3}{16}\right)$ hexagonal nut or screw head as measured across opposite flats. Sizes 6, 8, 10, 11, and 12 are most frequently used in electronics assembly. Sizes above $\frac{1}{2}$ inch are useful for the larger nuts associated with potentiometers and switches. The sizes of nutdrivers are often stamped into the shank, or colored handles are provided for quick identification. A shank length of 4 inches is suitable for most applications. Nutdrivers are available with hollow shanks for threading nuts onto long machine screws that extend more than $\frac{1}{4}$ to $\frac{3}{8}$ inch into the driver.

A screwdriver in conjunction with a nutdriver should be used to assemble and disassemble hardware whose fasteners require considerable force. This technique is shown in Fig. 25.11b.

## 25.5 WRENCHES

Wrenches are necessary for larger nuts found on some potentiometers and switches when a suitable nutdriver is not available or when accessibility does not permit its use. The most common types of wrenches used in electronic assembly are the *box, open-end,* and *adjustable* wrenches.

The box wrench, similar to a nutdriver, requires complete access over the nut since it encircles and grips the nut at its points. This wrench is a 12-point drive as opposed to the 6 points of a nutdriver. If used to secure a hexagonal nut, all 6 of the points will be contacted, virtually eliminating the possibility of shearing or rounding the points. The box wrench is preferable over other types of wrenches because it is much less likely to gouge the work surface. The box wrench should fit snugly onto the nut. All sizes listed for the nutdriver are also available for the box wrench, in addition to larger sizes that are not nor-

(a)

(b)

**Figure 25.11** Hexagonal nutdrivers: (a) nutdrivers are available with individual handles or interchangeable handles; (b) combined use of nutdriver and screwdriver.

mally required in electronic assembly. Lengths of 4 to 6 inches are suitable for most assembly applications. The proper use of a box wrench used to secure a potentiometer to a front panel is shown in Fig. 25.12. Masking tape may be applied to the panel where the edge of the wrench would contact the work surface around the nut to eliminate the possibility of damaging the finish.

Open-end wrenches are primarily intended for use with square nuts. The proper-size wrench can be determined when the entire length of the inside faces of the jaws fit snugly against any two opposite flats of the nut. This ensures maximum surface contact and min-

**Figure 25.12** Box wrench used in assembly.

imizes slippage and rounding of the nut. The jaws of this wrench are generally angled at 15 degrees to the handle center line. This design makes it possible to work in confined areas where the nut can be turned only partially. Open-end wrenches for electronic assembly are available in a complete range of sizes from $\frac{3}{16}$ to 1 inch and between $3\frac{1}{2}$ and 10 inches long. Although available in sizes much larger than 1 inch, these are seldom required in electronic applications.

Both the box and open-end wrenches require a complete set to accommodate all possible nut sizes that may be encountered in assembly. The adjustable wrench eliminates the need for complete sets since the distance between jaws can be changed to any size within its capacity. It does not, however, provide as secure a grip as either the box or open-end wrench. The adjustable wrench is similar to the open-end wrench with the 15-degree angle except that only one jaw is stationary. The second jaw can be moved by means of a thumb-operated *adjusting screw* located within the head of the wrench. To adjust the wrench for proper fit, the jaw is first opened sufficiently to allow both jaws to easily slip over opposite flats on the nut. The movable jaw is then closed until both jaws are gripping firmly against the flats. It is important that when positioned, the *pulling* force should be applied to the stationary jaw. If excessive force is applied to the movable jaw, the adjusting screw may be damaged or the wrench could slip, damaging the work surface or surrounding parts.

## 25.6 POP RIVETS

An alternative method of securing sheet metal or hardware by means other than machine screws and nuts or sheet metal screws is with the use of *pop rivets*. Pop riveting is rapid and requires little skill, and free access to only one side of the work is necessary. The only tool used for this process is a hand-type riveting gun.

A typical pop rivet, shown in Fig. 25.13a, consists of a *truss-style head, sealing plane, hollow shaft, stem,* and *button head*. The stem is tapered near the button head. When a tensile force is applied to the stem, the button head is drawn up into the hollow shaft. This action flares the end of the rivet against the metal, thus providing a positive fastening of the metal parts. When a pulling force of from 500 to 700 pounds is reached, the stem will break at the tapered section. The sheet metal or hardware to be fastened must be provided with clearance holes to accept the shaft of the rivet. A clearance of 3 to 5 thousandths of an inch is sufficient. The shaft is then pressed through the holes with the sealing plane flush against the metal surface. The pulling force is applied to the stem through the locking action of the pop rivet gun *nose piece* (stem collet). The nose piece is placed over the stem and pressed flush against the head of the rivet (Fig. 25.13b). As the operating lever of the gun is squeezed, the stem of the rivet is gripped, drawing the button head upward into the hollow shaft. Continued squeezing of the operating lever will create the desired flaring action. The stem is broken when sufficient pulling force is reached. The operating lever may have to be actuated more than once to complete this operation. When the lever is again released, the broken stem drops out of the nose piece. An installed pop rivet is shown in Fig. 25.13c.

Rivets are a permanent type of fastener intended for use if components, sheet metal, or hardware is not to be removed once fastened. This fact must be considered when using rivets, since they are difficult and time-consuming to remove. If it is necessary to remove a rivet, drilling through the rivet center with a drill bit size that is at least equal to the clearance hole is required.

Pop rivets of the style described are available in steel or aluminum with shaft diameters of $\frac{1}{8}$, $\frac{5}{32}$, and $\frac{3}{16}$ inch and shaft lengths of $\frac{1}{8}$ to $\frac{5}{8}$ inch. The clearance hole drill numbers for these three diameters are Nos. 30, 20, and 11, respectively. The stems of the three available shaft diameters require their own nose piece on the pop rivet gun. These nose pieces thread into the gun assembly and are easily removed with an open-end wrench.

**Figure 25.13** Pop rivet and its assembly: (a) pop rivet assembly and nomenclature; (b) pop rivet gun positioned against rivet head; (c) installed pop rivet.

## 25.7 PLIERS FOR HARDWARE ASSEMBLY

The two types of pliers most commonly used in electronic assembly are the *long-nose* and the *gas* pliers. These are shown in Fig. 25.14. Both styles consist of a *pivot, jaws,* and *handles*.

Long-nose pliers consist of a pair of long, narrow tapered jaws. The jaws are smooth or may have serrations near the tips for improved gripping. These pliers are designed to aid in assembly for holding or positioning small parts in tight spaces where fin-

**Figure 25.14** Pliers for hardware assembly: (a) long-nose pliers aid in assembling small parts; (b) gas pliers used to install power cord strain relief.

ger accessibility is limited. They are not intended to withstand severe gripping or twisting forces and should not be used for tightening machine nuts. Misusing the tool in this way could cause the jaws to spring out of alignment, rendering the pliers useless. Long-nose pliers holding a machine nut for assembly are shown in Fig. 25.14a.

Gas pliers have a much more rugged design than long-nose pliers and have coarse serrations over the complete inside surface of each jaw. These pliers are used primarily for holding round stock, such as inserting a power cord strain relief bushing, as shown in Fig. 25.14b. Excess gripping force on soft material should be avoided because the serrated jaws will mar the surface. For this reason, gas pliers should be used with extreme care. A minimum of two layers of masking tape wrapped around the stock will help prevent the jaws from marring.

Although many other styles of pliers are available, such as the *slip-joint* and *rib-joint* types, they are designed for holding heavy stock in which marring of the surface is not a factor. They should not be used in place of any of the drivers, wrenches, or pliers discussed in this chapter. Special pliers for cutting wire are considered in Chapter 26.

## 25.8 SPECIAL FASTENERS AND HARDWARE

Many instances arise in which the conventional fastening methods do not completely satisfy the assembly requirements because of many reasons, such as accessibility, orientation, frequent removal, and location. Many special fasteners and hardware that find wide application in electronic assembly are commercially available. Some of the more common of these will be discussed here.

The layout design may specify a single wire or cable to pass through a hole formed in the metal chassis element. For this requirement, a *grommet* should be used to prevent damage to the wire installation. A grommet is a rubber insulator that "cups" around the hole in the chassis and prevents wires from rubbing against the metal edges. A grommet being inserted into a chassis element is shown in Fig. 25.15a. An access hole must be provided in the chassis that is equal to dimension A in Fig. 25.15b. Some common size soft-rubber grommets together with the necessary size of access hole for each are listed in Table 25.4. Although dimension A is the grommet size designation, the selection of the proper-size grommet depends on the minimum required size of dimension B.

If the packaging design requires that a machine screw or other metal stud be insulated from the chassis, *fiber shoulder washers,* such as those shown in Fig. 25.16a, may

(a)          (b)

**Figure 25.15**  Grommets are inserted into holes to cover sharp edges: (a) grommet installation technique; (b) important dimensions for grommet installation.

**TABLE 25.4**
Rubber Grommet Dimensions

| Grommet Size Designation | A (in.) | B (in.) |
|---|---|---|
| $\frac{1}{4}$ | $\frac{1}{4}$ | $\frac{1}{8}$ |
| $\frac{5}{16}$ | $\frac{5}{16}$ | $\frac{3}{16}$ |
| $\frac{3}{8}$ | $\frac{3}{8}$ | $\frac{1}{4}$ |
| $\frac{1}{2}$ | $\frac{1}{2}$ | $\frac{3}{8}$ |
| $\frac{5}{16}$ | $\frac{5}{16}$ | $\frac{3}{16}$ |
| $\frac{3}{4}$ | $\frac{3}{4}$ | $\frac{9}{16}$ |
| $\frac{9}{16}$ | $\frac{9}{16}$ | $\frac{7}{16}$ |
| $\frac{5}{8}$ | $\frac{5}{8}$ | $\frac{1}{2}$ |
| $\frac{7}{8}$ | $\frac{7}{8}$ | $\frac{3}{8}$ |
| $\frac{7}{8}$ | $\frac{7}{8}$ | $\frac{5}{8}$ |
| $1$ | $1$ | $\frac{11}{16}$ |

be used. Two washers are necessary, as shown in Fig. 25.16b, to ensure complete electrical insulation of the stud from the surrounding metal edges of the access hole. The major diameter of the stud used determines the size of dimension $A$ in Fig. 25.16c, whereas dimension $B$ is the outside of the washer. Dimension $C$ is the outside diameter of the shoulder for which the access hole in the metal must be formed, and dimension $D$ corresponds to the shoulder height. If the height of the shoulder is more than one-half the thickness of the metal, a flat fiber washer may be necessary in place of a second shoulder washer to ensure flush mounting and total insulation. The appropriate size shoulder may be determined from Table 25.5.

The use of machine screws and nuts, as used to mount pc boards into a chassis, shown in Fig. 25.19, is one acceptable mounting arrangement. However, large areas of metal must be removed from the chassis. The sheet metal work could be reduced by the use of *standoffs* and *circuit board supports*, which often find application in prototype as well as production assemblies. This hardware is shown in Fig. 25.17. Straight standoffs are made of plastic for applications requiring electrical insulation and are available in lengths from $\frac{1}{8}$ to several inches. Standoffs, shown in Fig. 25.17a, are threaded at both ends for mounting and will accept a screw size range of No. 4 to No. 10. *Hinged* standoffs

**Figure 25.16** Fiber shoulder washers: (a) common shoulder washers; (b) recommended method of assembly; (c) important shoulder washer dimensions for installation.

*25.8 Special Fasteners and Hardware*

**TABLE 25.5**

Fiber Shoulder Washer Dimensions

| Screw Size | I.D.<br>A | O.D.<br>B | Shoulder Diameter<br>C | Shoulder Height<br>D |
|---|---|---|---|---|
| 3 | 0.110 | 0.052 | 0.187 | 0.031 |
| 4 | 0.136 | 0.052 | 0.187 | 0.031 |
| 5 | 0.136 | 0.312 | 0.187 | 0.031 |
| 6 | 0.140 | 0.375 | 0.237 | 0.031 |
| 8 | 0.172 | 0.375 | 0.246 | 0.031 |
| 10 | 0.196 | 0.375 | 0.308 | 0.031 |
| 12 | 0.250 | 0.500 | 0.312 | 0.028 |
| $\frac{1}{4}$ | 0.265 | 0.500 | 0.365 | 0.031 |
| $\frac{5}{16}$ | 0.375 | 0.750 | 0.500 | 0.031 |
| $\frac{3}{8}$ | 0.385 | 0.625 | 0.500 | 0.031 |

Dimensions $A, B, C, D$ are in inches.

are available in cadmium-plated brass. Both ends of this style standoff are threaded for mounting to accept screw sizes from 4-40 to 8-32.

By drilling holes into the four corners of the pc board, straight standoffs may be used between the board and chassis to support the board away from the chassis surface. The length of standoff is determined by the tallest projection or component extending from the side of the pc board that faces the chassis. A minimum of $\frac{1}{8}$ inch must be added to this dimension to ensure adequate clearance.

The hinged standoff is generally selected to make circuit testing or servicing easier. Two corners of the pc board are fastened to hinged standoffs and the two opposite corners are secured to standard standoffs. The height of the hardware must be selected to allow the board to rest parallel with the chassis surface with sufficient clearance. For servicing, the two screws at the corners of the board secured to the standard standoffs are unfastened. The board can then be lifted upward to an angle of 90 degrees by the pivoting action of the hinged standoffs, thereby exposing wiring and components for testing and repair. This method of mounting not only provides easy access for repair, but also can reduce the overall package size.

*Circuit board supports*, shown in Fig. 25.17b, are available in nylon and other plas-

(a)

(b)

**Figure 25.17**  Hardware for mounting printed circuit boards: (a) threaded standoffs; (b) nylon printed circuit board support.

tic materials. They consist of two snap-lock arrangements, one at each end of the support. For mounting, one end of the support snaps securely into a 0.187-inch-diameter hole drilled into the chassis. These supports provide another convenient method of mounting pc boards to chassis. The board is secured to the upper end of the support by a spring-type locking action. Support heights range from $\frac{3}{16}$ to $\frac{7}{8}$ inch. Circuit board supports can be mounted on all four corners of small boards. More may be used on larger boards for greater support. They have the advantage of allowing boards to be removed and quickly replaced and eliminate the need for removing large areas of metal from the chassis. These circuit board supports can be used in conjunction with connectors mounted for electrical interconnections between the board and chassis, which adds to the versatility of this method of mounting.

*Printed circuit board guides,* shown in Fig. 25.18, are grooved runners made of plastic and designed for use with edge connectors. These guides ensure alignment between the edge of a pc board and a connector and also provide additional support for the board. Generally fastened to the chassis with machine screws and nuts, pc board guides are also available with an adhesive backing for rapid attachment. The grooves in these guides are designed to accept all standard board thicknesses.

The final method of fastening to be considered is the use of adhesives, particularly *epoxy resins.* High-strength adhesives find wide application in electronic packaging since they provide a number of advantages over other methods of bonding. Some of these advantages are elimination of machine operations, provision of electrical insulation and low thermal resistance between parts joined, and uniform stress distribution over the entire bonded area. Although the use of adhesives requires thorough preliminary cleaning of parts and often heat and pressure during the curing period, the advantages of this method of bonding far outweigh the time necessary for its proper use.

Epoxy resins must generally be mixed with a *catalyst* (hardener) prior to use. The parts are measured by volume or weight. Manufacturers' mixing instructions must be consulted for the correct amounts for a specific epoxy since there is a wide variation among the many different adhesive types. The *working time* (time between mixing and initial hardening) must also be predetermined for the specific epoxy to avoid mixing more than can be used during the prescribed time.

For proper bonding, the surfaces must be thoroughly cleaned. Fine emery cloth to remove oxides may be used followed by degreasing with a solvent such as toluene or trichlorethylene. After the epoxy is properly mixed, a thin coating is applied to each of the mating surfaces and pressed together to squeeze out air bubbles and excess adhesive. Improved bonding to aluminum results if the surface is anodized prior to bonding. Depend-

**Figure 25.18** Printed circuit grooved guides.

**Figure 25.19** Completed mechanical assembly of all parts to a chassis and front panel element.

ing on the epoxy used, curing time will vary from 5 minutes to 24 hours at room temperature. These periods can be reduced if the bonded surfaces are baked in an oven, but care must be exercised not to exceed the temperature range of the epoxy. In addition, baking is not recommended when bonding devices to heat sinks because of the possibility of damaging components.

Epoxy is used for bonding when the stresses encountered will be only tensile or shear. Since the shelf life of many adhesives is often less than one year, the expiration date must not be exceeded for sound bonds. In addition, adhesives should be stored in a cool area to ensure longest possible shelf life.

Hardware and pc boards are shown assembled to a chassis and front panel element in Fig. 25.19 using materials and techniques discussed in this section.

After assembly is completed, wiring of the boards and chassis-mounted hardware is necessary to complete a system. In the next chapter we consider wire and wiring assembly techniques.

## EXERCISES

### A. Questions

25.1    List the four major factors in electronic packaging that are required for quality workmanship.

25.2    List the eight features that distinguish machine screws.

25.3    What is the advantage of the use of Phillips head screws over slotted head screws?

25.4    Define the term *fit* as it pertains to machine screws.

25.5    What is the major difference between thread-forming and thread-cutting screws?

25.6    For fastening two pieces of 18-gauge aluminum with a No. 6 sheet metal screw, determine the required clearance hole size and tap hole size. Refer to Appendix IV and Fig. 25.7.

25.7    What is one disadvantage of using pop rivets rather than screws and nuts?

25.8    What hole diameter is required to install a grommet having an inside diameter of $\frac{3}{16}$ inch? Refer to Fig. 25.15b and Table 25.4.

## B. True or False

Circle T if the statement is true, or F if any part of the statement is false.

25.1    Flat head machine screws are recommended for use on thin sheet metal.    **T    F**

25.2    The Allen drive configuration has a more positive tool positioning and gripping action than the Phillips head.    **T    F**

25.3    The UNC system has fewer threads per inch than the UNF system.    **T    F**

25.4    For most electronic assembly applications, the Class 3 machine threads are standard.    **T    F**

25.5    The tooth-type lock washer should not be reused when removed.    **T    F**

25.6    Split lock washers are more effective than any other type of lock washer.    **T    F**

25.7    Self-tapping screws are intended to be used for sheet metal fastening applications.    **T    F**

25.8    Pop rivets are used where a permanent type of fastener is required.    **T    F**

25.9    The diameter of a hole for the installation of a grommet is the same dimension as the grommet size designation.    **T    F**

25.10   Hinged standoffs are used to improve access to wiring and components.    **T    F**

## C. Multiple Choice

Circle the correct answer for each statement.

25.1    Some common machine screw head styles are binder, round, and (*oval, slotted*).

25.2    For equivalent screw sizes, (*UNC, UNF*) screws have more threads per inch.

25.3    Machine screws with a (*Class 1, Class 2, Class 3*) screw thread are considered as having a snug fit.

25.4    Pop rivets are a (*temporary, permanent*) type of fastener.

25.5    Shoulder washers are usually made from (*metal, fiber*) in order to provide electrical (*isolation, conduction*).

## D. Matching Columns

Match each item in column A to the most appropriate item in column B.

| COLUMN A | COLUMN B |
|---|---|
| 1.  Class 2 | a.  Wire protector |
| 2.  Hexagonal drive | b.  Washer |
| 3.  Shoulder washer | c.  Fine thread |
| 4.  Lock | d.  Allen |
| 5.  UNF | e.  Insulation |
| 6.  Split-joint | f.  Free fit |
| 7.  Grommet | g.  Pliers |

**Figure 25.20**

### E. Problems

25.1    Using an appropriate epoxy, bond the two halves of an ink eraser to an aluminum handle such as that shown in Fig. 25.20. The handle is fabricated from 16-gauge aluminum having blank dimensions of $\frac{3}{4}$ inch by 10 inches and bent around a $\frac{5}{8}$-inch-diameter dowel.

25.2    Select an appropriate epoxy and seal the four corners of the soldering iron tip cleaning sponge tray constructed in Problem 17.1.

25.3    Mount the lamp (L1), on–off switch (S8), and battery clip for the seven-fixed-interval UJT timer printed circuit designed and fabricated in Problems 6.2, 10.2, 16.2, and 17.2 onto the chassis fabricated in Problem 23.4. The pc board may be mounted onto the chassis with standoffs or by flush mounting using the appropriate size mounting hole. Consideration must be given to providing access to the pc board-mounted time interval switch. Wiring among chassis-mounted components and hardware and the pc board is done in Problem 26.5.

25.4    The meter for the tachometer printed circuit designed and fabricated in Problems 6.4, 10.4, 16.4, and 17.4 may be mounted onto a panel of the same design as that shown in Fig. 20.15. The pc board is to be mounted with proper length spacers and mounting screws to the rear of the meter, which in turn is secured to the panel. Wiring for the installation of the tachometer is done in Problem 26.6.

# Wire and Interconnection Techniques 26

*Upon completion of this chapter on hookup wire and interconnection techniques, the student should be able to*

- Know the composition, configurations, and size designations of common hookup wire.
- Calculate the resistance of a given length of wire.
- Measure the diameter of wire.
- Be familiar with the various types of wire insulation material and their characteristics.
- Select the proper type of wire insulation for a specific application.
- Know the common types of special wire configurations and their applications.
- Be familiar with the wire color-coding system.
- Use mechanical strippers to remove wire insulation.
- Prepare wire for soldering.
- Form proper wire connections on a variety of terminals.
- Use crimp-type terminals.
- Use manually operated wire-wrapping and dewrapping tools.
- Understand the use of semiautomatic and fully automatic wire-wrapping machines.
- Use proper bussing techniques on wire-wrapped terminals.

## 26.0 INTRODUCTION

One of the least emphasized and perhaps the most important aspect of packaging is the selection of the most suitable hookup wire for a specific assembly. From the standpoint of reliability, this selection must be based on such factors as *conductor material and plating, conductor size, type of insulation, flexibility, environmental factors,* and *current* and *voltage limits,* as well as *cost.* This chapter provides guidelines to aid in evaluating and selecting wire. Coupled with the problem of selecting are the methods for *interfacing interconnections.* Since there are a variety of standard interconnections (in addition to many special types), this chapter also includes pertinent information for evaluating the suitability of various connection methods.

Specific information also is provided here on common conductor materials and platings and American Wire Gauge (AWG) classifications of size, along with a comparison of solid wire with various types of stranded wire. Typical types of insulation are examined with respect to their application, including color-coding methods. Additional information is provided on current, voltage, and mechanical and environmental criteria to

**453**

allow for a total understanding of wire selection from the many types commercially available. Special wire types are also discussed.

Methods of interconnections are introduced, and the tools used for wire preparation and mechanical assembly are described, such as *strippers* and *diagonal cutters*. *Wrap-around* and *feed-through* connections as they apply to swaged solder-type terminals, eyelet-type terminals, and solder lugs are discussed and expanded to include special hardware connections, such as plugs and jacks. Solderless wiring techniques, such as solderless or crimp-type connectors and wire-wrap connections are set forth. Finally, information on cable ties, shrink tubing, and cable clamps is presented.

*Caution: Prior to performing any of the operations described in this chapter, approved safety glasses must be worn.* Also, the section on safety outlined at the beginning of the book should be reviewed for more specific information.

## 26.1 CONDUCTOR MATERIALS AND WIRE CONFIGURATIONS

Most of the wire used for electronic interconnections is fabricated from *annealed* (softened) copper. Copper used for this purpose is termed *Electrolytic Tough Pitch* (E.T.P.). Although other materials, such as aluminum, steel, and silver, are employed in wire construction, they are not common to electronics fabrication. Alloys of copper have been developed, however, that display superior characteristics when compared to pure copper. Although pure copper possesses excellent electrical conductivity characteristics as well as being malleable and ductile, it has the pronounced disadvantage of being highly susceptible to fractures under vibration and flexing conditions. Modern high-temperature annealing techniques, however, have made it possible to develop high-strength alloys with improved fatigue life. Some of these common high-strength alloys are *cadmium–copper, chromium–copper,* and *cadmium–chromium–copper.* One disadvantage in using these alloyed wires is that they are available only with coatings of silver or nickel, which are both expensive and difficult to solder. Tin-coated high-strength alloys are, unfortunately, not readily available.

Copper wire is usually coated, because pure copper oxidizes quickly when exposed to the atmosphere. *Tin* is the most common coating, although *silver* and *nickel* coatings are used in special applications. Tin-coated wire, recommended for applications wherein environmental temperatures will not exceed 150°C (302°F), is the least expensive of the three coatings mentioned. In addition, tin coating improves solderability. Silver coating, recommended for use in temperatures ranging from approximately 150 to 200°C (302 to 395°F) greatly improves solderability but tends to corrode at elevated temperatures. Silver is the most expensive coating of those mentioned. Nickel coating is capable of oxidation protection to temperatures of up to 300°C (575°F) but requires activated fluxes and high-temperature soldering techniques.

Tin-coated wire is by far the most commonly used in the electronics industry. Tin is applied to the bare wire by either *dipping* or *electroplating.* Dipping is the least expensive method and is used if uniform plating thickness is not critical. Electroplated tin should be specified if close tolerance automatic stripping is to be employed, since uniform plating is achievable by this method.

Wire used in the electronics industry can be divided into two groups: *solid* and *stranded.* Generally, solid wire that is comparable to a specific size of stranded wire is less expensive to manufacture and subsequently less expensive to purchase. Solid wire, however, does not possess the flexibility characteristics or the fatigue life of stranded wire. Whereas solid wire tends to fracture even under mild flexing, stranded wire remains highly flexible. Because much of the electronic equipment produced today is exposed to some form of vibration under normal use or has wiring flexibility requirements during assembly, stranded wire is the type most often specified.

Solid wire finds extensive application in the fabrication of leads for small components, such as resistors, capacitors, and solid-state devices. Hookup applications of solid wire are generally limited to conditions wherein flexibility is not a criterion or lead runs are to be short, rigid terminations such as the direct connection of two closely spaced swaged turret-type terminals on a pc board. Solid, uninsulated tinned copper wire used as a short jumper or as a common circuit point is termed *bus* wire.

When longer lengths of bus wire are needed, it is usually recommended that the wire be straightened prior to its use. The best way to straighten wire is to secure one end of it in a vise and the other end in the chuck of a hand drill (eggbeater). All that is usually required to straighten the wire is 8 to 10 complete turns of the drill in one direction followed by an equal number of turns in the opposite direction while applying continuous tension to the wire.

Stranded wire is classified as either *bunch* or *concentric*. Bunch-type stranded wire is constructed of several small-diameter solid wires "bunched" together without regard to symmetry. Although the cost of bunch wire is low and it is extremely flexible, it does not have a consistent diameter. Because of this disadvantage, it is not recommended, because modern insulation stripping tools may nick or break individual strands, which will weaken the wire as well as reduce current capacity.

Concentric-type stranded wire overcomes the disadvantages of bunch wire in that its consistent circular cross section makes it suitable to manual as well as automatic stripping techniques. *Counterdirectional* concentric stranded wire is a multilayer wire with alternate layers of strands rotating or twisting in an opposite direction to the next succeeding lower layer. *Unidirectional* concentric stranded wire is also a multilayer wire, but each layer of strands twists in the same direction but with a different degree of *pitch* (angle at which strands cross the axis of the overall wire) as they are twisted. Unidirectional stranded wire is more flexible than the counterdirectional type, but its diameter is not so precisely uniform.

Solid and stranded wire each have their specific size designation systems. Solid wire is designated in terms of its cross-sectional area and diameter by *gauge number*. The most commonly used designation is the *American Wire Gauge* (AWG) standard system. Appendix VII lists the AWG numbers of solid bare copper wire with its equivalent cross-sectional area and diameter for gauge Nos. 00 to 50.

To better understand the information provided in wire tables, certain relationships need to be examined. In the electrical field, characteristics of different wire gauges and lengths are, for convenience, compared to a standard so that different wire sizes may be readily compared. The accepted standard is a piece of solid annealed copper wire with a diameter of 0.001 inch (1 mil) and a length of 1 foot whose resistance is 10.37 ohms at 20°C. (This standard is often referred to as the *mil-foot*.) Since most electrical conductors have round cross sections, a system of circular measure has been established to reduce the tedious task of comparing conductor characteristics. The parameters involved in such a comparison are resistance, length, and cross-sectional area. These relationships are expressed mathematically as

$$\frac{\text{resistance wire I}}{\text{resistance wire II}} = \frac{\text{length wire I}}{\text{length wire II}} \times \frac{\text{area wire II}}{\text{area wire I}} \qquad (26.1)$$

The expression for determining the area of a circle is $A = \frac{\pi}{4}(\text{diameter})^2$. When the area of two conductors is compared, as above, $\frac{\pi}{4}$ is a constant for both wire I and wire II. Therefore, Equation (26.1) may be simplified as follows:

$$\frac{\text{resistance wire I}}{\text{resistance wire II}} = \frac{\text{length wire I}}{\text{length wire II}} \times \frac{(\text{diameter})^2 \text{ wire II}}{(\text{diameter})^2 \text{ wire I}} \qquad (26.2)$$

The term (*diameter*)$^2$ represents the cross-sectional area of a round wire. This term is designated as the *area* in *circular mills* (CM), where the area now is equal to the square of the diameter ($A = d^2$). All wire tables refer to diameters in mils and to cross-sectional areas in circular mils.

To determine the resistance of copper wire, the following relationship is used:

$$\text{resistance} = \frac{10.37 \times \text{length (feet)}}{\text{diameter}^2 \text{ (CM)}} \tag{26.3}$$

If, for example, it is necessary to find the resistance of 50 feet of AWG No. 18 wire, Appendix VII shows that the area of this wire is 1624 CM. Substituting these known values into Equation (26.3) yields

$$R = \frac{(10.37)(50)}{1624} = 0.32 \text{ ohm} \tag{26.4}$$

The wire diameter in mils can be converted to inches merely by moving the decimal point three places to the left (multiply by 0.001). For example, AWG No. 18 wire has a diameter of 40.30 mils, which is equal to 0.0403 inch.

Because one of the most important criteria for selecting wire is its current-carrying capacity, wire tables provide the cross-sectional areas of each wire gauge expressed in CM. The fourth column in Appendix VII lists the maximum recommended current capacity in amperes for common wires used in electronics. These values were obtained by the "rule of thumb" relationship that each 500-CM cross section will safely accommodate 1 ampere of current. For example, AWG No. 10 wire, having a cross section of approximately 10,000 CM, can handle 20 amperes. AWG No. 13 wire, which is 3 gauge numbers higher, reduces the cross-sectional area by one-half to approximately 5000 CM and therefore is capable of handling one-half the current, or 10 amperes. Again, AWG No. 23 (10 gauge numbers higher than No. 13), with its cross section of approximately 500 CM, is capable of handling 1 ampere. From these relationships, the following guidelines can be used: For each *decrease* or *increase* of three gauge numbers, the CM area doubles or halves, respectively, as does its current-carrying capacity. Correspondingly, a gauge number decrease or increase by a factor of 10 will increase or decrease the cross section by 10 or $\frac{1}{10}$, respectively, as will its current capacity. (Refer to Appendix VII.) Common gauges of hookup wire used in the electronics industry range from AWG No. 10 to No. 26, and their current-carrying capacities, from approximately 20 amperes to $\frac{1}{2}$ ampere.

The AWG standard numbering system was developed principally for bare solid-copper wire. This system is also used to designate stranded wire together with an additional number to specify the number of strands composing the wire. For example, in the designation 7/34, the first number specifies that there are a total of 7 strands of wire and the second number indicates that each strand is AWG No. 34. To determine the equivalent AWG solid-wire size to a specific stranded wire, the cross-sectional area of each strand is multiplied by the number of strands in the overall wire. For example, to determine the equivalent solid-wire size of No. 7/34 stranded wire, the cross-sectional area of No. 34 is found to be 40 CM (from Appendix VII). The 7 strands constitute a total cross section of $7 \times 40$, or 280 CM. Appendix VII reveals that its closest equivalent is AWG No. 26 solid wire. Therefore, both No. 7/34 and AWG No. 26 have a maximum recommended current-carrying capacity of approximately 0.5 ampere. Table 26.1 lists some of the common stranded-wire types with their equivalent AWG numbers.

To determine the AWG wire size of either solid or stranded wire, a *wire gauge,* such as that shown in Fig. 26.1a, may be used. Each slot in the gauge is accompanied by the corresponding AWG number size and a relief hole having a diameter greater than the slot width. The purpose of the relief hole is to allow for proper measurement "feel" and

**TABLE 26.1**

Stranded-Wire Configurations
to AWG Numbers

| Equivalent AWG Number | Stranding |
|:---:|:---:|
| 10 | 37/26 |
| 12 | 19/25 |
| 14 | 19/27 |
| 14 | 37/29 |
| 16 | 19/29 |
| 16 | 27/30 |
| 18 | 7/26 |
| 18 | 19/30 |
| 18 | 27/32 |
| 20 | 7/28 |
| 20 | 19/32 |
| 22 | 7/30 |
| 22 | 19/34 |
| 24 | 7/32 |
| 24 | 19/36 |
| 26 | 7/34 |
| 26 | 19/38 |

also for damage-free removal of the wire from the gauge after measurement. The proper method of positioning the gauge and wire to be measured is shown in Fig. 26.1b. Notice from the figure that when measuring stranded wire, the insulation is not completely removed as it is for solid wire. Rather, a section of insulation is separated to expose the conductor. This section should not initially be cut closer than 1 inch from the end of the wire to prevent the individual strands from separating or flattening as the wire is placed in the gauge. When removing the section of insulation, the underlying wires must not be twisted, for the diameter of the conductor would be altered and an erroneous reading would result.

When measuring stranded wire, the wire gauge reading will be slightly larger than the corresponding AWG number for solid wire because of the slight increase in diameter

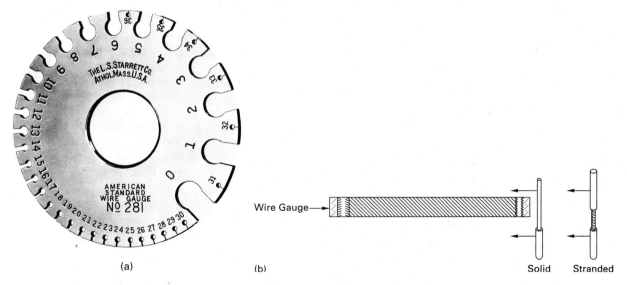

(a)          (b)          Solid   Stranded

**Figure 26.1**   Wire gauge and measuring technique: (a) wire gauge to measure AWG sizes from No. 0 to No. 36, *courtesy of the L.S. Starrett Company, Athol, MA;* (b) perpendicular gauge and wire positioning for accurate measurement in gauge slot.

caused by the spaces between strands. Therefore, when measuring stranded wire, the gauge slot into which the wire fits will be one AWG number *less* than the actual size and will also yield an *odd*-numbered AWG value. For example, a stranded AWG No. 20 wire will properly fit into the gauge slot marked AWG No. 19. Therefore, this rule of thumb indicates that the correct size of this wire is AWG No. 20. (It may be well to emphasize at this point that *odd* AWG number sizes are not commonly stocked and, for this reason, should not be specified.) An optional method of measuring stranded wire is to measure *one strand* in the gauge, count the number of strands in the total conductor, and refer to Table 26.1 to determine the equivalent AWG number.

To provide additional help in selecting wire, we use the following example to determine the most suitable type of wire plating, type (stranded versus solid), and size.

*Example.* Power amplifier requirements:

*Current.* 1.6 amperes dc maximum

*Operating temperature range.* 70°C (see Sec. 26.2)

*Flexibility requirements.* pc board is rigidly supported—no vibration problems—minimum flexibility required for assembly

*Solution.* Since flexibility is not a prime factor, solid wire is selected because it is the least expensive. The temperature requirements are minimal, thus allowing for the use of tin coating that will further reduce costs and ensure availability. The size of the wire will now be selected based on the current requirements of the circuit. Most of the interconnections are signal lines in which current demands are very low. The only wires that will be expected to handle 1.6 amperes are those originating from the power supply. Appendix VII indicates that AWG No. 20 wire will handle 2 amperes. All signal wires could be a great deal smaller than this; but for the sake of uniformity and simplicity, a realistic choice would be AWG No. 20 wire for *all* interconnections. Although the cost of wire is proportional to size, the savings realized by the choice of a smaller wire are negligible.

## 26.2 INSULATION MATERIAL

Hookup wire used in making interconnections in electronic equipment must be provided with an insulation to prevent short circuits and to protect the conductor. The only exception to this rule involves the use of bus wire, which, because of rigidity and short length, poses no insulation problems.

The selection of the most suitable insulation for a specific application is based on three criteria: *electrical, mechanical,* and *environmental.* Each of these criteria is discussed.

Common wire insulations and insulation coatings are divided into three main categories: *rubbers, vinyls,* and *special high-temperature* or *special-application materials.* A comparative tabulation of various insulating materials is given in Table 26.2. This table compares the following mechanical and physical properties:

1. Voltage breakdown
2. Insulation resistance
3. Low dissipation factor
4. Abrasion resistance
5. Flexibility
6. Cut-through resistance
7. Water absorption resistance
8. Flame resistance
9. Acid resistance
10. Useful temperature range

**TABLE 26.2**

Comparative Ratings for Wire Insulation

| Family | Material | Common Designation | Voltage Breakdown | Insulation Resistance | Low Dissipation Factor | Abrasion Resistance | Flexibility | Cut-Through Resistance | Water Absorption Resistance | Flame Resistance | Acid Resistance | Useful Temperature Range (°C) |
|---|---|---|---|---|---|---|---|---|---|---|---|---|
| | Silicone rubber | Silicone | Exc. | Exc. | Good | Fair/poor | Exc. | Poor | Fair | Fair | Fair | − 60 to 200 |
| Rubber | Polychloroprene rubber | Neoprene | Fair | Poor | Fair | Exc. | Exc. | Good | Good | Good | Good | − 30 to 80 |
| | Butyl rubber | Butyl | Good | Good | Good | Good/exc. | Exc. | Fair | Good | Poor | Good/exc. | − 40 to 85 |
| | Polyvinyl chloride | PVC | Good | Good | Fair | Good | Good | Good | Good | Good | Good/exc. | − 55 to 105 |
| Polyolefin (vinyl) | Polyethylene | PE | Exc. | Exc. | Good | Fair/good | Fair | Good | Exc. | Poor | Good/exc. | − 60 to 80 |
| | Polypropylene | Poly | Good/exc. | Exc. | Good | Good/exc. | Poor | Good | Exc. | Fair/poor | Exc. | − 20 to 125 |
| | Polytetrafluoroethylene | Teflon* (TFE) | Good | Good | Exc. | Exc./good | Good | Fair | Exc. | Exc. | Exc. | − 70 to 250 |
| | Monochlorotrifluoroethylene | Kel-F† | Good | Exc. | Fair | Good/exc. | Good | Good | Exc. | Exc. | Exc. | − 80 to 200 |
| Fluorocarbon | Polyvinylidine fluoride | Kynar‡ | Exc. | Fair | Good | Exc. | Good | Exc. | Good | Good | Good | − 65 to 125 |
| | Polyurethane | Urethane | Good | Poor | Fair | Exc. | Exc. | Good | Poor | Poor | Fair | − 55 to 80 |
| | Polyamide | Nylon | Poor | Poor | Fair | Exc. | Poor | Good | Fair/poor | Good/fair | Fair | − 40 to 120 |
| | Polyester film | Mylar* | Exc. | Exc. | Poor | Good | Fair | Exc. | Good | Fair/poor | Good | − 65 to 120 |

Ratings are in decreasing order: excellent (exc.), good, fair, poor.

*Trademark of DuPont.

†Trademark of Minnesota Mining & Manufacturing Company.

‡Trademark of Pennsalt Chemical Corp.

The *voltage breakdown* and *insulation resistance* are the two factors that are of key importance in high-voltage applications. Insulations designated *excellent* are the only ones that should be considered for this purpose. As seen from Table 26.2, silicone, polyethylene (PE), and Mylar possess high corona resistance and are commonly used for high-voltage cables. Neoprene, urethane, and nylon insulations would not be suitable for this application because of their poor rating. The amount of vibration that a wire is exposed to is directly related to its insulation voltage breakdown rating. Therefore, this factor needs to be considered in the choice of insulation.

Signal attenuation at high frequencies is a function of the insulation's *dissipation factor*. Generally, insulations with a low dissipation factor are preferable. Teflon has the lowest dissipation factor among those listed in Table 26.2.

Mechanical considerations, such as *abrasion resistance, flexibility,* and *cut-through resistance,* represent a compromise between cost versus installation ease. Abrasion and cut-through resistance are generally directly related—the higher the value of one, the higher the value of the other. Of the insulations listed in Table 26.2, Kynar is the toughest, having excellent abrasion and cut-through resistance characteristics. The insulation flexibility becomes an extremely important consideration, especially in high-density packaging for ease of installation. The *rubber* family (silicon, neoprene, and butyl) exhibits the highest flexibility and the *vinyls* (PE and poly) the lowest.

*Water-absorption resistance* is an environmental consideration. Rubber and urethane materials are more susceptible to moisture than the vinyl materials. Teflon materials are the least susceptible to water absorption.

*Flame-resistant* materials find application in electronic equipment in which explosions or flammable conditions may be encountered. Teflon and Kel-F are the materials specified for this application, since they do not support combustion. All the other insulation materials listed in Table 26.2 will support combustion in varying degrees.

Insulations that are highly *acid resistant,* such as butyl rubber, polypropylene, and Teflon, find wide application in the space industry. Urethane, nylon, and silicone rubber are inferior in this property.

The useful temperature range for PVC, nylon, and PE is narrow, whereas for Teflon and Kel-F it is wide. In considering this property, it is important to use good judgment in terms of cost. For example, it would not be advisable to choose Teflon (upper limit 250°C) based on its superior temperature characteristics if the temperature range of PE (upper limit of 80°C), as well as its other characteristics, will satisfy the design requirements because Teflon is considerably more expensive. From the wide range of materials available, it is evident that the selection of the most appropriate wire insulation for a specific application is a compromise between specifications and cost.

To illustrate the use of Table 26.2, the following problem is considered:

*Circuit.*   stereo power amplifier

*Electrical specifications*
    Voltage 36 volts dc maximum
    Current 1.6 amperes dc maximum
    Frequency 20 to 20 kilohertz (kHz)

*Mechanical specifications*
    Stationary
    No expected high-level vibrations
    Low-density packaging
    All leads pass through grommets

*Environmental specifications*
    Temperature range +50 to 90°F (+10 to 32°C)
    Humidity 40% typical

These specifications appear to present no critical circuit requirements insofar as wire insulation selection is concerned. Since voltage and current requirements are low, any insulation with a *voltage breakdown* and *insulation resistance* comparison rating of *good* and *excellent* is acceptable. This stipulation eliminates neoprene, Kynar, urethane, and nylon from further consideration.

With a frequency range specified at 20 hertz to 20 kHz, any insulation with a dissipation factor rating of *good* or *excellent* will be suitable. Low-dissipation-loss factors become critical only at higher frequencies.

Because all wires will be passed through rubber grommets, the possibility of abrasion is nonexistent and the quality of *abrasion resistance* need not be considered in this evaluation even if a *fair* or *poor* rating is assigned a specific material. Therefore, for this consideration, any of the wire insulations listed are suitable.

*Flexibility* also is not critical because of the low-density packaging of the circuit and the resultant uncomplicated wiring. Therefore, no insulation material need be rejected because of its flexibility rating. To reduce assembly time, especially if large-scale production is considered, an insulation with a comparison rating for *cut-through resistance* of *good* or *excellent* should be considered. This eliminates silicone, butyl, and Teflon from further consideration. Since the use of this type of power amplifier is usually in the home or other location wherein temperature and humidity are controlled at levels that exclude critical environmental problems, any insulation with a *water-absorption resistance* rating of *good* or *excellent* is acceptable. *Flame* and *acid resistance* is, of course, of no consideration in the intended environment. Therefore, no insulation, even with a *fair* rating for either of these categories, is omitted from consideration.

Finally, the expected ambient temperature of the unit is approximately +10 to 32°C, with a factor of 2 used in derating on each end of the temperature range to allow for extreme conditions (the inside case temperature will increase under full power output). The range of derated temperatures now considered becomes 5 to 64°C. Table 26.2 shows that all the insulations listed have *useful temperature ranges* that are well within these specifications. As a result, no insulation material will be eliminated because of temperature limitations.

An evaluation chart is provided in Table 26.3 to show more clearly the procedure for eliminating insulation materials that are below minimum ratings. The *Analysis Results* column of Table 26.3 indicates that either polyethylene (PE) or polypropylene (poly) plastic vinyl insulation is suitable for the example problem. The final selection between these two materials now becomes a matter of cost and availability.

For purposes of identification, hookup wire insulation is provided with a color code. This provides for rapid identification of wires, which is an indispensible aid when working with cables and harnesses. Color-coding is extremely useful in both assembly and in troubleshooting. Military and industrial standards have been established for color-coding insulation. These standards are listed in Table 26.4. Similar to the resistance color code adopted from EIA (Electronic Industries Association) standards, the solid color of the insulation is designated by numbers 0 through 9 and represents the first digit of the color-code number. With 10 basic colors, 10 numerical possibilities are available to identify individual wires. For complex wiring involving more than 10 leads that must be identified, insulation is also available with a broad tracer band that is a different color from the body color and spirals the length of the wire. These tracer colors now extend the available different color combinations to 100. When more than 100 wires are involved, as in the case of complex harnesses, insulation is available with a *second* tracer that is narrower than the first. Both tracers are closely spaced, running parallel with each other, and spiral the entire length of the wire. The use of the second tracer extends the numerical coding capabilities to 910 possible combinations. To illustrate the numbering system, a wire with a *red* body color, broad *green* tracer, and narrow *orange* tracer (red/green/orange) can be classified by the three-digit number 253.

**TABLE 26.3**

Sample Evaluation Chart for the Selection of Wire Insulation

| Material | Voltage Breakdown | Insulation Resistance | Low Dissipation Factor | Abrasion Resistance | Flexibility | Cut-Through Resistance | Water Absorption Resistance | Flame Resistance | Acid Resistance | Useful Temperature Range | Analysis Results* |
|---|---|---|---|---|---|---|---|---|---|---|---|
| Silicone | ✓ | ✓ | † | ✓ | ✓ | † | † | ✓ | ✓ | ✓ | 2 |
| Neoprene | † | † | † | ✓ | ✓ | ✓ | ✓ | ✓ | ✓ | ✓ | 3 |
| Butyl | ✓ | † | ✓ | ✓ | ✓ | † | ✓ | ✓ | ✓ | ✓ | 1 |
| PVC | ✓ | ✓ | † | ✓ | ✓ | ✓ | ✓ | ✓ | ✓ | ✓ | 1 |
| PE | ✓ | ✓ | ✓ | ✓ | ✓ | ✓ | ✓ | ✓ | ✓ | ✓ | 0—suitable |
| Poly | ✓ | ✓ | ✓ | ✓ | ✓ | ✓ | ✓ | ✓ | ✓ | ✓ | 0—suitable |
| Teflon (TFE) | ✓ | ✓ | † | ✓ | ✓ | † | ✓ | ✓ | ✓ | ✓ | 1 |
| Kel-F | ✓ | † | ✓ | ✓ | ✓ | ✓ | ✓ | ✓ | ✓ | ✓ | 1 |
| Kynar | ✓ | † | † | ✓ | ✓ | ✓ | ✓ | ✓ | ✓ | ✓ | 1 |
| Urethane | † | † | † | ✓ | ✓ | ✓ | † | ✓ | ✓ | ✓ | 3 |
| Nylon | † | † | † | ✓ | ✓ | ✓ | ✓ | ✓ | ✓ | ✓ | 3 |
| Mylar | ✓ | ✓ | † | ✓ | ✓ | ✓ | ✓ | ✓ | ✓ | ✓ | 1 |

*Total number of insulation characteristics that do not meet minimum required ratings

✓Insulation characteristic meets minimum required rating.

†Insulation characteristic below example problem minimum required rating.

TABLE 26.4

Color-Coding Standards for Wire Insulation

| Color Code | | Numeric Wire Code | |
|---|---|---|---|
| | | Numerical Possibilities | Description |
| 0 | Black | 10 | Solid color insulation |
| 1 | Brown | 90 | Solid color with one |
| 2 | Red | | broad trace |
| 3 | Orange | | Solid color with one |
| 4 | Yellow | 810 | broad trace and |
| 5 | Green | | one narrow trace |
| 6 | Blue | | |
| 7 | Violet | | (910 total numeric combinations) |
| 8 | Gray | | |
| 9 | White | | |

Standards also have been established that do not employ a numbering system but rather designate wire color with respect to circuit function. Two such standards are the Government Standard MIL-STD-122 color code for chassis wiring and the EIA color code for power transformers. These classifications are shown in Table 26.5.

## 26.3 SPECIAL WIRE CONFIGURATIONS

There are several readily available wire configurations that are used for special applications. *Shielded wire,* shown in Fig. 26.2a, is a variation of hookup wire previously discussed. It consists of an insulated length of hookup wire enclosed in a conductive envelope in the form of a braided wire shield. The shield is formed from fine strands of tinned, annealed copper wire. The strands are woven into a braid that provides approximately 85% coverage of the underlying insulated conductor. Shielded wire is also available with an outer insulation covering to prevent shorts when running the wire between termination points. Another variation of shielded wire has *two* inner conductors twisted around each other. This twist improves unwanted signal cancellation.

Shielded wire is employed in audio circuits at the input stage where the signal level

**TABLE 26.5**

Wire Color Identification Related to Circuit Function

| Functional Color Coding Chassis Wiring | | | Color Code for Power Transformers | | |
|---|---|---|---|---|---|
| Color | Code Number | Function | Winding | | Color Code |
| Black | 0 | Grounds, grounded elements, and returns | Primary { Untapped | | Black |
| Brown | 1 | Heaters, or filaments not connected to ground | Tapped | | Common black, tap black/yellow |
| Red | 2 | Power supply, $+V_{cc}$ or B+ | High-voltage plate | | Red |
| Orange | 3 | Screen grid | High-voltage center tap | | Red/yellow |
| Yellow | 4 | Transistor emitters or cathodes | Rectifier filament | | Yellow |
| Green | 5 | Transistor bases and control grids | Rectifier filament center-tap | | Yellow/blue |
| Blue | 6 | Transistor collectors, anodes for semiconductor elements and tube plates | Filament (No. 1) | | Green |
| | | | Filament (No.1) center tap | | Green/yellow |
| Violet | 7 | Power supply, $-V_{cc}$ or B− | Filament (No. 2) | | Brown |
| Gray | 8 | Ac power lines | Filament (No. 2) center tap | | Brown/yellow |
| White | 9 | Miscellaneous, above-ground returns | Filament (No. 3) | | Gray |
| | | | Filament (No. 3) center tap | | Gray/yellow |

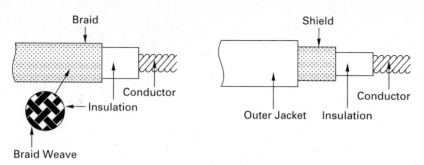

**Figure 26.2**  (a) Shielded wire; (b) coaxial cable.

is low and undesired noise (typically 60-hertz hum) or feedback from the output stage of a high-gain amplifier could produce unwanted signals. The shield is either connected to the chassis at the circuit input or it may be connected to the chassis at both ends to provide the most effective method of minimizing undesired signal pickup.

*Coaxial cable* (coax), shown in Fig. 26.2b, is used exclusively in rf circuits. It is similar in construction to shielded wire, with the outer insulating jacket isolating the shield from ground. The major difference between coaxial cable and shielded wire is in their electrical characteristics. Coax is designed specifically to transmit rf energy, from one point to another, with minimum loss (attenuation). Insulation material and thickness is controlled with extreme accuracy during manufacture to produce cables possessing 50, 75, or 95 ohms characteristic impedance for proper matching. Shielded wire, on the other hand, is intended for low-frequency applications in which its impedance is not critical.

Losses from *skin effect* at high frequencies must be minimized when using coaxial cable. Skin effect is the term used to describe the type of loss or attenuation of the signal caused by the current traveling along the surface (skin) of the conductor and partly into the adjacent insulation. Because of its outstanding insulating characteristics at high frequencies, nylon is the type of insulation most often used in the manufacture of coaxial cables.

Another special wire configuration is *Litzendraht* (Litz) wire, which is used in the manufacture of tuning coils. At short-wave frequencies, this type of wire has low skin effect, owing to its configuration. Litz wire is a stranded conductor containing 25 or more extremely fine wire strands. Each strand is independently insulated with cotton insulation and wound from the center to the outer surface of the overall conductor. This technique of winding tends to equalize the skin effect problem.

## 26.4  TOOLS FOR WIRE PREPARATION AND ASSEMBLY

To form a sound electrical connection between the hookup wire used for interconnections and the terminal to which it is to be soldered, *all* insulation must be completely removed from the end of the wire to expose the conductor. In addition, it may become necessary to remove contaminants, such as dirt, finger oils, or oxidation, from the bare wire. (Oxidation will occur if the bare conductor has been exposed to the atmosphere for a prolonged time before soldering.) It is vitally important when preparing to make a solder connection that all surfaces to be soldered (alloyed) are absolutely free of all contaminants.

Mechanical wire strippers, such as the two styles shown in Fig. 26.3, are among the most common tools used to remove wire insulation. Each has its advantages and disadvantages with regard to resulting quality and ease of operation. The *step-adjustable* stripper is shown in Fig. 26.3a. All that is required to set the jaws is to rotate the *selector step wheel* until the desired gauge number is oriented with the *stop pin*. Its obvious disadvantage is that it is capable of stripping only eight gauge wire sizes (AWG No. 12 to AWG

(a)

(b)

**Figure 26.3** Mechanical wire strippers: (a) step adjustable; (b) insulation removal with automatic wire stripper.

No. 26). To minimize the possibility of conductor damage when using this stripper, it should be held perpendicular to the wire when pulling off the insulation.

The most popular mechanical stripper is the *automatic* style shown in Fig. 26.3b. This stripper has six marked cutting positions on the jaws to accommodate various gauge sizes. The wire to be stripped is placed into the appropriate cutting jaw position and between the gripping jaws. With the wire held in position, the handles are gently squeezed together. This causes the movable upper cutting jaws to first press the insulation down into the lower stationary cutting jaws. Additional squeezing of the handles causes the insulation to be severed about the conductor while the upper movable gripping jaw firmly grasps the wire with the lower stationary gripping jaw. Further squeezing of the handle causes both sets of jaws, while engaged, to separate, thereby removing the insulation from the end of the wire (Fig. 26.3b). As pressure on the handles is removed, both sets of jaws disengage, thus releasing the stripped wire. These strippers are also available with a stop that, when set to the desired length of strip, will consistently remove the same length of insulation from wires.

*Thermal* wire strippers, although not popular in prototype construction, are another means of removing the insulation from wires. They consist of a heating element and two pivoted handle grips having electrodes that serve as the stripping jaws. The wire to be stripped is positioned between the electrodes, which are then clamped together around the wire. Squeezing the insulated handles activates a switch that causes the electrodes to heat the insulation to its melting point. With the electrodes energized, the wire is rotated so that the insulation melts uniformly. When the insulation has melted, the pressure on the handles is released, the wire is removed, and the insulation is pulled off of the conductor.

Thermal strippers will not nick or cut conductors. They are more expensive than mechanical strippers, however, and may also cause irritating fumes as the insulation is melted.

Whichever method of wire stripping is employed, more wire than is necessary to make the connection should be stripped. The reasons for this are that (1) in the case of stranded wire where tinning (discussed later in this section) of the strands is required, solder buildup usually occurs at the conductor end and can be removed simply by cutting; and (2) cutting and stripping each wire to its exact required dimension will not provide the necessary "play" that the technician may need to adjust wire positions during wire operations.

As mentioned previously, it may be necessary to further clean the exposed wire prior to soldering. This can be quickly accomplished with the use of the lead cleaner shown in Fig. 25.20. Designed primarily for use on solid wire (including component leads), this eraser-style cleaner will remove contaminants such as dirt and oxides from the surface of the wire.

To strip enamel or any type of film insulation from solid wire, an *abrasive* lead cleaner may be used. It is made of braided wire formed about a piece of aluminum. Details for constructing this type of cleaner are given in Problem 26.7.

The final step prior to interconnection is the *tinning* of stranded wire. This process is necessary to allow for the formation around a terminal in a manner that will result in as neat a joint as obtainable with the use of solid wire. Approximately $\frac{1}{2}$ inch of insulation is first removed from the wire and the individual strands are firmly twisted together in a direction to return to their original twist orientation that is disturbed by the stripping action. This twisting also returns the wire to its original diameter, which is important to avoid difficulties in making connections to small eyelet terminals or the center prong on phono plugs.

The exposed and twisted wire is then dipped into a liquid flux such as Kester 1544 or 1589 flux. Allowing the flux to contact the insulation should be avoided so that cleaning up after the tinning process will be faster.

The tinning of stranded wire can be effectively accomplished using one of two methods. The first is with the use of a soldering iron, as shown in Fig. 26.4a. The insulated portion of the wire is gently held between the jaws of a bench vise with the fluxed conductor extended. The hot soldering iron is brought into contact with the bottom of the wire and the solder is applied to the top directly onto the strands. Heat from the iron will cause the solder to melt and be drawn through the strands since melting solder follows the

(a)

(b)

**Figure 26.4** Tinning stranded wire: (a) using a soldering iron; (b) dipping the wire into liquid flux and then into molten solder.

direction of heat flow. Drawing the iron and the solder along the length of the wire from the end to the insulation and back again will ensure solid, uniform bonding of the individual strands.

The second method of tinning is with the use of a *solder pot.* This is shown in Fig. 26.4b. The pot contains a pool of molten solder maintained at a temperature between 600 and 650°F. To begin the tinning process, the dross formation on the surface of the molten solder is skimmed away with a small wooden spatula, such as a tongue depressor. The fluxed wire is then slowly placed into the solder so that the insulation is just above the surface. This slow entry allows the individual strands to arrive at a sufficient temperature so that proper wetting action and bonding results. Withdrawing the wire is again done slowly so that the surface tension of the solder will draw off any excess solder on the wire. (*Caution:* Safety glasses should be worn when working with molten solder, to avoid injury.)

Effective tinning results when sufficient solder has wetted all conductor surfaces forming a solid bond between them. Excessive solder should be avoided. The contour of each strand should be plainly visible.

After the wire has been tinned, the flux residue should be removed. This is done by placing the wire on a paper towel and scrubbing the end with a stiff acid brush using isopropyl alcohol. The wire is now ready to be wrapped around any type of connection without concern for any strands becoming separated from the group.

After the tinning operation, the excess wire length along with the accumulation of solder that usually builds up at the wire end can be removed with a pair of *diagonal cutters.* Selecting the most suitable type of diagonal cutters (dykes) depends primarily on the size of the wire being used as well as accessibility during assembly. Because of the variety of cutting situations encountered, many styles of diagonal cutters are available. A selection of common styles used for electronic wiring is shown in Fig. 26.5. The size of these diagonal cutters ranges from 6 inches long for heavy-gauge wire (AWG No. 14 maximum) to 4 inches long for use with smaller wire such as AWG No. 24 and smaller.

The cutting edges of the diagonal cutters can be classified as *regular, semi-flush,* or *full-flush.* Regular cutters, used for general-purpose applications, leave a "point" at the end of the wire after cutting. Semiflush cutters leave a less pronounced point. Full-flush

**Figure 26.5** Various styles of wire-cutting tools.

cutters, intended for use only on annealed copper wire, leave no point or chamfer but rather a smooth, flat end perpendicular to the wire axis. Since the regular and semiflush cutters distort the end of a conductor, this could create a problem when attempting to insert the wire into a close tolerance hole such as those associated with small eyelets and holes in pc board terminal pads. In addition, the sharp points these cutters leave are undesirable when protruding from terminal connections. For these reasons, the full-flush cutter is recommended.

The jaws of most diagonal cutters produce a slight "shock" as they come together when cutting a wire. When used on hookup wire, this presents no problem. However, sensitive semiconductor devices could be damaged by this shock. For this reason, *shear-type* cutters are recommended for delicate work. The jaws of these cutters pass each other similar to a pair of scissors, thereby minimizing the cutting shock.

Diagonal cutters are available in *rounded, tapered,* and *offset* noses to fulfill the requirements of a variety of cutting needs. The most appropriate cutting jaws, shapes, and edge types are selected after the work to be performed is evaluated. When heavier-gauge wires are to be cut and close cutting tolerances and accessibility are not critical, less delicate cutters may be employed. Intricate wiring using finer-gauge wire will require more delicate cutters.

Rubber jaw inserts are available on some styles of cutters, which grip the end of the wire to be removed. These inserts prevent the cut end from flying out of the jaw when the wire is cut. There is a serious safety hazard involved in using diagonal cutters not equipped with rubber inserts. It is always a good practice to point the conductor end *downward* to prevent the flying end from causing injury. Cutters with plastic grips are preferable to cutters with no grips because they not only provide more positive gripping, but also reduce hand fatigue.

## 26.5  WIRE CONNECTIONS

One of the most common types of terminal used to make interconnections between pc boards and chassis-mounted components is the *swaged turret terminal* discussed in Chapter 16. To obtain a good electrical connection, it is vitally important that a solid mechanical connection first be made. The wire to be soldered to a turret-type terminal should be formed as tightly as possible around the turret without scraping or nicking the conductor surface or deforming the wire contour. With the insulation positioned no more than $\frac{1}{8}$ inch away from the terminal, the conductor is first tightly formed 180 degrees around the turret with long-nose pliers (Fig. 26.6a). The excess length of wire is then cut at point *A*, which leaves a sufficient length to continue forming a complete 360-degree *wraparound* connection. This wrap being formed with long-nose pliers is shown in Fig. 26.6b. Although 180-degree wraps are often used in electronic wiring, the 360-degree wrap provides the most secure mechanical connection. Long-nose pliers with smooth jaws should be used when making mechanical connections to any terminal configuration to avoid nicking and deforming the conductor surface.

Soldering to a turret terminal is performed using the soldering techniques discussed in Chapter 17 for soldering leads to pc board terminal pads. A small amount of solder is first applied to the soldering iron tip (solder bridge). The tip is next brought into contact with the wire and shoulder of the terminal, as shown in Fig. 26.6c. Positioning the tip directly opposite from the insulation will minimize heat damage to plastic and rubber insulation. Solder is applied immediately when the connection has reached the proper temperature (i.e., when the solder of the bridge or the solder used in tinning stranded wire begins to flow). Only a *small* amount of solder is necessary to make a sound electrical connec-

**Figure 26.6** Wiring to a turret-type terminal: (a) initial 180° wrap; (b) forming the complete 360° wrap; (c) correct iron tip and solder position; (d) properly soldered terminal.

tion. If properly soldered, the contour of both the wire and the terminal will be plainly visible. A properly wired and soldered turret terminal connection is shown in Fig. 26.6d.

Wraparound connections also find wide application when wiring to any eyelet-type terminal, such as those associated with *terminal strips, switches, potentiometers,* and *pc board connectors.* Wiring to these terminals is shown in Fig. 26.7. Wire forming about an eyelet-type terminal lug begins by positioning the insulation no more than $\frac{1}{8}$ inch from the eyelet opening. The wire wrap is then formed in a similar manner to the 360-degree wrap for turret terminals, but since the cross section of the eyelet is a very thin rectangular shape, the wrap is more difficult to form. It is recommended that the first two 90-degree contours, as shown in Fig. 26.7a, be preformed before hooking the conductor through the eyelet. Once positioned into the eyelet, the excess lead length is cut at point *A* and the final 90-degree bend is made to complete the wrap (Fig. 26.7b). It is not good practice to make more than three wraparound connections to a single eyelet or to solder them until all have been mechanically formed.

When mechanical strength requirements are minimal, wires can be soldered more easily into small-diameter holes, such as those found on solder lugs, by means of a *feed-through* connection, shown in Fig. 26.8. For this type of connection, the diameter of the wire should be approximately equal to that of the hole in the lug. As with other types of connections, the insulation is positioned at a maximum of $\frac{1}{8}$ inch from the terminal, with the prepared wire end passing through the lug hole. The excess wire is trimmed to within approximately $\frac{1}{16}$ inch from the terminal. Because the diameter of the wire approximates that of the hole, soldering will properly alloy both circumferences. The resulting connection is mechanically satisfactory and electrically sound, although feed-through connections should be avoided in equipment that will be subjected to severe shock and vibration.

Wires may be *dressed* between connections by either *point-to-point* or *square-corner* wiring techniques. Point-to-point wiring, as the name implies, is a wiring tech-

(a) Preformed wire   Pull   Point A

(b) Finished wrap   ⅛" Maximum

**Figure 26.7**   Wire forming for eyelet terminals.

nique used primarily in high-frequency circuits or any economical wiring application. Wiring interconnections are made with regard to the shortest routing path possible to minimize lead length, which is critical at high frequencies. This wiring technique is quicker and uses less wire than the other methods discussed here. Point-to-point wire dress usually does not add to the neatness of the package. However, with careful wiring positioning and neat soldering connections, the resulting wiring will be reliable and will display quality workmanship.

Square-corner wiring is a wire dress technique in which all wires installed in the package run either parallel or perpendicular to one another. When any wire path direction changes, the change is made with a 90-degree bend. Although not used in high-frequency work, this technique imparts a very neat appearance to the package. The only disadvantage to square-corner wiring is the additional amount of wire and time necessary to complete a package.

## 26.6 SPECIAL WIRE ASSEMBLIES AND TECHNIQUES

A common yet difficult wiring connection involves the preparation and installation of shielded wire to phono plugs and jacks. The function of the braid of the wire is to cover and thereby isolate the inner (signal) conductor. There should be minimum exposure of the signal wire insulation after the shield and signal wire connections are made to the connectors. If the shielded wire has an outer insulating jacket, approximately $1\frac{1}{2}$ inches of the jacket should be removed. The shield is then pushed back from the end so as to "loosen it." A pick (scriber) is then carefully worked into the braid without damaging any individual strands to form a hole large enough through which the insulated signal wire can be pulled. This technique is shown in Fig. 26.9a. The pick is then passed into the hole between the braid and the signal wire insulation so that it protrudes from the other side.

$^1/_{16}$" Maximum   $^1/_8$" Maximum

**Figure 26.8**   Feed-through connection.

(a)

(b)

**Figure 26.9** Technique for preparing shielded wire for soldering: (a) scribe used to provide access hole in shield; (b) removing signal lead from shield.

Doubling the shielded wire back at the breakout point and lifting the pick will remove the signal wire from the hole in the braid. This technique is shown in Fig. 26.9b. The end of the braid is then tinned and will serve as the ground lead.

The prepared shield and signal wire ready for soldering to a connector are shown in Fig. 26.10a. Soldering the signal wire to a plug requires careful preparation. The hollow center prong of the plug is first filled with solder. This step can best be done by gently gripping the outer shell of the plug in a bench vise. The correct soldering iron tip position and application of solder is shown in Fig. 26.10b. When solder begins to appear at the tip of the prong, the iron is removed. Solder must not be allowed to flow in the outer surface of the prong or it may prove to be impossible to plug it into a jack. To install the wires onto the plug, at least $\frac{3}{4}$ inch of signal wire is stripped and tinned, leaving only $\frac{3}{16}$ inch of insulation between the stripped signal conductor and the braid. The braid is also tinned. The center prong of the plug is reheated and the signal wire is inserted into the top of the plug until the braid is even with the top of the ferrule. Some of the signal conductor may protrude from the end of the prong. This is removed flush with diagonal cutters. The braid is next wrapped 360 degrees around the ferrule and then soldered, using the technique shown in Fig. 26.10c. Proper installation of shielded wire to a phono jack is shown in Fig. 26.10d.

## 26.7 SOLDERLESS WIRING TECHNIQUES

*Solderless-* or *crimp-*type terminals provide a strong metal-to-metal (terminal-to-wire) bond that is suitable for applications using either stranded or solid wire. These terminals are particularly desirable in some production assemblies because they eliminate soldering. Two common configurations of solderless-type (crimp) connectors are the *ring tongue* and *fork tongue* styles shown in Fig. 26.11.

Selecting the most suitable style depends on the application. If rapid connection and disconnection are not required but a higher degree of mechanical security is necessary, the

(a)

Solder

Solder Droplet Indicates
Sufficient Solder

Soldering Iron Tip

(b)

(c) Cutaway view showing correct assembly of wire and shield

(b)

**Figure 26.10**   Preparation and soldering of shielded wire to plugs and jacks: (a) shielded wire prepared for soldering; (b) preparation of prong with solder; (c) proper position of iron tip and solder; (d) correctly prepared jack.

Barrel Insulation

Metal Barrel — | — Tongue

**Figure 26.11**  Crimp terminals.

Ring Tongue

Fork Tongue

ring tongue should be used. If, however, rapid connecting and disconnecting requirements are specified with no critical mechanical requirements, the fork tongue style is preferable.

The *barrels* of crimp terminals are available in uninsulated or insulated forms. The insulator is usually polyvinyl chloride (PVC) and offers a degree of strain relief for the wire if strong vibration is expected. These insulators are commonly available in colors so that the range of AWG sizes that each barrel will accept can be quickly identified (see Table 26.6).

Solderless terminals in common use in electronics have barrels that are capable of accepting wire sizes from AWG No. 10 to AWG No. 22. A No. 10 terminal will accept both AWG Nos. 10 and 12 wire; a No. 14 terminal will accept AWG Nos. 14 and 16 wire; and a No. 18 terminal will accept AWG Nos. 18, 20, and 22 wire. The terminals are available to accept screw sizes from No. 4 to $\frac{3}{8}$ inch.

Conductors are assembled into solderless terminals with a *crimping tool* such as the one shown in Fig. 26.12. The wire insulation is stripped to a length equal to the metal barrel length. The conductor is then inserted into the terminal until the insulation seats against the end of the metal barrel. The barrel is then placed into the appropriate jaw crimping position of the tool. The split side of the barrel should be placed against the smooth portion of the jaw opposite the crimping tooth. When the handles are firmly squeezed together, as shown in Fig. 26.12, the barrel will be crimped against the conductor and will make a sound mechanical and electrical connection.

## 26.8 WIRE-WRAP CONNECTIONS

An alternative method of providing randomly located metal studs to which external connections can be made *without* the necessity of soldering wires is through the use of *wire-wrapping* techniques. Wire-wrap terminals are square or rectangular pins having at least two sharp corners that are mechanically secured to the pc board by staking (press-fitted) or soldering. Lead connections are made to these pins with mechanical or pneumatic wire-wrapping tools using solid wire. The resulting connection is both mechanically secure and electrically sound.

Two types of wire-wrap pins are shown in Fig. 26.13, each having its own configuration and mounting arrangement. The most basic wire-wrap pin is shown in Fig. 26.13a. It consists simply of a 0.045-inch-square wire from $\frac{3}{4}$ to 2 inches long with both ends tapered for ease of initial insertion into the pc board. No special equipment is required for their installation since the pin is simply pressed into a predrilled hole. The recommended drill hole has a diameter of 0.050 inch with a 0.002-inch tolerance. Although commer-

**TABLE 26.6**

Color Code for Insulated Crimp Connector Barrel Sizes

| Terminal Number | Barrel Insulation Color | Accepts AWG Sizes |
|---|---|---|
| No. 18 | Red | 22 to 18 |
| No. 14 | Blue | 16 to 14 |
| No. 10 | Yellow | 12 to 10 |

**Figure 26.12** Technique for installing crimp terminals.

cially available vibrating tables and templates are employed to position the pins into the holes, the cost of such equipment is prohibitive in small volume or prototype work. These pins can be inserted easily by hand. The tapered end of the pin is first inserted into the drill hole of the board using finger pressure. This procedure is sufficient to temporarily hold the pin in a staking position. Similar to the staking of solder-type terminals, an arbor press is used with an anvil and flat-faced setting tool (see Sec. 16.1). The guide hole in the anvil should be large enough to accept the square configuration of the pin with sufficient clearance. The arbor press then is used to force the pin through the pc board, allowing it to extend to any desired height on the opposite side of the board. Although held in position only by the contact between the corners of the pin and the inside edge of the hole, it is sufficiently gripped in place until it can be soldered to the terminal pad. Connecting wires are then wrapped along the length of the pin extending above either side of the board.

Another wire-wrap terminal that provides for uniform staking height is shown in Fig. 26.13b. It consists of double-ended square stock with either 0.025-inch or 0.045-inch flats made of brass and plated with gold alloy. Dimension $E$ in Fig. 26.13b is generally selected to be equal to the thickness of the pc board insulating material. To insert this pin, a hole is drilled that is approximately 0.002-inch smaller than the maximum diameter of the *reverse barb*. For the 0.025-inch square pin, dimension $L$ is 0.052 inch (hole size $\approx$ 0.050 inch) and for the 0.045-inch square pin it is 0.096 inch (hole size $\approx$ 0.094 inch). The pin is positioned in such a manner as to allow its crown to contact the foil side (terminal pad) of the board. The pin is then secured with an anvil and setting tool in an arbor press. The re-

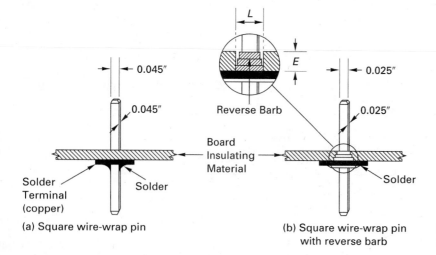

(a) Square wire-wrap pin

(b) Square wire-wrap pin with reverse barb

**Figure 26.13** Styles of wire-wrap terminals.

verse barb prevents the pin from turning or backing out of the insulating material. Soldering between the crown and terminal pad will make a reliable electrical connection.

Mechanical and electrical connection for wire-wrap terminals is achieved by wrapping, under tension, solid wire around a post terminal that has at least two sharp edges. These edges, as well as the wire itself, are deformed when tightly wrapped. The average pressure between the wire and the sharp contact points of the post is in the order of 30,000 psi, which is more than sufficient to produce a sound mechanical and electrical connection. Two properly formed types of wire-wrap connections are shown in Fig. 26.14.

Wire-wrap connections are formed by manual operation and automatic or semiautomatic wire-wrapping machines. Manually operated wire-wrap tools, such as the one shown in Fig. 26.15a, are available to quickly form wire-wrap connections. Pneumatic wire-wrap tools are also available, but are usually employed with semiautomatic machines in mass-production applications.

To use the tool shown in Fig. 26.15a, solid wire of the desired gauge is first stripped and then inserted into the outermost hole at the end of the wire wrapping tip (Fig. 26.15b). The tip is then positioned over the terminal and the trigger is quickly squeezed. The nose of the gun spins the wire around the terminal to form the desired connection.

The number of turns of wire wrapped is determined by the specifications provided in Table 26.7. The amount of insulation that must be stripped from the end of the wire can be quickly determined by trial and error.

Should it be necessary to unwrap a connection that is not readily accessible (on high-density packaging, wrapped terminals may be within 0.1 inch away from each other) *de-wrapping* tools such as the one shown in Fig. 26.15c are available. To remove a wrapped connection, the center hole in the nose is positioned over the terminal. Downward pressure of the tool against the wire will quickly unwrap the wire from the post. The tool must be kept vertical during the entire operation to avoid bending the terminal post.

When the number of terminal connections to be made becomes impossible to manually wire wrap at a cost-effective rate and without excessive errors, semiautomatic and fully automatic wire-wrapping machines are employed. One application of dense wiring includes that of large backplanes having thousands of wiring terminals on one side and heavily populated on the other side with printed circuit board or cable connectors. Another example is on panels requiring the discrete wiring of vast numbers of integrated circuit sockets.

**Figure 26.14** Two styles of wire wrapping.

**Figure 26.15** Hand operated wire-wrapping tools and wrapping method: (a) hand operated wire-wrap tool; (b) wire-wrap connection process; (c) de-wrapping tool.

Semiautomatic wire wrapping is performed on a machine having a vertical work-table to which the panel to be wired is clamped. Data-controlled carriages move the table in both *x* and *y* directions. The table movement is such that each new pin to be wired positions itself in front of the operator. This feature aids in reducing operator fatigue. A terminal pointer with a tool guide (mounted wrapping tool) traverses the *x* axis and stops after each table repositioning in front of the next terminal to be wrapped. The operator feeds the specified size of wire into the tool guide and, with slight pressure, pushes the tool onto

**TABLE 26.7**

Recommended Wiring Specifications for 0.025- and 0.045-Inch-Square
Wire-Wrap Terminals

| | Number of Turns | |
|---|---|---|
| AWG Number | Type 1 Connection | Type 2* Connection |
| 18 | 4 | 5 |
| 20 | | |
| 22 | 5 | 6 |
| 24 | | |
| 26 | 6 | 7 |
| 28 | | |
| 30 | 7 | 8 |

*Type 2 indicates one turn of insulated conductor contacting the terminal.

the pin to make each connection. Rates of between 450 and 600 wraps per hour are typical with semiautomatic machines.

When higher wiring densities and speeds exceed the capabilities of semiautomatic systems, a fully automatic wire-wrapping machine, such as that shown in Fig. 26.16a, is employed. This machine has an operator's console which includes the following: manual operating switches and data input, fault monitoring system, cycle counter, cycle interrupt, emergency stop, repeat wire operation, and sequence number readout.

The panel to be wired is clamped vertically inside the machine. Unlike the semiautomatic machines in which the board moves only in the $x$ and $y$ directions, the automated systems also provide for 90-degree board rotation. Two sets of programmable wire-wrapping tools and dressing fingers are mounted on two vertical carriages which are also data controlled. The dressing fingers form the specified contour of the wire as it is routed above selected channels between closely spaced terminal locations. After the wire-pattern excursion is completed between the two terminals, the machine automatically

(a)        (b)

**Figure 26.16** (a) 14FV vertical wire-wrapping machine, *courtesy of EPE Technology;* (b) basic wire patterns.

strips and cuts the wire to length. The tool bits then lower the wire ends over the two terminals to be connected and simultaneously form the wire wraps. One wrapping bit rotates clockwise while the other end of the wire is wrapped in a counterclockwise direction. The basic wire patterns formed are shown in Fig. 26.16b. The symbols at the bend locations represent the use of a dressing finger to form the pattern. The data also include the $z$ levels to obtain the specified height of the wire wrap on the terminal. This ensures proper spacing between multiple wraps on a single terminal. Because wire wrapped in opposite directions may result on the same terminal, the spacing between these wraps is critical to prevent a wire being installed from disturbing one previously formed.

The automatic wire-wrapping machine shown in Fig. 26.16a will accommodate a panel size of up to 26 by 26 inches with a wrap area of 22 by 22 inches. It is made to wrap AWG Nos. 26 and 30 wire at a rate as high as 1100 to 1200 wraps per hour.

Several terminals properly wrapped with solid wire are shown in Fig. 26.17. The type of wire normally used is copper-alloy solid wire with a tin, silver, or gold plating. Sizes of wire for this application range from AWG (American Wire Gauge) No. 18 to No. 26. The size of the wire will determine the exact number of turns to make a properly wrapped connection. This information for two types of interconnections is given in Table 26.7. The type 1 interconnection requires that the wire be wrapped around the terminal with its insulation approximately $\frac{1}{16}$ inch from the first turn; type 2 interconnections require that at least one turn of insulation also be formed around the terminal. This second classification utilizes the flexibility of the wire insulation to reduce stress encountered during vibrations of the wire at the point where it enters into the first wrap on the terminal. In addition, quality workmanship dictates that no more than three independent lead wraps be made on a single terminal pin and that (1) turns from one independent wrap do not overlap any portion of another wrap, (2) adjacent turns on the same wrap touch but do not overlap, (3) the sum of all gaps should not exceed one wire diameter, excluding the first and last turns (spiral and open wrap), (4) there are not insufficient turns, and (5) the wire end does not protude from the terminal (end tail). The appearance of unacceptable wire wraps is shown in Fig. 26.18a.

It is often the case that several terminals must be connected together or *bussed*. To avoid later servicing problems, bussed pairs of terminals to be connected should be provided with level 1 wraps first, followed by level 2 wraps to complete the string. Alternately wrapping terminals from level 2 to level 1 is unacceptable. If a wire that has its end wrapped at different levels must be relocated, the entire string may have to be removed to replace only one wire (see Fig. 26.18b).

**Figure 26.17** Wire-wrap assembly. *Courtesy of Augat, Inc.*

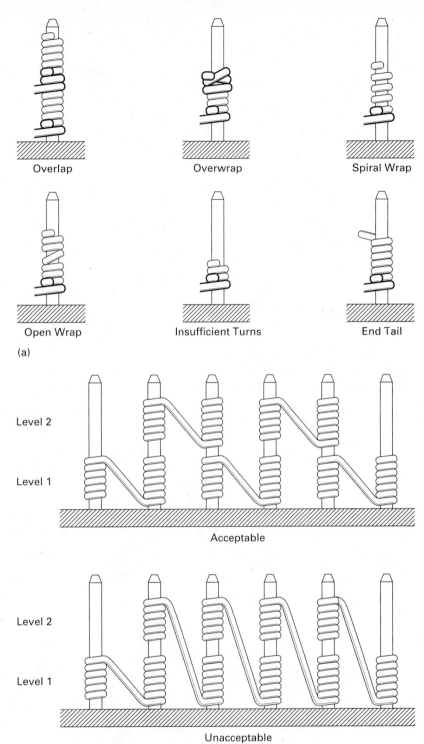

Overlap

Overwrap

Spiral Wrap

Open Wrap

Insufficient Turns

End Tail

(a)

Level 2

Level 1

Acceptable

Level 2

Level 1

Unacceptable

(b)

**Figure 26.18** Wire-wraps and bussing technique: (a) unacceptable wire wraps; (b) bussing.

Proper locations of both wire wraps and routing is essential to optimum circuit performance, especially in high-speed circuitry. If a signal wire in this type of design must be relocated because of stray inductance or capacitance problems, it is important that proper wire-wrap levels are adhered to throughout the design. Also, to reduce the number of parallel wires, advantage of the machine's data-controlled pallet-rotation ability should be taken. The software can be used to spread out the wires and result in a marked reduction of density buildup, in addition to a reduction in the number of parallel wires.

## 26.9 CABLE TIES

*Cable ties* are used when two or more wires are bundled together to form a harness or cable. They consist of a nylon strap with a locking eyelet. Small, closely spaced grooves are on the inside surface of the strap perpendicular to the length of the strap. These grooves will be in contact with the wires when assembled. The cable tie is first looped around the section to be bundled. The tapered end of the strap is then fed through the eyelet and pulled tightly. The strap is held securely by the locking eyelet, which engages the grooves and prevents the strap from slipping. Excess strap length is cut off with diagonal cutters close to the eyelet. The proper procedure for bundling leads with cable ties is shown in Fig. 26.19.

Special hand-operated cable tie assembly tools are available to facilitate the use of these bundling aids. One such tool is shown in Fig. 26.20. It consists of a gripping and cutting *head,* a *tension control,* and an *actuating trigger.* The tension control, roughly gauged from *loose* to *tight,* is set for the desired tension. The specific setting is determined by "feel," using the trial-and-error method. With the cable tie looped around the wires, the tapered end of the strap is fed through the eyelet and into the gripping head of the assembly tool. Each time the trigger is actuated, the head grips the strap and pulls it for a short distance through the eyelet. It may be necessary to actuate the trigger several times to reach the amount of tension that will firmly grip the leads. When the preset tension setting is reached, the cutting head shears the strap close to the eyelet. The use of this tool allows for rapid and uniformly controlled cable tie assembly.

Cable ties are positioned at regularly spaced intervals along the harness and at each breakout point. At breakout points, a single cable tie is looped in a *crossed* fashion around the breakout (see Fig. 26.21).

(a)

(b)

**Figure 26.19** Cable ties are used to bundle leads: (a) insertion of tab through locking eyelet; (b) trimming excess tab length.

Figure 26.20  Hand-operated cable tie assembly tool.

## 26.10 SHRINKABLE TUBING

Cable ties for harness construction is the most popular method used. Another technique of harnessing employs plastic and elastomer materials classified as *heat-shrinkable tubing*. This tubing is available in at least 12 different materials, each having its unique properties.

Manufacturers of shrinkable tubing employ special mechanical and thermal means to expand the molecules of the tubing and "fix" them in a strained state. When heated, the strains established during the manufacturing process are relieved, thus returning the tubing diameter to its original size. In addition, the wall thickness, which is reduced during manufacture, returns to its original thickness when heated. *Shrink temperatures* for some of the common tubing materials range from 175 to over 600°F (79 to 316°C).

In its expanded state, shrinkable tubing is easily slipped over the leads to be bundled after being cut to the required length (Fig. 26.22a). Shrink temperature is achieved with the use of a portable electric *heat gun,* shown in Fig. 26.22b, which concentrates heated air completely around the tubing, thus shrinking it and forming a tight fit. This

Figure 26.21  Technique for securing breakout points.

(a)

(b)

**Figure 26.22** Installation of shrinkable tubing: (a) initial placement of tubing sections; (b) air deflector allows for uniform heating.

process requires only one or two practice trials to determine the optimum length of time for the application of heat to shrink the tubing to the desired size.

Shrinkable tubing exhibits several desirable characteristics that are not attainable with cable tie methods. Shrinkable tubing provides a dust-free, waterproof, and abrasive and oil-resistant covering over the harness as well as the ends of the connectors. Short lengths of shrinkable tubing may also be used on a cable-tied harness where abrasion at a particular point could be a problem. The one disadvantage of this tubing is that it can be used only where accessibility will permit.

## 26.11 CABLE CLAMPS

When the harness has been completed, it is initially installed into the system. If connectors have been preassembled onto the harness, they are first mounted. All leads requiring soldered connections are then assembled onto the subassemblies. The orientation of the installed harness must be such as to protect the wire insulation from abrasion. In addition, the harness must not contact high-temperature components.

The harness must finally be rigidly fastened to the chassis to prevent movement or

**Figure 26.23**  Typical cable clamps.
*Courtesy of Weckesser Company, Inc.*

undue strain on terminal connections. Commercially available *cable clamps*, such as those shown in Fig. 26.23, are used for this purpose. These clamps are constructed of nylon or Teflon and are provided with either a clearance hole for typical machine screw sizes or with an adhesive backing for securing the clamp to the chassis. They are available in sizes to accept harness diameters from $\frac{1}{16}$ inch to several inches and provide an excellent means of securing a harness rigidly to chassis elements.

## EXERCISES

### A. Questions

26.1   What is the reason that copper wire is tin-coated?

26.2   What is the advantage of stranded wire over solid wire?

26.3   Define the term *bus wire*.

26.4   Why is concentric-type wire preferred over bunch-type wire?

26.5   What does the abbreviation AWG represent?

26.6   Explain the terms *mil-foot* and *circular mil*.

26.7   Determine the resistance of 3 feet of AWG No. 22 copper wire.

26.8   What is the relationship between cross-sectional area and AWG number of copper wire?

26.9   What is the relationship between current-carrying capacity and AWG number?

26.10   What three criteria determine the most suitable wire insulation for a specific application?

26.11   Place the following insulation materials in order of poor, fair, good, and excellent for insulation resistance: Teflon (TFE), nylon, Mylar, and Kynar.

26.12   Describe the body and tracer colors of hookup wire in a harness assembly that is assigned the code number 528.

26.13   What is Government Standard MIL-STD-122?

26.14   What is the primary difference between coaxial cable and shielded wire?

26.15   How is standard wire prepared prior to making a connection to a terminal?

26.16   What is the number and barrel color of a crimp terminal that will accept AWG wire numbers 22 to 18?

## B. True or False

Circle T if the statement is true, or F if any part of the statement is false.

26.1　Unidirectional stranded wire is more flexible than the counterdirectional type, but its diameter is not as precisely uniform.　　　T　F

26.2　In the AWG system, as the numbers increase, the corresponding wire diameters decrease.　　　T　F

26.3　Wire resistance increases by a factor of 4 when the diameter is doubled.　　　T　F

26.4　The term *circular mil* refers to the diameter of wire.　　　T　F

26.5　When measuring the size of stranded wire with a wire gauge, the slot that it snugly fits into will be one AWG number more than the gauge reading due to the space between strands.　　　T　F

26.6　Shielded wire is made to be used on low-frequency applications where its impedance is not critical.　　　T　F

26.7　When tinning stranded wire with a soldering iron, the wire is positioned between the solder and the iron's tip.　　　T　F

26.8　When forming wraparound connections, the wire insulation should touch the terminal to prevent the possibility of shorts between adjacent terminals.　　　T　F

26.9　A crimp connector having a red barrel will accept a larger-diameter wire than will a connector with a yellow barrel.　　　T　F

## C. Multiple Choice

Circle the correct answer to each statement.

26.1　Wire having an AWG number of 22 is (*smaller*, *larger*) in diameter than one having an AWG number of 16.

26.2　AWG No. 22 wire has a (*smaller*, *larger*) current-carrying capacity than AWG No. 14 wire.

26.3　The most common material used for coating copper wire is (*tin*, *nickel*).

26.4　The AWG wire numbering system was developed primarily to designate the size of (*stranded*, *solid*) wire.

26.5　(*Solid*, *stranded*) wire is used to make electrical connections to wire-wrap terminals.

## D. Matching Columns

Match each item in column A to the most appropriate item in column B.

| COLUMN A | | COLUMN B |
|---|---|---|
| 1. Diagonal cutters | a. | Insulation |
| 2. AWG | b. | Cross-sectional area |
| 3. Wire-wrap | c. | Full-flush |
| 4. Stranded wire | d. | Rf circuitry |
| 5. Kel-F | e. | American Wire Gauge |
| 6. Coaxial cable | f. | Solderless |
| 7. CM | g. | Concentric |

**Figure 26.24**

### E. Problems

26.1 Construct a pair of test leads such as those shown in Fig. 26.24. The test leads are to be made from 36-inch lengths of plastic insulated No. 18 AWG stranded wire. Strip $\frac{1}{2}$ inch of insulation from all four lead ends and twist and tin the stranded conductors. Slip the rubber insulated grips onto the leads before soldering the alligator clips to the lead ends. After soldering the clips, slide the rubber insulators into position.

26.2 Construct a set of interconnecting leads for a stereo system by first centering a 2-foot length of shrinkable tubing about two 3-foot lengths of insulated shielded conductor cables. Slide 2-inch lengths of shrinkable tubing over all four ends of the cables. Remove sufficent outer insulation from all cable ends to prepare the copper braid and the inner insulated conductors for soldered connections to RCA phono plugs. Once the phono plugs have been soldered, slip the shrinkable tubing onto the plugs until the end of the tubing is even with the leading edges of the outer portion of the connector. Finally, shrink the tubing about the cables and plugs.

26.3 Construct four 6-foot speaker leads from plastic insulated No. 16 AWG stranded wire. Slide $1\frac{1}{2}$-inch lengths of shrinkable tubing over all lead ends. Strip and tin all lead ends to a length of $\frac{1}{2}$ inch. Position the uninsulated barrels of solderless fork tongue connectors onto all prepared conductor ends and crimp. Slide the shrinkable tubing over the entire barrel length and shrink into place.

26.4 Using the technique discussed in this chapter for straightening bus wire, prepare a $32\frac{1}{2}$-inch length of tinned No. 14 AWG bus wire to construct a UHF antenna. Form the wire into a loop with a $7\frac{3}{4}$-inch diameter. Adjust the loop so that both loop ends overlap equally. Crimp solderless spade lugs onto both loop ends and form two 90-degree bends on each end for attachment to terminal lugs on the rear of a television set. The first 90-degree bends are to be made 4 inches from the end of the spade lugs and perpendicular to the circumference of the loop. The remaining 90-degree bend is to be located 2 inches from the ends of the lugs for the most convenient attachment.

26.5 Install plastic insulated No. 20 AWG stranded wire between the chassis-mounted components and hardware and the pc board for the seven-fixed-interval UJT timer designed, fabricated, and assembled in Problems 6.2, 10.2, 16.2, 17.2, and 25.3.

26.6 Install interconnection wires to the tachometer designed, fabricated, and assembled in Problems 6.4, 10.4, 16.4, 17.4, and 25.4. All leads will be plastic-insulated No. 18 AWG stranded wire. The power lead insulation color is to be red, with lead length sufficient for connection to the battery side of the fuse block. The ground lead insulation color should be black. This lead is to be terminated with a solderless spade lug and attached to any convenient screw on the car chassis. Se-

**Figure 26.25**

lect green for the insulation color for the remaining lead to be connected to the distributor side of the coil. A solderless ring tongue connector of sufficient size for the coil connection should be crimped onto the end of the green lead.

26.7   Fabricate a lead cleaner such as that shown in Fig. 26.25. Use a $\frac{5}{8}$- by 10-inch strip of 16-gauge aluminum, two 5-inch lengths of $\frac{5}{8}$-inch-wide copper braid, and two 1-inch lengths of shrinkable tubing.

# Unit
# SEVEN
## Student Projects

# Projects

## 27.0 INTRODUCTION

The projects presented in this chapter have been selected to allow students to practice the skills discussed in this book. These projects illustrate varying degrees of construction complexity. While the cost of constructing these projects was a consideration, primary emphasis was placed on the application of current technology as well as their usefulness.

The authors are aware of the necessity to substitute parts and devices because of availability considerations. Many devices and components are not critical, and where indicated, equivalencies may be substituted. For this reason, and also to familiarize the student further with the selected circuit, it is strongly recommended that the circuit be breadboarded and tested before fabrication. This approach also provides an opportunity for the advanced student to modify a project.

The presentation of each project begins with a general discussion of the circuit operation to acquaint the student with its function. This is followed by a circuit schematic and a complete parts list. Finally, construction hints are provided to avoid packaging difficulties.

The finished packages shown pictorially are to be considered as typical. The student, of course, is not restricted to the design shown and should feel free to employ ingenuity in fabricating a package that will reflect his or her own requirements and desires.

The following projects are presented in this chapter:

- Darkroom Timer
- TTL Logic Probe
- 1.2- to 20-Volt Laboratory Power Supply With Digital Readout
- Portable Fahrenheit/Celsius Thermometer With LCD Readout
- Portable Sportsman's Barometer With LCD Readout
- Automobile Security Alarm System
- Microcontroller-Based Refrigeration Alarm System

As an aid in construction, pin configurations for all of the devices used in these projects are provided in Fig. 27.1.

## 27.1 DARKROOM TIMER

This darkroom timer circuit employs the popular 555 timer IC, which is readily available from electronic parts suppliers. The timer features four *fixed time-delay* intervals, a *timing cycle in progress* indicator, a *timed-out* indicator, and separate *start* and *reset* controls.

Diodes

Light – Emitting Diodes (LEDs)

Anode

Cathode
IN914
IN4001

(a)

Flat

Anode →          ← Cathode
(+)                        (−)

(b) Red, Green, Yellow

Seven – Segment Display

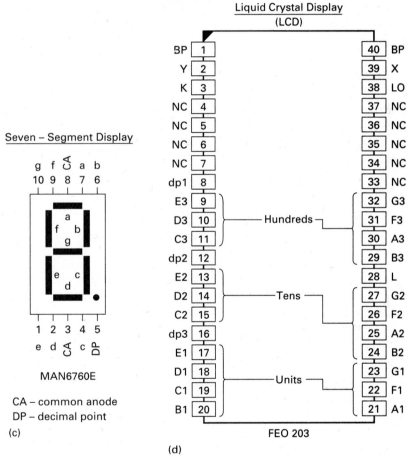

| g | f | CA | a | b |
|---|---|---|---|---|
| 10 | 9 | 8 | 7 | 6 |

| 1 | 2 | 3 | 4 | 5 |
|---|---|---|---|---|
| e | d | CA | c | DP |

MAN6760E

CA – common anode
DP – decimal point

(c)

Liquid Crystal Display (LCD)

| BP | 1 | | 40 | BP |
| Y | 2 | | 39 | X |
| K | 3 | | 38 | LO |
| NC | 4 | | 37 | NC |
| NC | 5 | | 36 | NC |
| NC | 6 | | 35 | NC |
| NC | 7 | | 34 | NC |
| dp1 | 8 | | 33 | NC |
| E3 | 9 | Hundreds | 32 | G3 |
| D3 | 10 | | 31 | F3 |
| C3 | 11 | | 30 | A3 |
| dp2 | 12 | | 29 | B3 |
| E2 | 13 | | 28 | L |
| D2 | 14 | Tens | 27 | G2 |
| C2 | 15 | | 26 | F2 |
| dp3 | 16 | | 25 | A2 |
| E1 | 17 | | 24 | B2 |
| D1 | 18 | Units | 23 | G1 |
| C1 | 19 | | 22 | F1 |
| B1 | 20 | | 21 | A1 |

FEO 203

(d)

**Figure 27.1**  Device configurations for projects: (a) diodes; (b) light-emitting diodes (LEDs): red, green, yellow; (c) seven-segment display, CA—common anode, DP— decimal point; (d) liquid crystal display (LCD);

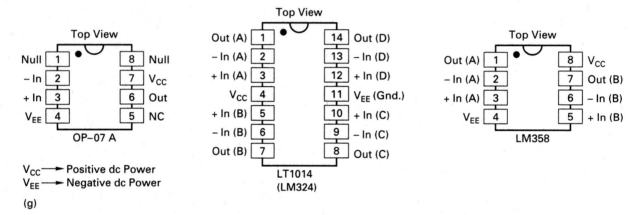

$V_{CC}$ ⟶ Positive dc Power
$V_{EE}$ ⟶ Negative dc Power

(g)

Digital Logic

Figure 27.1 (Cont.) (e) transistors; (f) silicon-controlled rectifier (SCR); (g) operational amplifiers; (h) digital logic;

**Figure 27.1** (Cont.) (i) regulator; (j) regulator; (k) timer; (l) temperature sensor; (m) absolute pressure sensor; (n) transducer;

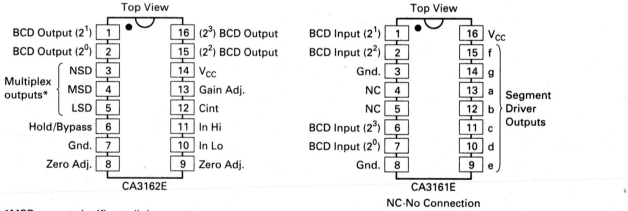

*MSD – most significant digit
 NSD – next significant digit
 LSD – least significant digit

(q)

**Figure 27.1** (Cont.) (o–q) converters.

Although this timer circuit was designed to aid in photographic film developing, it is sufficiently flexible to be adapted to many purposes where a portable timer is necessary.

**Circuit Operation.** Power is applied to the timer circuit shown in Fig. 27.2 by closing switch $SW_1$. The voltage at pin 3 of $IC_1$ is at approximately 0 volts. This forces the *timing cycle in progress* indicator $LED_2$ off and the *timed-out* indicator $LED_1$ to be on. One of four time delays can be selected with the time-selector switch $SW_2$. The time delay is set by $C_1$ and the two-resistor combinations functioning with $SW_2$. The timer shown in Fig. 27.2 is designed for delays of 10 seconds, 30 seconds, 1 minute, and 3 min-

(a)

Parts List

$IC_1$ – NE555 timer
$LED_1$ – green LED
$LED_2$ – yellow LED
$R_1, R_3$ – 1-MΩ, trim potentiometer
$R_2$ – 220-kΩ, $^1/_4$-W, 5% resistor
$R_4$ – 1-MΩ, $^1/_4$-W, 5% resistor
$R_5, R_7$ – 1-MΩ, trim potentiometer
$R_6$ – 1.8-MΩ, $^1/_4$-W, 5% resistor
$R_8$ – 6.8-MΩ, $^1/_4$-W, 5% resistor
$R_9, R_{10}$ – 22-kΩ, $^1/_4$-W, 5% resistors
$R_{11}$ – 100-kΩ, $^1/_4$-W, 5% resistor
$R_{12}, R_{13}$ – 680-Ω, $^1/_2$-W, 10% resistors
*$C_1$ – 22-$\mu$F, 16-WVDC tantalum electrolytic capacitor
$C_2$ – 0.01-$\mu$F, 50-WVDC ceramic capacitor
$SW_1$ – SPST
$SW_2$ – four-position wafer switch
$SW_3, SW_4$ – momentary contact switches
$B_1$ – 9-V transistor battery and connector
Assorted hardware and chassis
*See text.

(b)

**Figure 27.2** Circuit schematic for the darkroom timer.

utes. Potentiometers $R_1$, $R_3$, $R_5$, and $R_7$ are used to adjust the individual time delays. A description on setting these potentiometers for the desired delays is given in the circuit calibration section.

Momentarily depressing the start switch $SW_3$ forces pin 2 of $IC_1$ to about 0 volts and begins the selected time delay. Pin 3 goes to a positive voltage, lighting $LED_2$ and shutting off $LED_1$. At the completion of the delay time, pin 3 returns to about 0 volts, turning on $LED_1$ (*timed-out* indicator) and shutting off $LED_2$ (*timing cycle in progress*).

Normally, $R_9$ holds pin 4 of $IC_1$ high at about 9 volts. If, however, the *reset* switch $SW_4$ is depressed during a timing cycle, pin 3 of $IC_1$ will go immediately to about 0 volts, which activates $LED_1$ and cancels the cycle. A new delay time cycle can then be initiated with the start switch $SW_3$. It is important that the timing capacitor $C_1$ have a good-quality tantalum or polystyrene dielectric.

**Circuit Calibration.** The time delays for the circuit of Fig. 27.2 are calculated using the relationship $T \cong 1.1RC_1$, where $T$ is in seconds, $C_1$ is in farads, and $R$ is the two-resistor combination $(R_1 + R_2)$ or $(R_3 + R_4)$ or $(R_5 + R_6)$ or $(R_7 + R_8)$ in ohms, determined by the position of the time selector switch $SW_2$.

The setting of a 10-second time delay will be used to demonstrate the calibration of the circuit. Resistor $R_1$ is first adjusted to one end of its range. The circuit is cycled and the delay timed. $R_1$ is then adjusted to the other end of its range and again the cycle is timed. These two readings will provide the range of the time delays possible for the two-resistor combination of $R_1$ and $R_2$. To obtain an exact 10-second delay time, $R_1$ is adjusted within the range and the time cycle recorded. This fine tuning is repeated until the exact 10-second delay results. To set and fine tune for the other three time-delay periods, the calibration procedure described is repeated for each position of the selector switch $SW_2$.

**Construction Hints.** There are no special construction problems with this project. As shown in Fig. 27.3, a suggested package is a small, commonly available chassis. Note that $LED_1$ and $LED_2$ are positioned to be prominently visible. Since the *start* and *reset*

**Figure 27.3**  Suggested package for the darkroom timer.

switches will require extensive use, $SW_3$ is positioned on the top and $SW_4$ on the front face of the package. The power switch $SW_1$ is placed along the side.

## 27.2 TTL LOGIC PROBE

Digital circuitry plays a major role in today's electronics industry. While new families of devices are introduced to keep up with the demands of changing technology, TTL (transistor–transistor–logic) devices maintain their dominance due in part to their ready availability and simplicity of use. A technician working with and troubleshooting digital circuits using these devices finds a time-saving instrument in a *TTL logic probe*. The probe described in this project uses readily available parts, is easy to build, and provides effective results.

**Circuit Operation.** The TTL logic probe circuit, shown in Fig. 27.4, is designed around a LM358 low-power dual op amp capable of single-supply operation. Power for

Parts List

IC$_1$ – LM358 low-power dual op amp
$R_1$ – 4.7-kΩ, $^1/_4$-W, 10% resistor
$R_2$ – 2.5-kΩ, 10-turn trim pot
$R_3$ – 2.7-kΩ, $^1/_4$-W, 10% resistor
$R_4$ – 5-kΩ, 10-turn trim pot
$R_5$, $R_6$ – 120-Ω, $^1/_2$-W, 10% resistor
$R_7$ – 3.9-MΩ, $^1/_4$-W, 10% resistor
LED$_1$ – light-emitting diode (RED)
LED$_2$ – light-emitting diode (GREEN)
D$_1$ – IN4001 diode ($I_0$ = 1A, PIV = 50 V)
Two alligator clips and assorted hardware
Case – see text

**Figure 27.4** TTL logic probe schematic.

the probe is applied when the red alligator clip, connected to $J_1$, is attached to the +5-volt power supply bus of the digital circuit to be tested. The probe ground is made when the black alligator clip, connected to $J_2$, is attached to the digital circuit's ground bus. The purpose of diode $D_1$ is to prevent damage to the probe circuitry if these power supply connections are inadvertently reversed.

With power applied, the probe is calibrated as follows: The positive lead of a dc voltmeter is connected to test point 1 ($TP_1$) and the negative lead to $J_2$. Potentiometer $R_2$ is adjusted until a reading of exactly 0.8 volt appears at $TP_1$. This is the maximum allowable voltage level for a TTL logic 0. The positive lead of the voltmeter is next connected to test point 2 ($TP_2$), and potentiometer $R_4$ is adjusted for a voltage reading of 2.0 volts, which is the minimum allowable TTL logic 1 level.

The circuit can be tested by first contacting the probe to a known logic 0 (or to digital ground). The output of op amp B at pin 7 will go to about +3.0 volts and the green light-emitting diode ($LED_2$) will light to indicate a logic 0 state. The red diode ($LED_1$) will remain off. The probe is next contacted to a known logic 1 (or to the digital power supply's +5-volt bus). This will cause the output of op amp A to go to about +3.0 volts, lighting $LED_1$, which indicates a logic 1 level. Resistors $R_5$ and $R_6$ limit the LED "on" current to about 10 mA to conserve power.

The purpose of $R_7$ is to maintain the probe voltage above 0.8 volt and below 2.0 volts, which is called the "forbidden" area for TTL logic. Thus, when the probe contacts neither the logic 0 nor logic 1 state, both of the LEDs will be off.

**Construction Hints.** Construction of the TTL logic probe requires no special considerations. Fabrication of a pc board is recommended but is not essential. The probe package shown in Fig. 27.5 was constructed using a modified probe case. The sharp metal probe is isolated from the case and wired to terminal $J_3$ on the pc board with a short length of hookup wire.

Power (red) and ground (black) leads are AWG No. 24 stranded wire fitted with miniature alligator clips and extend into the probe case. All probe circuitry shown with the digital ground symbol ▽ must be electrically connected together and to the black external ground lead $J_2$.

A final consideration in the packaging of the probe is in the composition of the case. If it is metal, care must be exercised to ensure that none of the circuitry touches it electrically. Any type of round hollow material, such as a short length of PCV piping or a hard plastic tube, may be used as the probe case.

## 27.3 1.2- TO 20-VOLT LABORATORY POWER SUPPLY WITH DIGITAL READOUT

One of the most useful projects the experimenter can construct is a laboratory-type dc power supply. Applications of this type of circuit are almost unlimited. However, for a power supply to be most useful, it should have the following features: (1) adjustable output-voltage capability over a wide range, (2) high-load-current capability, (3) excellent voltage regulation at all load currents, (4) current limiting with short-circuit protection to prevent damage to the supply when accidental overloads or shorts are applied to the output terminals, and (5) metering to monitor output voltage.

*The variable-voltage-*regulated power supply circuit, shown in Fig. 27.6, will meet all of the foregoing requirements with the following specifications:

*Output voltage:* adjustable from 1.2 to 20 volts dc
*Output current:* 0 to 1 ampere
*Voltage regulation:* 0.1%

**Figure 27.5** TTL logic probe package.

*Including:* Load-current limiting
Thermal-overload protection
Safe-area-device protection

The requirements listed for a basic laboratory-type power supply are easily met with the use of an integrated regulator chip having only three external terminals: input (in), output (out), and adjustment (adj). These are shown in Fig. 27.6. The resulting circuit is (1) low in overall parts count, (2) inexpensive, and (3) simpler to package than a comparable-type discrete power supply.

**Circuit Operation.** For purposes of analysis, the power supply circuit schematic diagram, shown in Fig. 27.6, can be divided into two basic sections: an unregulated power supply and a voltage regulator section. The unregulated power supply consists of transformer $T_1$, diodes $D_1$ to $D_4$ in a full-wave bridge configuration, and the filter capacitor $C_1$.

The basic regulator section consists of the LM317T regulator chip and support components $R_2$, $R_3$, $R_4$, $C_2$, and $C_3$. The LM317T regulator chip requires that the output of the unregulated supply, across filter capacitor $C_1$, be limited to a maximum of 40 volts dc. This specification is easily met by $T_1$, since its theoretical peak output voltage under no-load current conditions is approximately 34.5 volts dc. Under full-load current conditions of 1 ampere, the dc output voltage of the unregulated supply is approximately 27 volts dc.

All of the electronics necessary to produce an adjustable, positive regulated power supply are contained in one three-terminal integrated circuit package, available in several standard package styles. The TO-220 plastic case was selected for this project. An internal reference voltage of 1.2 volts (constant) is established between the output (out) and the adjustment (adj) terminals of the LM317T regulator. This reference voltage is applied directly across the 240-ohm resistor ($R_2$) establishing a constant reference current of 5 mA flowing through $R_2$. This reference current also flows through the potentiometer ($R_3$) and parallel padding resistor $R_4$. An increase in the resistance of $R_3$ results in larger output voltages while a decrease in $R_3$ results in lower voltages. The minimum dc output voltage can never be lower than the reference voltage of 1.2 volts.

An ac ripple voltage is present on the input terminal of the regulator chip. At full-load currents of 1 ampere, this ripple can be as high as 5 volts peak to peak. On the output terminal of the regulator, there is essentially no ripple voltage (5 mV or less). The ripple voltage that is present at the input terminal is absorbed by the regulator.

Capacitor $C_2$ is a 0.1-μF disk required if the lead length from the unregulated supply terminal is excessive (2 inches or longer). It is, however, good practice to always include this capacitor. Capacitor $C_3$ (10-μF tantalum) is connected from the output terminal of the regulator to common to improve the transient response of the regulator.

**Parts List**

$IC_1$ – LM317T — positive adjustable regulator (TO–220 package)
    with #6 fiber shoulder washer and thermalloy
    #43-77-2 thermofilm mica washer

$D_1$ through $D_4$ — 1N4001 diodes 1A @ 50 PIV or equivalent

$T_1$ – transformer, secondary 25.2-Vac, 2-A Triad F41-X
    or equivalent

$F_1$ – fuse, 400 mA, 250 Vac, slow-blo

$SW_1$ – SPST slide switch

$J_1$ through $J_3$ – banana jacks

$L_1$ – NE5I

$R_1$ – 220-k$\Omega$, $^1/_2$-W, 10% resistor
$R_2$ – 240-$\Omega$, $^1/_2$-W, 5% resistor
$R_3$ – 10-k$\Omega$, 2 -W, linear taper potentiometer
$R_4$ – 5.6-k$\Omega$, $^1/_2$-W, 10% resistor
$R_6$ – 24-k$\Omega$, $^1/_2$-W, 5% resistor
$C_1$ – 1000-$\mu$F, 50-WVDC electrolytic capacitor
$C_2$ – 0.1-$\mu$F, 100 -WVDC disc capacitor
$C_3$ – 10-$\mu$F, 35-WVDC tantalum electrolytic capacitor
$DPM_1$ digital voltmeter readout PCB

**Figure 27.6** Schematic diagram of variable-voltage-regulated power supply.

498

**Figure 27.7** Schematic diagram of the digital panel meter readout for the regulated power supply project.

Parts List

IC$_{101}$ – 7805 – positive 5-V regulator (TO-220 package) or equivalent
IC$_{102}$ – CA3162E analog-to-digital converter (16-pin DIP)
IC$_{103}$ – CA3161E BCD-to-seven segment decoder/driver (16-pin DIP)
LED$_1$, LED$_2$, LED$_3$ – MAN6760E seven-segment LED displays
    (common anode) or equivalent
D$_{101}$ – IN4001 diode 1 A at 50 PIV or equivalent
Q$_{101}$, Q$_{102}$, Q$_{103}$ – 2N3906 pnp transistors or equivalent
R$_{101}$, R$_{102}$ – 56-Ω, 2-W, 10% resistors

R$_{103}$ – 100-kΩ, 1% resistor
R$_{104}$ – 1-kΩ, 1% resistor
R$_{105}$ – 50-kΩ, trim potentiometer
R$_{106}$ – 10-kΩ, trim potentiometer
R$_{107}$ – 100-Ω, $1/2$-W, 10% resistor
C$_{101}$, C$_{102}$, C$_{103}$ – 10-μF, 25-WVDC tantalum capacitors
C$_{104}$ – 0.33-μF, 100-WVDC disk capacitor

499

The dc output voltage appearing between terminals $J_1$ and $J_2$ is monitored with a digital panel meter (DPM), which is described here in detail.

**Digital Panel Meter.** Recent advances in the development of integrated circuit technology have made it possible to replace older, less reliable analog panel meters with digital readout at lower costs and moderate increases in circuit complexity. To illustrate this, the power supply described in this project will be provided with a digital panel meter instead of the traditional analog meter. This DPM is designed around two 16-pin dual-in-line packaged ICs and support components as shown in Fig. 27.7. The core of this DPM system is the RCA CA3162E integrating analog-to-digital (A/D) converter and its companion, the CA3161E BCD-to-seven segment decoder/driver IC.

Dc power for the DPM is provided from the unregulated output section at $J_4$ of Fig. 27.6. This voltage ($V_{in}$) is then converted by $IC_{101}$ (Fig. 27.7) to +5.0 volts.

The output voltage of the power supply to be monitored is between $J_1$ and $J_2$ of Fig. 27.6. This voltage is connected between the + and − input of the DPM (see Fig. 27.7). The voltage is divided by $R_{103}$ and $R_{104}$ and feeds into the A/D converter ($IC_{102}$), which converts the analog voltage into a binary-coded decimal (BCD) equivalent that appears at pins 2, 1, 15, and 16. $IC_{103}$, in turn, takes this BCD code and converts it into a seven-line output that represents a decimal number. These seven-segment outputs are multiplexed to $LED_1$, $LED_2$, and $LED_3$. The digits are selected by pins 3, 4, and 5 of $IC_{102}$, which provide base drive to turn on $Q_{101}$, $Q_{102}$, or $Q_{103}$. These transistors then feed current to the anodes of the appropriate LED readouts.

The DPM assembled on a double-sided pc board is shown in Fig. 27.8. The DPM requires calibration prior to being integrated into the power supply project. This is done with the calibration setup shown in Fig. 27.9. A power supply of approximately 30 volts is connected between $V_{in}$ and − of the DPM. To zero the meter, the + and − inputs are

**Figure 27.8** Printed circuit board assembly of the digital panel meter (DPM) section of the power supply project.

**Figure 27.9** Calibration setup for the digital panel meter.

shorted and $R_{105}$ is adjusted for an output reading of 00.0. To calibrate the meter, the short is removed and a precise voltage, for example, 20 volts, is applied between the + and − inputs. Potentiometer $R_{106}$ is then adjusted for a reading of exactly 20.0 on the DPM.

**Construction Hints.** For optimum performance, the power supply circuit should be packaged in a well-ventilated enclosure. The LM317T regulator package will require heat sinking to achieve the required specifications. A metal chassis element can be used as a suitable heat sink provided that it consists of at least 25 square inches of $\frac{1}{16}$-inch-thick aluminum. The regulator chip must be electrically insulated from the chassis. This is done with the use of a silicon-greased mica washer and insulating shoulder washer (Fig. 27.10). Note that the leads are bent at 90 degrees onto a pc board. The pc board is used to assemble all the small fragile circuit components. A three-pronged plug is essential for electrical grounding of the chassis. The basic power supply package is shown in Fig. 27.11.

## 27.4 PORTABLE FAHRENHEIT/CELSIUS THERMOMETER WITH LCD READOUT

The accurate measurement of temperature requires a sensor that is linear and its readings repeatable. To satisfy both of these requirements, this electronic thermometer project is designed around the popular AD590 linear integrated circuit from Analog Devices, Inc. This device is a practical current source producing an output current that is directly proportional to temperature in degrees Kelvin (°K). Because temperature readings are more commonly read in degrees Fahrenheit (°F) and degrees Celsius (°C), additional electronic circuitry is required for conversion.

The thermometer designed for this project is portable, accurate, and measures temperature in the ranges of 0 to 199.9 degrees Fahrenheit and 0 to 100 degrees Celsius. Its

**Figure 27.10** Assembly diagram of LM317T regulator package to chassis.

**Figure 27.11** Basic package configuration for the power supply project.

electronics can be divided into three main subdivisions: the power supply circuit, the temperature sensor with its associated signal-conditioning circuitry (SCC), and an integrating-type analog-to-digital converter (ADC) with its liquid crystal display (LCD) readout. Each of those circuits is described in detail.

**Universal Power Supply.** A common design specification for portable hand-held equipment is that all of the electronics be powered from a single energy source such as a 9-volt transistor battery. While standard CMOS digital logic can operate at this voltage level, much of the current analog devices, such as operational amplifiers, require split (dual) voltage sources. The power supply circuit shown in Fig. 27.12 provides a regulated $\pm 5$-volt source from a 9-volt battery. When $SW_1$ (READ) is closed, $IC_2$ converts the +9-volt battery level into a precise +5.000 volts reference at pin 6 of $IC_2$. This reference at $J_1$ is thus made independent of reductions in battery voltage due to age and use. A negative nonregulated output of $-9$ volts is then developed with $IC_1$ and two noncritical 10-$\mu$F tantalum capacitors, $C_1$ and $C_2$. $IC_3$ regulates this output to $-5$ volts at $J_2$.

**Signal-Conditioning Circuit.** The power supply circuit shown in Fig. 27.12 provides a precise +5-volt reference for both the AD590 temperature sensor and its associated bridge circuitry. This is shown in Fig. 27.13. Resistor $R_{101}$ converts the sensor's temperature-dependent current into a temperature-dependent voltage that is applied to the + input of the instrumentation amplifier (IA). The $-$ input of the IA is returned to the null (zero offset) circuitry of the bridge reference through the scale selector switch $SW_2$. When $SW_2$ is *down,* the offset voltage at $TP_3$ is applied to the IA to zero the Fahrenheit scale. When $SW_2$ is *up,* the voltage at $TP_3$ is returned to the $-$ input of the IA to zero the Celsius scale. The three-op amp IA, consisting of $IC_{102}$ A, B, and C and its associated circuitry, performs two major functions. First, it provides a single-ended output voltage to drive the input of the ADC voltmeter and second, it applies the correct amplification so that the output voltage of $TP_4$ is scaled to $(10 \text{ mV}/^\circ\text{F})T_F$ or $(10 \text{ mV}/^\circ\text{C})T_C$, depending on the position of $SW_2$.

**Circuit Calibration—Fahrenheit Scale.** With power supply switch $SW_1$ closed and $SW_2$ down ($^\circ$F position), a digital voltmeter is connected between $TP_1$ and circuit ground. At room temperature of approximately 77.5$^\circ$F, the voltage across $R_{101}$ should be about 293 mV. By pinching the AD590 sensor between the index finger and thumb, the

**Figure 27.12** Power supply electronics to develop ±5 volts for the thermometer project.

Parts List

$B_1$ – 9 V transistor battery
$SW_1$ – SPST pushbutton switch
$IC_1$ – ICL7660 voltage converter
$IC_2$, $IC_3$ – 5-V precision voltage reference ICs REF-02
$C_1$, $C_2$ – 10-$\mu$F, 25-WVDC tantalum capacitors
$R_1$ – 2.2-k$\Omega$, $^1/_4$-W, 10% resistor

body temperature will increase this voltage at $TP_1$ by about 7 mV for a reading in the vicinity of 300 mV. With this basic test completed, the calibration procedure continues. The *offset adjustment* is made by connecting a digital multimeter (DMM) to $TP_2$ and adjusting $R_{103}$ for a voltage reading of precisely 255.4 mV. *Span adjustment* (or gain adjustment) requires that the DMM be connected to read the voltage of $TP_4$. The room temperature is then measured, in °F, with a high-quality laboratory thermometer, and the output voltage reading at $TP_4$ corresponding with this temperature is calculated using the relationship $V_0 = (10 \text{ mV/°F}) \, T_F$. For example, if the room temperature measures 77.5°F, then $V_0 = (10 \text{ mV/°F}) \, 77.5°F = 77.5 \text{ mV}$. For this example, $R_{109}$ would be adjusted for a reading at $TP_4$ of 775 mV. With a fixed decimal point positioned between the 7 and the 5, the digital voltmeter would provide a direct reading in °F (i.e., 77.5).

**Circuit Calibration—Celsius Scale.** The *offset adjustment* is made by first placing $SW_2$ in the *up* position for the °C scale. A DMM is then connected to $TP_3$ and $R_{106}$ is adjusted for a voltage reading of exactly 273.2 mV. (No span adjustment is required on the Celsius scale.) The DMM is moved to $TP_4$ and again, room temperature is precisely measured in °C. The corresponding voltage reading of $TP_4$ for the temperature is calculated with the equation $V_0 = (10 \text{ mV/°C}) \, T_C$. For example, if the room temperature is 24.5°C, then $V_0 = (10 \text{ mV/°C}) \, 24.5°C = 245 \text{ mV}$, which corresponds to 24.5°C.

**Circuit Description—$3\frac{1}{2}$-Digital Readout System.** The readout system for the thermometer project is built around the familiar ICL7106 dual-slope integrating-type A/D converter and a 0.6-inch high-character LCD (see Fig. 27.14). Both the ADC and the LCD are housed in 40-pin dual-in-line packages. The supporting components have been selected to optimize circuit performance for a 0- to 199.9-mV full-scale input range and a corresponding output display.

Output signals from the SCC and $TP_4$ of Fig. 27.13 are applied to the A/D converter

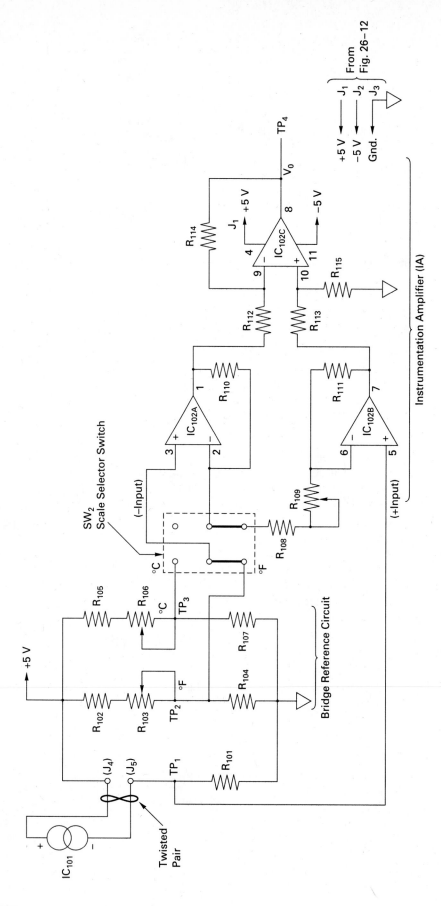

## Parts List

IC$_{101}$ – AD590 two-terminal temperature sensor from Analog Devices
*IC$_{102}$ – LT1014 quad precision op amp from Linear Technology
R$_{101}$ – 1-kΩ, $^1/_8$-W, 1% resistor
R$_{102}$, R$_{105}$ – 20-kΩ, $^1/_8$-W, 1% resistors

R$_{103}$, R$_{106}$, R$_{109}$ – 25- kΩ, 10-turn potentiometers
R$_{104}$, R$_{107}$ – 2.49-kΩ, $^1/_8$-W, 1% resistor
R$_{108}$ – 12- kΩ, $^1/_8$- W, 5% resistor
R$_{110}$, R$_{111}$, R$_{112}$, R$_{113}$, R$_{114}$, R$_{115}$ – 10-kΩ, $^1/_8$- W, 1% resistors
SW$_2$ – DPDT miniature toggle switch

* The LM324 quad op amp from National Semiconductor is a substitute

**Figure 27.13**   Sensor and signal conditioning circuitry (SCC) for the thermometer project.

IC$_{201}$ – ICL7106 integrating-type A/D converter
IC$_{202}$ – FE0203 liquid crystal display (LCD)
IC$_{203}$ – CA4030 quad exclusive OR gate
R$_{201}$, R$_{206}$, R$_{207}$, R$_{208}$ – 1-MΩ, $^1/_8$-W, 10% resistors
R$_{202}$ – 47-kΩ, $^1/_8$-W, 10% resistor
R$_{203}$ – 100-kΩ, $^1/_8$-W, 10% resistor
R$_{204}$ – 1-kΩ, 10-turn potentiometer

R$_{205}$ – 24-kΩ, $^1/_8$-W, 5% resistor
C$_{201}$ – 0.01-μF, 50-WVDC disk capacitor
*C$_{202}$ – 0.22-μF capacitor
*C$_{203}$ – 0.47-μF capacitor
*C$_{204}$ – 0.1-μF, capacitor
C$_{205}$ – 100-pF, 50-WVDC disk capacitor
SW$_3$ – SPST, PC mount, switch block

*See text.

**Figure 27.14**  Circuit schematic of the ICL7106 A/D converter and LCD readout.

through a low-pass filter ($R_{201}$ and $C_{201}$) having a cutoff frequency of approximately 15.9 Hz. The input signals that are passed are processed by the dual-slope method of conversion employed by the 7106 IC. This conversion method samples input voltage and converts it into a current that charges capacitor $C_{202}$ for a fixed interval. At the end of this charging time interval, the input signal is replaced by a precise reference voltage which is developed across $C_{204}$ and established externally by adjusting $R_{204}$. This reference voltage allows $C_{202}$ to be discharged at a constant rate until its charge is completely removed. The time for this discharge is therefore directly proportional to the value of the input voltage level and is measured with an internal counter operating from an internal clock.

To test that the internal clock is operating correctly, an oscilloscope (or counter) is connected between pins 37 and 32 (GND) of $IC_{201}$. With a 10 × probe, a clock frequency of 48 kHz ± 20% is an acceptable reading.

All digital decoding and drive necessary to interface to a low-power LCD is provided on the 7106 IC. The decimal-point position for the LCD is selected by depressing one of the three switches labeled *dp* on $SW_3$. For this project, $dp_3$ should be the choice so that the output of the SCC will be interpreted directly into units of °F or °C.

As a final note, it is recommended that capacitors $C_{202}$, $C_{203}$, and $C_{204}$ be manufactured with a polypropylene or Mylar dielectric.

**Final Circuit Calibration.** To complete the final calibration of the digital readout system, a DMM is connected between pins 36 and 32 (GND). Potentiometer $R_{204}$ is adjusted until an exact reading of 100.0 mV is obtained. The system is then ready to accept the input signals from the SCC circuit.

**Construction Hints.** The electronic thermometer shown in Fig. 27.15 is housed in a readily available plastic instrument box. All of the signal-conditioning circuitry was assembled on a 2- by 3-inch single-sided pc board that is piggybacked with a 3- by 5-inch double-sided board on which is packaged the ADC and LCD display. The universal power supply circuit is assembled on a 1- by 2-inch pc board. The AD590 temperature sensor, soldered to a 3-foot length of twisted pair AWG No. 24 wire, is external to the chassis, as shown in Fig. 27.15a.

To complete the package, a 1- by 2-inch cutout was formed in the plastic case to expose the LCD. A small DPDT switch ($SW_2$, the scale selector switch) is mounted on the front of the case and labeled. Last, the pushbutton SPST switch ($SW_1$) is also mounted on the case front. Depressing this button applies power and displays the temperature. Upon releasing the button, power is removed in order to conserve battery life. These are shown on the completed thermometer package in Fig. 27.15b.

## 27.5 PORTABLE SPORTSMAN'S BAROMETER WITH LCD READOUT

The sportsman's barometer project, shown in Fig. 27.16, is built around a solid-state absolute pressure sensor (type SCX15ANC from Sensym Corporation). The rugged construction of this sensor provides high reliability at moderate cost.

A portable barometer has significant advantages to those who spend a great deal of time outdoors hiking, skiing, fishing, and so on, where changing weather conditions may alter one's plans suddenly. A drop in barometeric pressure signifies that stormy weather is approaching, while a rising pressure indicates fair-weather conditions ahead. A steady pressure denotes unchanging weather patterns.

This portable barometer outputs absolute pressure digitally in units of inches of mercury (in. Hg) in a range from 28.0 to 32.0 in. Hg. As an aside, fishermen have found that the best fishing occurs when sea-level pressure is between 29.6 and 30.4 in. Hg and rising.

Note in Fig. 27.16 that both the universal power supply section and the $3\frac{1}{2}$-digit

(a)

**Figure 27.15** Portable Fahrenheit/ Celsius thermometer with liquid crystal display: (a) major pcb assemblies; (b) completed project.

(b)

readout sections are similar to those described in the section on the thermometer project. They are shown here for purposes of completeness, but their operation will not be repeated.

**Circuit Operation—Pressure Sensor and Signal-Conditioning Circuit.** Pressure sensor PS100 outputs a differential voltage between pins 5 and 3 that is directly proportional to absolute pressure. This voltage is applied to the instrumentation amplifier consisting of $IC_{102}$, $IC_{103}$ and resistors $R_{104}$, $R_{105}$, $R_{106}$, $R_{107}$, and $R_{108}$. $R_{105}$ is a precision 1% 3.24-k$\Omega$ resistor which sets the IA's gain and scales the output at pin 6 of $IC_{103}$ into units of in. Hg. The output voltage of $IC_{103}$ is applied to the readout section for the digital display. Dc power for the three op amps as well as for the sensor is obtained from the universal power supply.

**Circuit Calibration.** Prior to calibrating the sensor SCC, the reference voltage for the digital readout section should be adjusted just as it was in the thermometer project. A DMM is connected between pins 36 and 32 (GND) and $R_{204}$ is adjusted for an exact dc

**Parts List: Universal Power Supply Section**

$B_1$ – 9-V transistor battery
$SW_1$ – SPST switch
$IC_3$ – ICL7660 voltage converter
$IC_1$, $IC_2$ – 5-V precision voltage reference ICs REF-02
$C_1$, $C_2$ – 10-$\mu$F, 25-WVDC tantalum capacitors
$R_1$ – 2.2-k$\Omega$, $^1/_4$-W, 10% resistor

(a)

**Figure 27.16**   Complete circuit schematic and parts list for the sportsman's barometer project: (a) universal power supply section;

voltage reading of 1.00 volt. The readout section is now ready to accept input signals. To calibrate the barometer, it is first necessary to obtain the barometric pressure by calling the local airport or weather station. Closing $SW_1$ to apply power to the system, $R_{102}$ is adjusted until the digital readout displays the correct pressure.

**Construction Hints.** The completed sportsman's barometer project is shown in Fig. 27.17. Note that it is packaged in the same type of plastic case as the previous project on the thermometer. For this reason, all of the construction hints presented in that section also apply to this project.

## 27.6 AUTOMOBILE SECURITY ALARM SYSTEM

Although no alarm system is foolproof, they do serve as a deterrent to automobile thefts, especially those thefts committed by the amateur. The most positive types of automobile alarm systems are those that employ mechanical door, hood, and trunk switches. Although this type of system is more difficult to install than the voltage- or current-sensing type of alarm systems, it is less susceptible to false triggering due to climate changes, extreme variations in temperature, or outside electrical interference. The most effective alarm systems are those that are completely concealed inside the automobile, thus giving

Parts List: Pressure Sensor and SCC Section

PS100 – SCX15ANC absolute pressure sensor from Sensym Corporation

IC$_{101}$, IC$_{102}$, IC$_{103}$ – OPO7A op amp
R$_{101}$, R$_{103}$ – 100-kΩ, $^1/_8$-W, 1% resistors
R$_{102}$ – 10-kΩ, 10-turn potentiometer
R$_{104}$, R$_{106}$, R$_{107}$, R$_{108}$ – 10-kΩ, $^1/_8$-W, 1% resistors
R$_{105}$ – 3.24-kΩ, $^1/_8$-W, 1% resistor

3$^1/_2$-Digit Readout Section

Pressure Sensor and Signal-Conditioning Circuit Section

(b)

Parts List: 3$^1/_2$ – Digit Readout Section

IC$_{201}$ – ICL7106 integrating-type A/D converter
IC$_{202}$ – FEO203 liquid crystal display (LCD)
IC$_{203}$ – CA4030 quad exclusive OR gate
R$_{201}$ – 1-MΩ, $^1/_8$-W, 10% resistor
R$_{202}$ – 470-kΩ, $^1/_8$-W, 10% resistor
R$_{203}$ – 100-kΩ, $^1/_8$-W, 10% resistor
R$_{204}$ – 25-kΩ, 10-turn potentiometer

R$_{205}$ – 24-kΩ, $^1/_8$-W, 5% resistor
C$_{201}$ – 0.01-μF, 50-WVDC disk capacitor
*C$_{202}$ – 0.22-μF capacitor
*C$_{203}$ – 0.047-μF capacitor
*C$_{204}$ – 0.1-μF capacitor
C$_{205}$ – 100-pF, 50-WVDC disk capacitor

*Use polypropylene or Mylar capacitors for best results

**Figure 27.16** (Cont.) (b) pressure sensor and signal conditioning circuit section.

509

**Figure 27.17** Sportsman's barometer.

potential thieves a false sense of security. The automobile security alarm system and parts list, shown in Figs. 27.18 and 27.19, is a reliable, relatively inexpensive system that will provide a high degree of security for an automobile.

**Circuit Operation.** An automobile security alarm system is usually installed in the trunk of the automobile. All the required electrical connections are then made from this location.

Switch $SW_1$ is a DPDT slide or toggle switch that serves two functions: It "arms" or "disarms" the alarm system. In the "armed" position, the switch acts to connect the distributor side of the ignition points to ground, thus preventing the car from being started. This arm/disarm switch is usually mounted under the dashboard of the car, hidden from view but accessible to the driver.

The alarm system includes (1) an exit delay, (2) an entrance delay, (3) alarm and re-cycle, and (4) an "instant on" protection feature for both the hood and trunk. To activate the system, the arm/disarm switch $SW_1$ is moved from the "off" or "disarmed" position to the "armed" position. After the switch has been activated, the operator has a preset exit delay time of approximately 17 seconds to leave the car and lock the doors. (If closing the doors takes longer than 17 seconds, the horn will sound.) The exit delay time interval is set by the timing capacitor $C_3$ and resistor $R_2$. At the end of the exit delay cycle, the output of $IC_2$, at pin 3, goes from approximately 0 volts to 6 volts, applying power to the reset terminals of $IC_3$ and $IC_4$ at pin 4. The system is now armed and ready to detect an intruder entering any protected door. If desired, $LED_1$ can be connected between jacks $J_2$ and $J_3$ as shown in Fig. 27.18. $LED_1$ will light at the end of the 17-second exit delay inter-

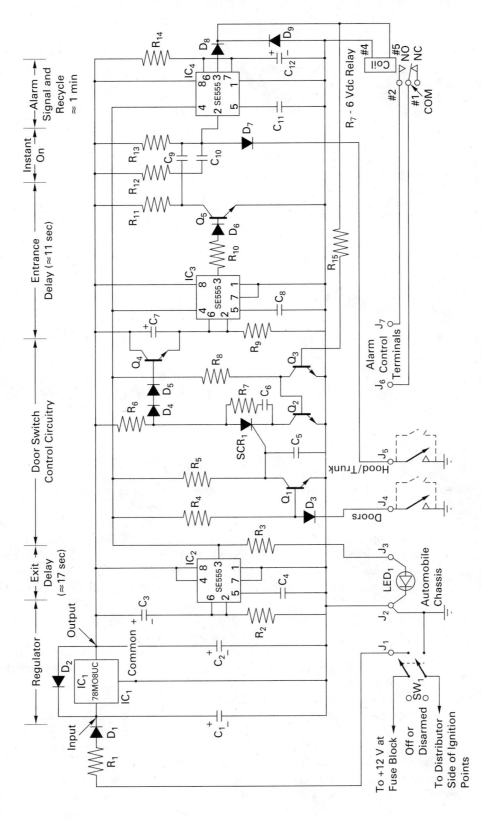

**Figure 27.18**   Circuit schematic for the automobile security alarm system.

511

**Figure 27.19** Parts list for the automobile alarm project of Fig. 27.18.

val, indicating that the system is armed. This light can be mounted on the dashboard of the car.

Once the circuit is armed, the door switch control circuitry, consisting of R$_4$, R$_5$, Q$_1$, D$_3$, SCR$_1$, R$_6$, D$_4$, D$_5$, Q$_4$, and Q$_2$, is also powered. As long as all door switches are open (door switches are open when car doors are closed since the door position depresses the switch plunger), transistor Q$_4$ remains on and acts as a short circuit across C$_7$, preventing IC$_3$ from operating. An intruder opening a door causes the switch plunger to release and closes the switch, which causes the collector of Q$_1$ to go high, firing the gate of SCR$_1$. The base of Q$_4$ will go to approximately 0 volts, causing its collector-to-emitter to act as an open circuit across C$_7$. The timing capacitor C$_7$ (associated ith IC$_3$) begins to charge. The intruder now has 11 seconds to find the arm/disarm switch and return it to the "off" position. If the switch is not thrown within the preset entrance delay time of 11 seconds, the charge developed across C$_7$ (through resistor R$_9$) will cause the output of IC$_3$ (pin 3) to go high, thus triggering IC$_4$. With IC$_4$ triggered, its pin 3 goes high, turning on diode D$_8$, which places 6 volts across the coil of the relay, Ry. The relay switches from its normally open condition to cause a closed condition at points J$_6$ and J$_7$. As can be seen in Fig. 27.20a, b, or c, when J$_6$ and J$_7$ are shorted together by the switching action of relay Ry, they complete the alarm circuit. This causes the alarm siren or automobile horn to sound for a duration of approximately 1 minute, this time interval being set by R$_{14}$ and C$_{12}$. After the alarm sounds for 1 minute, the system will recycle. If, for example, during the 1-minute sound interval, the intruder is frightened off and closes the door which he initially opened, the system will recycle. Even though the door has been returned to its closed position, the sound will complete its 1-minute cycle, after which the system is set and ready to detect another door intrusion. If, on the other hand, the intruder leaves without closing the door which he initially opened, the alarm will sound for 1 minute, shut down for an 11-second interval, and then sound again for 1 minute. This 1 minute on/11 seconds off cycle will continue until the opened door is finally closed.

(a) One-Wire horn system

(b) Two-Wire horn system

**Figure 27.20** Horn and siren wiring diagrams.

(c) Optional siren hook-up

Since it is not advisable to have any delay on either the hood or trunk switches, the "instant-on" feature accomplished by $C_{10}$, $R_{12}$, and $D_7$ is included. Opening either the hood or trunk after the system has been armed will instantly sound the alarm for the same preset 1-minute interval as established by $R_{14}$ and $C_{12}$.

Diode $D_1$ has been added to prevent damage to any of the components in the system should the correct polarity be reversed when wiring the system into the automobile. In addition, an on-board voltage regulator circuit has been added. This circuit consists of $IC_1$, $C_1$, $C_2$, and $D_2$, and its purpose is to minimize any voltage spikes from the car's electrical system being transferred to the IC timers, causing damage or false triggering.

**Construction Hints.** Because of the complexity of the circuit, it is strongly recommended that all components be assembled onto a pc board. The pc board should be mounted onto a chassis and protected by a sturdy enclosure. This unit is best secured in a safe position in the trunk.

Jack $J_2$ should be wired to the automobile chassis frame close to where the circuit enclosure is mounted. At least AWG No. 18 wire should be used and the connection tested to ensure that a good electrical contact has been made.

It is generally unnecessary to buy switches for the door, trunk, or hood, since these are often standard equipment on many cars. All existing switches should be checked for perfect working order. If a switch must be installed at any of the locations, the plunger type is most suitable for this application. A newly installed switch must be wired in parallel with either those using exit and entrance delay (door switches) or instant-on (trunk or hood). For example, hood and trunk switches should be wired in parallel with each other but not in parallel with the door switches (Fig. 27.18). Since diode $D_3$ isolates the alarm circuit from the 12-volt source and all door switches are in parallel, only a single wire connection is necessary between $J_4$ and the terminal connection on one door switch. In like manner, only one wire is necessary between $J_5$ and either the trunk or hood switch.

**Caution.** Never attempt to switch any automobile horn directly with the pc on-board relay. These horns draw in excess of 3 amperes each and should be controlled only from the car's horn relay. Depending on the type of automobile horn system, wiring according to Fig. 27.20a or b is recommended. As seen in Fig. 27.18, $J_6$ and $J_7$, have been left uncommitted for controlled switching of any particular type of horn circuit. This allows for convenient wiring connections to be made between $J_6$ and $J_7$ and any of the circuits shown in Fig. 27.20a, b, or c. If the car has no horn relay or if a siren is desired, an automobile horn relay must be installed. Sirens for alarm systems for use on a 12-volt dc source can draw in excess of 20 amperes. A horn relay capable of handling at least this much current must be selected. The siren, +12 volts, and ground connections should be made with a minimum of AWG No. 10 wire.

The finished security alarm package is shown in Fig. 27.21.

## 27.7 MICROCONTROLLER-BASED REFRIGERATION ALARM SYSTEM

This project illustrates an industrial trend in the area of data acquisition (DA) and process control, that is, the development of small, computer-based systems to first measure physical parameters, such as temperature, barometric pressure, and weight, and then input the measured analog signals into either a microprocessor ($\mu P$) or microcontroller ($\mu C$) for further data processing and/or control. Within the processor, the analog signal is immediately converted into a digital equivalent code that is stored in memory. It is this digital code that represents the physical measurement and is a necessary first step to create any computer-based system.

Microcontroller-based data acquisition and process control systems are commonplace today. They are found in such diverse fields as automotive testing, biomedical engineering, and the meteorological sciences, to name just a few.

The software needed to run a microcontroller-based process control system is called an *application program* and most application programs begin with software code, called a *driver,* that performs the actual measurement and stores the data in memory. Therefore, the driver acts as the software interface between input hardware (sensor and analog inter-

(a)

(b)

**Figure 27.21** Completed alarm project: (a) alarm electronics mounted on a double-sided pcb; (b) chassis for alarm.

face circuitry) and the code to be written to control the output devices and solve the engineering problem.

Application programs can be written to take the stored digital code and display it on a video monitor, send it to a printer, mathematically manipulate it to perform scaling, compare it against a previously inputted set-point value to make a control decision, such as turning an output device *on* or *off,* or any number of other operations limited only by the engineering problem to be solved and the programmer's imagination. It is the application software, running on the processor, controlling and manipulating the stored data, that provides the tremendous versatility within a microcontroller-based system. Often, changes to software are all that are needed to completely redefine a control systems entire function.

**Statement of the Problem.** The object is to steadily monitor the internal temperature of a standard home refrigerator that is typically set below 38°F (3.3°C) for normal operation. When working correctly, a green status light must be *on* continuously. If a fault occurs and the refrigerator's temperature rises above the normal set-point temperature, the green status light should begin blinking and an audible alarm should be activated to alert the user to a potential problem—the possibility of food spoilage—if the difficulty is not immediately corrected.

**A Basic Overview of the Solution.** As with any measurement and control problem, this project begins with the selection of an appropriate sensor to measure a physical quantity, in this case the normal refrigerator storage temperature of 38°F (3.3°C). A thermistor or *therm*ally sensitive res*istor* is a negative *temp*erature *co*efficient (TEMPCO) device and a good choice to measure temperature in this range and then convert it into an analog voltage. This voltage is then buffered and applied directly to one input of the on-chip analog-to-digital converter (ADC) of the microcontroller.

Two separate microcontroller output control lines (TTL compatible) will be needed to complete the project. One line will drive the green status indicator (LED) circuitry and the second will control a low-voltage, low-current, solid-state piezoelectric buzzer. In addition, the entire system will be powered by a single 5-volt dc voltage regulator.

Finally, an assembly-language application program is included to both acquire the temperature information and then solve the monitor/alarm problem and complete the project. The application program for this project compares the measured input data (in hexadecimal form) with the hexadecimal equivalent of 38°F (3.3°C). When the input voltage drops below the set-point value (temperature exceeds 38°F), the LED must begin to flash at a rate of approximately $\frac{1}{2}$ hertz and the alarm should sound at the same rate.

Of course many other applications that solve useful engineering problems can be developed for the data acquisition and process control hardware presented in this project. The uses are virtually unlimited.

**Circuit Operation.** Figure 27.22 shows the complete block diagram and circuit schematic for this refrigerator temperature monitor and alarm project, which was designed around motorola's low-cost M68HC11 evaluation board (EVBU). The system hardware, external to the EVBU board, includes (1) a linear voltage regulator for system power, (2) sensor and interface circuitry to monitor temperature, and (3) status lamp and audio alarm. What follows is a detailed description of each hardware element group.

**Single 5-Volt dc Power Supply Circuitry.** Dc power is supplied to all functions from a single three-terminal linear regulator, $IC_1$. Diode $D_1$ is added to prevent damage if power is applied incorrectly and the output of the +5-volt dc regulator is wired directly to connector $P_2$ on the EVBU printed circuit board. $R_1$ diverts a small amount of power (HEAT) from the IC regulator package.

**Sensor and Analog Interface Circuit.** Figure 27.22b is the complete schematic needed to interface the output signal of a UUA41J1 thermistor ($T_1$) into one ADC input of the microcontroller. This circuit consists of a stable 100-μA current source ($IC_2$) thermistor sensor and a single-supply operational amplifier ($IC_3$) configured as a voltage-follower with a gain of +1.

$IC_2$ delivers a precise 100 μA to the grounded sensor, thermistor $T_1$. This drive current $I$ is constant in value and independent of power supply variations. $IC_3$ buffers thermistor output voltage $V_T$, and presents a single-ended voltage ($V_o$) to the ADC input within its 5-volt range. To check operation of the sensor and analog interface circuit shown in Fig. 27.22b, apply dc power and connect a good-quality digital multimeter to pin 1 of $IC_3$ to measure $V_o$. Temporarily replace the thermistor with a decade-resistor box set to

(a) The main components and wiring for the refrigerator alarm

(b) Temperature sensor and analog-to-digital converter wiring

**Parts List**

PCB — M68HC11 Evaluation Board microcontroller (EVBU) from Motorola
   Corporation
$IC_1$ — 7805 positive +5 Vdc linear voltage
   regulator (TO-220 package)
$IC_2$ — REF-200 - 100-$\mu$A IC current source
   from Burr-Brown Corporation
$IC_3$ — $^1/_2$ LM358 — single-supply operation
   amplifier
$IC_{4A}$, $IC_{4B}$ — 74HO7 TTL buffer/driver
$Q_1$ — 2N3904 – NPN transistor (TO-92
   package)
$D_1$ — IN4001 – diode ($I_F$ = I A, PIV = 50 V)
$T_1$ — UUA41J1 – thermistor from Fenwell
   Corp.
$LED_1$ — Light-emitting diode (green)
$BZ_1$ — Audio alarm
$R_1$ — 47 $\Omega$, $^1/_2$ W, 10%
$R_2$ — 220 $\Omega$, $^1/_4$ W, 10%
$R_3$ — 4.7 k$\Omega$, $^1/_4$ W, 10%
$R_4$ — 1 k$\Omega$, $^1/_4$ W, 10%
$C_1$, $C_2$ — 10 $\mu$F, 35 WVDC electrolytic
$C_3$ — 1 $\mu$F, 15 WVDC electrolytic

**Figure 27.22** Microcontroller-based refrigerator alarm system.

27,600 ohms (thermistor's resistance at 38°F) and measure $V_o$. The voltage $V_o$ should equal $V_T = (I) R_s = (100 \ \mu A) (27.6 \ K) = 2.76$ volts. Replace the decade box with the thermistor when proper circuit operation is verified.

$R_4$ acts as a pull-up resistor, applying a 5-volt reference to the ADC's $V_{RH}$ pin. (Pin 52 on connector $P_4$.) Capacitor $C_3$ is connected between the ADC's $V_{RH}$ and $V_{RL}$ pins and provides filtering. See the ADC wiring detail shown in Fig. 27.22b.

Access to the 8-bit ADC (via Port E) on the M68HC11 microcontroller chip is made available to the user at connector $P_4$ of the EVBU board. See the pin assignments for $P_4$ given in Fig. 27.23a. While there are 8 input channels on the M68HC11 chip, we will use only one, $PE_1$ (pin 45 on connector $P_4$), for this project.

**Green Status LED and Audio Alarm.** A medium-size green light-emitting diode ($LED_1$), used as a status indicator, is shown in Fig. 27.22a with accompanying drive circuitry. Output control line $PB_0$, accessible at pin 42 of connector $P_4$, is wired to the base of $Q_1$ through a 4.7-k$\Omega$ current-limiting resistor, and the LED's anode is connected to $Q_1$'s emitter. Under software control, when $PB_0$ is driven high (Logic 1), $Q_1$ conducts and the LED lights. A Logic 0 on $PB_0$, will cause the LED to be off.

In a similar fashion, port B's line $PB_4$ is used to control the audio alarm $BZ_1$. However, for this circuitry, the alarm will sound when $PB_4$ goes low to a Logic 0 and will be off when $PB_4$ goes high to a Logic 1. To supply 40 mA of drive current for $BZ_1$, two 7407 buffer/driver gates, $IC_{4A}$ and $IC_{4B}$, are connected in parallel as shown in Fig. 27.22a.

**Communicating with the M68HC11.** Note that Fig. 27.22 also shows the EVBU board connected to a personal computer's serial input through a ribbon cable connected to connector $P_2$. The pin assignments for connector $P_2$ are shown in Fig. 27.23b. For communication between the EVBU and the PC, the authors used a software communication program called KERMIT. It should be noted here also that once the applications software has been assembled into the EEPROM area of the microcontroller, the PC can be disconnected from the EVBU. This refrigerator monitor and alarm system is a stand-alone project requiring the PC only for software assembly, not for actual project usage.

**Construction Hints.** Assembly of all the analog interface and output drive circuitry, shown in Fig. 27.22, is best accomplished on the small 3-inch by 3-inch wire-wrap area on the EVBU printed circuit board. Use the wire-wrap techniques discussed in Sec. 26.8 for component assembly and wiring. Using wire-wrap sockets for $LED_1$, driver $Q_1$, current source $IC_2$, buffer $IC_3$, and alarm buffer/driver $IC_4$ will greatly facilitate troubleshooting and repair. Power supply connections complete the wiring of the analog interface circuit to the microcontroller evaluation board. Once wired, the M68HC11 pcb is mounted in a commercially purchased case using standoffs, and small stick-on rubber feet complete the package. Figure 27.24 illustrates the complete system assembly. Note that no provision was made to make connector $P_2$ accessible from the case. The $P_2$ connector is needed only for software assembly, not for system utility.

**The Applications Program.** The applications program needed to complete this project must first condition the ADC to accept data, then acquire data from the sensor and analog interface circuit, store the measured value in a memory location, and then perform a comparison to determine if the measured temperature is below 38°F (normal operation) or above the set point temperature. If the measured temperature is above 38°F, the LED must begin flashing and the audio alarm must sound. Figure 27.25 (see pg. 521) shows the flowchart to accomplish these tasks, and the actual assembly-language code is given in Fig. 27.26 (see pg. 522). In a later section, specific information will be given on how to assemble the application program code into the M68HC11 microcontroller's EEPROM area of memory.

Starting at EEPROM memory location $B600, the first two lines of code send the

(a) Pin assignments for I/O connector, P4 on the EVBU PCB (top view)

**Figure 27.23** Transducers are interfaced to the ADC of the M68HC11 microcontroller at port E located on the $P_4$ connector. Port B output lines are also accessible at the $P_4$ connector. Communication between the μC and the PC is serial, using connector $P_2$.

| | | | | |
|---|---|---|---|---|
| GND | 1 | | | |
| TXD | 2 | | 14 | NC |
| RXD | 3 | | 15 | NC |
| NC | 4 | | 16 | NC |
| CTS | 5 | | 17 | NC |
| DSR | 6 | | 18 | NC |
| SIG-GND | 7 | | 19 | NC |
| DCD | 8 | | 20 | DTR |
| NC | 9 | | 21 | NC |
| NC | 10 | | 22 | NC |
| NC | 11 | | 23 | NC |
| NC | 12 | | 24 | NC |
| NC | 13 | | 25 | NC |

(b) Pin designation for the P2 communications terminal on the EVBU board

**Figure 27.24** The completed package for the microcontroller refrigeration alarm system.

correct control word to the µC's *options* register, instructing it on power up to activate the on-chip ADC. Then lines 3 and 4 turn on $LED_1$, indicating normal operation.

The flow chart in Fig. 27.25 shows the *driver* software beginning at line 5. The driver must first properly initialize the ADC control register and then provide a necessary 65-microsecond time delay to perform the analog-to-digital conversion. Note that the M68HC11 microcontroller's one-line assembler interprets all numbers as hexadecimal. Note also that we use the $ symbol here in the text to denote hexadecimal numbers, but the one-line assembler needs no such prefix. Figure 27.26 shows the code properly written without the $ prefix.

Lines 5 and 6 of the program immediately load the control word $01 (0000 0001) into accumulator A and then store it in the ADC control/status register at location $1030. The $01 control word sets the ADC for a single measurement at port $PE_1$. Next, lines 7, 8, and 9 establish a time delay of about 65 microseconds, giving the ADC sufficient time to make the conversion and then transfer the data to an internal data storage register at location $1031.

The hex equivalent of 38°F ($80) is next loaded in accumulator A and then temporarily stored at location $40 where it is compared with the measured value. The compare and branch instructions (lines 13 and 14) returns the operation back to location $B60A for another measurement if the previous measurement was below 38°F or to line 15 if the temperature was above 38°F.

Oscillation of the LED and audio alarm follows. Lines 15 to 22 of this applications program switch off the LED and turn on the buzzer for approximately $\frac{1}{2}$ second. Lines 23 to 30 switch the LED on and the buzzer off for approximately $\frac{1}{2}$ second.

Line 31 completes the oscillation loop by returning the program to address $B61F to repeat the oscillation cycle.

**Loading the Applications Program.** To assemble an applications program into the microcontroller, connect your PC's serial communications port to connector $P_2$ of the EVBU, and use a communications software package like KERMIT. Then with power off,

**Figure 27.25** Flow chart details the applications program for the refrigerator project.

insure that jumper $J_1$ is connected to the MONITOR position (pins 1 and 2 wired together) as shown in Fig. 27.22a.

We will assume that KERMIT is loaded in a subdirectory called MS-Kermit on your personal computer, and that you wish to load the application program into the M68HC11's EEPROM memory. Change to the KERMIT directory by typing the underscored information below:

C:\>_KERMIT<CR>_

Then at the KERMIT prompt, type

MS-Kermit>_SET PORT 1<CR>_

This instructs the software as to which serial communication port on your computer you have connected to $P_2$ of the EVBU. Port 1 is shown but port 2 should be substituted if it applies to your hardware. Then set the baud rate in software to match the pin selector rate on the EVBU. Again at the KERMIT prompt, type

```
B600    LDAA    #93     \   Initialize ADC at power up          ;line  1
        STAA    1039    /                                       ;line  2
        LDAA    #01     \   Set PB0 line high and PB4 low and   ;line  3
        STAA    1004    /   turn LED on, and alarm off          ;line  4
B60A    LDAA    #01     \                                       ;line  5
        STAA    1030    |   Measure refrigerator                ;line  6
        LDAB    #1A     |   temperature                         ;line  7
B611    DECB            |                                       ;line  8
        BNE     B611    /                                       ;line  9
        LDAA    #C0     \   Store 38 degrees (HEX $C0)          ;line 10
        STAA    40      /   in RAM memory                       ;line 11
        LDAA    1031    \                                       ;line 12
        CMPA    40      |   Compare set point against           ;line 13
        BLS     B60A    /   measurement                         ;line 14
B61F    LDAA    #01     \   Set PB0 low and PB4 high            ;line 15
        STAA    1004    /   turn LED off and buzzer on          ;line 16
        LDAA    #FF     \                                       ;line 17
        LDAB    #FF     |                                       ;line 18
B628    DECA            |   1/2-Hertz oscillator                ;line 19
        BNE     B628    |                                       ;line 20
        DECB            |                                       ;line 21
        BNE     B628    /                                       ;line 22
        LDAA    #00     \   Set PB0 high and PB4 low            ;line 23
        STAA    1004    /   turn LED off and buzzer on          ;line 24
        LDAA    #FF     \                                       ;line 25
        LDAB    #FF     |                                       ;line 26
B637    DECA            |   1/2-Hertz oscillator                ;line 27
        BNE     B637    |                                       ;line 28
        DECB            |                                       ;line 29
        BNE     B637    /                                       ;line 30
        BRA     B61F        Repeat osc. until reset             ;line 31
        SWI                 Software interrupt                  ;line 32
```

**Figure 27.26**   The assembly language code needed to monitor refrigerator temperature and sound an alarm if temperature exceeds 38°F.

```
MS-Kermit>SET BAUD 9600<CR>
```

A baud rate of 9600 is shown, but rates of 4800 and 2400 can also be used, depending on which pins are strapped together on the EVBU. Finally, to connect to the M68HC11, type

```
C:>\Kermit>C<CR>
```

Then pressing the *Reset* button on the EVBU will display the BUFFALO message:

```
BUFFALO 2.5 (ext) -
Bit User Fast Friendly Aid to Logical Operation
```

Next key in a carriage return <CR> to display the BUFFALO prompt (>). The EVBU is supplied with monitor/debugger firmware (in EPROM memory) called BUFFALO, which also includes a one-line assembler/disassembler. These features allow the user to program the µC. The **ASM** command allows the user to access the one-line assembler and enter assembly-language instructions, like the application program shown in Fig. 27.26. To load the software, use the ASM command below followed by the starting address location you wish for the program:

```
>ASM B600<CR>
```

You will see the specified address, followed by the current instruction or information at that location, and then the BUFFALO prompt on the next line. As an example,

```
B600 SBCA, $AA,X
        >
```

might be the current information at $B600. Refer again to Fig. 27.26 and load the instructions, following each with a carriage return. Note that the descriptive information and line numbers are not part of the assembly language program, but are presented as documentation only. For example, the first line of the driver program is loaded at location $B600, after the prompt as

```
>LDAA #93<CR>
```

After the program is assembled, you are ready to exit the one-line assembler by holding down the **CTRL** key and simultaneously pressing the **A** key. The BUFFALO prompt is once again displayed.

**System Operation.** With power off, reconnect jumper $J_1$ to the EEPROM position as shown in Fig. 27.22. When power is once again applied, the system will immediately jump to the starting EEPROM memory location at $B600 and begin measuring temperature.

# Appendices

Comparison of Aluminum and Steel Sheet Metal Gauges

| Gauge Number | Aluminum (American or Brown & Sharpe) | Steel (U.S. Standard) |
|---|---|---|
| 0000000 | — | 0.500 |
| 000000 | 0.5800 | 0.46875 |
| 00000 | 0.5165 | 0.4375 |
| 0000 | 0.4600 | 0.40625 |
| 000 | 0.4096 | 0.375 |
| 00 | 0.3648 | 0.34375 |
| 0 | 0.3249 | 0.3125 |
| 1 | 0.2893 | 0.28125 |
| 2 | 0.2576 | 0.265625 |
| 3 | 0.2294 | 0.25 |
| 4 | 0.2043 | 0.234375 |
| 5 | 0.1819 | 0.21875 |
| 6 | 0.1620 | 0.203125 |
| 7 | 0.1443 | 0.1875 |
| 8 | 0.1285 | 0.171875 |
| 9 | 0.1144 | 0.15625 |
| 10 | 0.1019 | 0.140625 |
| 11 | 0.09074 | 0.125 |
| 12 | 0.08081 | 0.109375 |
| 13 | 0.07196 | 0.09375 |
| 14 | 0.06408 | 0.078125 |
| 15 | 0.05707 | 0.0703125 |
| 16 | 0.05082 | 0.0625 |
| 17 | 0.04526 | 0.05625 |
| 18 | 0.04030 | 0.05 |
| 19 | 0.03589 | 0.04375 |
| 20 | 0.03196 | 0.0375 |
| 21 | 0.02846 | 0.034375 |
| 22 | 0.02535 | 0.03125 |
| 23 | 0.02257 | 0.028125 |
| 24 | 0.02010 | 0.025 |
| 25 | 0.01790 | 0.021875 |
| 26 | 0.01594 | 0.01875 |
| 27 | 0.01420 | 0.0171875 |
| 28 | 0.01264 | 0.015625 |
| 29 | 0.01126 | 0.0140625 |
| 30 | 0.01003 | 0.0125 |
| 31 | 0.008928 | 0.0109375 |
| 32 | 0.007950 | 0.01015625 |
| 33 | 0.007080 | 0.009375 |
| 34 | 0.006305 | 0.00859375 |
| 35 | 0.005615 | 0.0078125 |
| 36 | 0.005000 | 0.00703125 |
| 37 | 0.004453 | 0.006640625 |
| 38 | 0.003965 | 0.00625 |
| 39 | 0.003531 | — |
| 40 | 0.003145 | — |

All decimal values in inches.

Twist Drill Sizes and Decimal Equivalents of Number (Wire Gauge), Letter, and Fraction Size Drills

| Drill Size | Decimal Equivalent (in.) | Drill Size | Decimal Equivalent (in.) | Drill Size | Decimal Equivalent (in.) |
|---|---|---|---|---|---|
| 80 | 0.0135 | 40 | 0.0980 | 2 | 0.2210 |
| 79 | 0.0145 | 39 | 0.0995 | 1 | 0.2280 |
| $\frac{1}{64}$ in. | 0.0156 | 38 | 0.1015 | A | 0.2340 |
| 78 | 0.0160 | 37 | 0.1040 | $\frac{15}{64}$ in. | 0.2344 |
| 77 | 0.0180 | 36 | 0.1065 | B | 0.2380 |
| 76 | 0.0200 | $\frac{7}{64}$ in. | 0.1094 | C | 0.2420 |
| 75 | 0.0210 | 35 | 0.1100 | D | 0.2460 |
| 74 | 0.0225 | 34 | 0.1110 | E | 0.2500 |
| 73 | 0.0240 | 33 | 0.1130 | $\frac{1}{4}$ in. | 0.2500 |
| 72 | 0.0250 | 32 | 0.1160 | F | 0.2570 |
| 71 | 0.0260 | 31 | 0.1200 | G | 0.2610 |
| 70 | 0.0280 | $\frac{1}{8}$ in. | 0.1250 | $\frac{17}{64}$ in. | 0.2656 |
| 69 | 0.0292 | 30 | 0.1285 | H | 0.2660 |
| 68 | 0.0310 | 29 | 0.1360 | I | 0.2720 |
| $\frac{1}{32}$ in. | 0.0312 | 28 | 0.1405 | J | 0.2770 |
| 67 | 0.0320 | $\frac{9}{64}$ in. | 0.1406 | K | 0.2810 |
| 66 | 0.0330 | 27 | 0.1440 | $\frac{9}{32}$ in. | 0.2812 |
| 65 | 0.0350 | 26 | 0.1470 | L | 0.2900 |
| 64 | 0.0360 | 25 | 0.1495 | M | 0.2950 |
| 63 | 0.0370 | 24 | 0.1520 | $\frac{19}{64}$ in. | 0.2969 |
| 62 | 0.0380 | 23 | 0.1540 | N | 0.3020 |
| 61 | 0.0390 | $\frac{5}{32}$ in. | 0.1562 | $\frac{5}{16}$ in. | 0.3125 |
| 60 | 0.0400 | 22 | 0.1570 | O | 0.3160 |
| 59 | 0.0410 | 21 | 0.1590 | P | 0.3230 |
| 58 | 0.0420 | 20 | 0.1610 | $\frac{21}{64}$ in. | 0.3281 |
| 57 | 0.0430 | 19 | 0.1660 | Q | 0.3320 |
| 56 | 0.0465 | 18 | 0.1695 | R | 0.3390 |
| $\frac{3}{64}$ in. | 0.0469 | $\frac{11}{64}$ in. | 0.1719 | $\frac{11}{32}$ in. | 0.3438 |
| 55 | 0.0520 | 17 | 0.1730 | S | 0.3480 |
| 54 | 0.0550 | 16 | 0.1770 | T | 0.3580 |
| 53 | 0.0595 | 15 | 0.1800 | $\frac{23}{64}$ in. | 0.3594 |
| $\frac{1}{16}$ in. | 0.0625 | 14 | 0.1820 | U | 0.3680 |
| 52 | 0.0635 | 13 | 0.1850 | $\frac{3}{8}$ in. | 0.3750 |
| 51 | 0.0670 | $\frac{3}{16}$ in. | 0.1875 | V | 0.3770 |
| 50 | 0.0700 | 12 | 0.1890 | W | 0.3860 |
| 49 | 0.0730 | 11 | 0.1910 | $\frac{25}{64}$ in. | 0.3906 |
| 48 | 0.0760 | 10 | 0.1935 | X | 0.3970 |
| $\frac{5}{64}$ in. | 0.0781 | 9 | 0.1960 | Y | 0.4040 |
| 47 | 0.0785 | 8 | 0.1990 | $\frac{13}{32}$ in. | 0.4062 |
| 46 | 0.0810 | 7 | 0.2010 | Z | 0.4130 |
| 45 | 0.0820 | $\frac{13}{64}$ in. | 0.2031 | $\frac{27}{64}$ in. | 0.4219 |
| 44 | 0.0860 | 6 | 0.2040 | $\frac{7}{16}$ in. | 0.4375 |
| 43 | 0.0890 | 5 | 0.2055 | $\frac{29}{64}$ in. | 0.4531 |
| 42 | 0.0935 | 4 | 0.2090 | $\frac{15}{32}$ in. | 0.4687 |
| $\frac{3}{32}$ in. | 0.0937 | 3 | 0.2130 | $\frac{31}{64}$ in. | 0.4844 |
| 41 | 0.0960 | $\frac{7}{32}$ in. | 0.2187 | $\frac{1}{2}$ in. | 0.5000 |

Machine Screw Sizes and Decimal Equivalents

| Gauge No. / Fraction | Maximum Diameter* (G) | Minimum Diameter* (T) | Threads per Inch | |
|---|---|---|---|---|
| | | | UNC | UNF |
| 0 | 0.0600 | 0.0438 | — | 80 |
| 1 | 0.0730 | 0.0527 | 64 | |
| | | 0.0550 | | 72 |
| 2† | 0.0860 | 0.0628 | 56 | |
| | | 0.0657 | | 64 |
| 3 | 0.0990 | 0.0719 | 48 | |
| | | 0.0758 | | 56 |
| 4† | 0.1120 | 0.0795 | 40 | |
| | | 0.0849 | | 48 |
| $5\frac{1}{8}$ | 0.1250 | 0.0925 | 40 | |
| | | 0.0955 | | 44 |
| 6† | 0.1380 | 0.0974 | 32 | |
| | | 0.1055 | | 40 |
| 8† | 0.1640 | 0.1234 | 32 | |
| | | 0.1279 | | 36 |
| 10† | 0.1900 | 0.1359 | 24 | |
| | | 0.1494 | | 32 |
| 12† | 0.2160 | 0.1619 | 24 | |
| | | 0.1696 | | 28 |
| $\frac{1}{4}$ | 0.2500 | 0.1850 | 20 | |
| | | 0.2036 | | 28 |
| $\frac{5}{16}$ | 0.3125 | 0.2403 | 18 | |
| | | 0.2584 | | 24 |
| $\frac{3}{8}$ | 0.3750 | 0.2938 | 16 | |
| | | 0.3209 | | 24 |
| $\frac{7}{16}$ | 0.4375 | 0.3447 | 14 | |
| | | 0.3726 | | 20 |
| $\frac{1}{2}$ | 0.5000 | 0.4001 | 13 | |
| | | 0.4351 | | 20 |
| $\frac{9}{16}$ | 0.5625 | 0.4542 | 12 | |
| | | 0.4903 | | 18 |
| $\frac{5}{8}$ | 0.6250 | 0.5069 | 11 | |
| | | 0.5528 | | 18 |
| $\frac{3}{4}$ | 0.7500 | 0.6201 | 10 | |
| | | 0.6688 | | 16 |
| $\frac{7}{8}$ | 0.8750 | 0.7307 | 9 | |
| | | 0.7822 | | 14 |
| 1 in. | 1.0000 | 0.8376 | 8 | |
| | | 0.9072 | | 14 |

*Diameters in inches.

†Sizes most frequently used in electronics.

## Clearance Hole Sizes and Hole Tolerance Requirements for Machine Screws

| Screw Size | Clearance Drill | *Decimal Equivalent of Drill (in.) | *Hole Tolerance (in.) |
|---|---|---|---|
| 0–80 | 51 | 0.0670 | 0.0070 |
| 1–64 | $\frac{5}{64}$ | 0.0781 | 0.0081 |
| 1–72 | | | |
| 2–56 | 42 | 0.0935 | 0.0075 |
| 2–64 | | | |
| 3–48 | 37 | 0.1040 | 0.0050 |
| 3–56 | | | |
| 4–40 | 31 | 0.1200 | 0.0080 |
| 4–48 | | | |
| 5–40 | 29 | 0.1360 | 0.0110 |
| 5–44 | | | |
| 6–32 | 27 | 0.1440 | 0.0060 |
| 6–40 | | | |
| 8–32 | 17 | 0.1730 | 0.0090 |
| 8–36 | | | |
| 10–24 | 8 | 0.1990 | 0.0090 |
| 10–32 | | | |
| 12–24 | 1 | 0.2280 | 0.0120 |
| 12–28 | | | |
| $\frac{1}{4}$–20 | G | 0.2610 | 0.0110 |
| $\frac{1}{4}$–28 | | | |
| $\frac{5}{16}$–18 | $\frac{21}{64}$ | 0.3281 | 0.0156 |
| $\frac{5}{16}$–24 | | | |
| $\frac{3}{8}$–16 | $\frac{25}{64}$ | 0.3906 | 0.0156 |
| $\frac{3}{8}$–24 | | | |
| $\frac{7}{16}$–14 | $\frac{29}{64}$ | 0.4531 | 0.0156 |
| $\frac{7}{16}$–20 | | | |
| $\frac{1}{2}$–13 | $\frac{33}{64}$ | 0.5156 | 0.0156 |
| $\frac{1}{2}$–20 | | | |
| $\frac{9}{16}$–12 | $\frac{37}{64}$ | 0.5781 | 0.0156 |
| $\frac{9}{16}$–18 | | | |
| $\frac{5}{8}$–11 | $\frac{41}{64}$ | 0.6406 | 0.0156 |
| $\frac{5}{8}$–18 | | | |
| $\frac{3}{4}$–10 | $\frac{25}{32}$ | 0.7812 | 0.0312 |
| $\frac{3}{4}$–16 | | | |
| $\frac{7}{8}$–9 | $\frac{29}{32}$ | 0.9062 | 0.0312 |
| $\frac{7}{8}$–14 | | | |

*All dimensions in inches.

# Dimensional Information for Hexagonal Machine Screw Nuts

| Machine Screw Nut Size | Width Across Flats (in.) (W) | Thickness (in.) (T) |
|---|---|---|
| 0–80 | $\frac{5}{32}$ | $\frac{3}{64}$ |
| 1–64 | | |
| 1–72 | $\frac{5}{32}$ | $\frac{3}{64}$ |
| 2–56 | $\frac{5}{32}$ | $\frac{1}{16}$ |
| 2–56 | | |
| 2–64 | $\frac{3}{16}$ | $\frac{1}{16}$ |
| 3–48 | | |
| 3–56 | $\frac{3}{16}$ | $\frac{1}{16}$ |
| 4–40 | $\frac{3}{16}$ | $\frac{1}{16}$ |
| 4–40 | | |
| 4–48 | $\frac{1}{4}$ | $\frac{3}{32}$ |
| 5–40 | $\frac{1}{4}$ | $\frac{3}{32}$ |
| 5–40 | | |
| 5–44 | $\frac{5}{16}$ | $\frac{7}{64}$ |
| 6–32 | $\frac{1}{4}$ | $\frac{3}{32}$ |
| 6–32 | | |
| 6–40 | $\frac{5}{16}$ | $\frac{7}{64}$ |
| | $\frac{1}{4}$ | $\frac{3}{32}$ |
| 8–32 | $\frac{5}{16}$ | $\frac{7}{64}$ |
| | $\frac{11}{32}$ | $\frac{1}{8}$ |
| 8–36 | $\frac{11}{32}$ | $\frac{1}{8}$ |
| 10–32 | $\frac{5}{16}$ | $\frac{1}{8}$ |
| | $\frac{11}{32}$ | $\frac{1}{8}$ |
| | $\frac{3}{8}$ | $\frac{1}{8}$ |
| 10–24 | $\frac{3}{8}$ | $\frac{1}{8}$ |
| 12–24 | $\frac{7}{16}$ | $\frac{5}{32}$ |
| $\frac{1}{4}$–20 | | |
| $\frac{1}{4}$–28 | $\frac{7}{16}$ | $\frac{3}{16}$ |
| $\frac{5}{16}$–18 | $\frac{9}{16}$ | $\frac{7}{32}$ |
| $\frac{5}{16}$–24 | | |
| $\frac{3}{8}$–16 | $\frac{5}{8}$ | $\frac{1}{4}$ |
| $\frac{3}{8}$–24 | | |

| Screw Size | Flat | | | Lock Washers Internal | | | External | | | Split | |
|---|---|---|---|---|---|---|---|---|---|---|---|
| | O.D. | I.D. | Thick. | O.D. | I.D. | Thick. | O.D. | I.D. | Thick. | W | T |
| 0 | $\frac{5}{32}$ | $\frac{1}{16}$ | 0.020 | — | — | — | — | — | — | 0.022 | 0.022 |
| 1 | $\frac{3}{16}$ | $\frac{5}{64}$ | 0.016 | — | — | — | — | — | — | 0.022 | 0.022 |
| | | | | | | | | | | 0.030 | 0.015 |
| | | | | | | | | | | 0.030 | 0.020 |
| | $\frac{7}{32}$ | $\frac{3}{32}$ | 0.018 | | | | | | | | |
| 2 | $\frac{1}{4}$ | $\frac{3}{32}$ | 0.016 | 0.200 | 0.095 | 0.015 | 0.275 | 0.095 | 0.019 | 0.030 | 0.015 |
| | $\frac{1}{4}$ | $\frac{3}{32}$ | 0.031 | | | | | | | 0.035 | 0.020 |
| | $\frac{7}{32}$ | $\frac{7}{64}$ | 0.022 | | | | | | | | |
| 3 | $\frac{1}{4}$ | $\frac{7}{64}$ | 0.016 | 0.225 | 0.109 | 0.015 | — | — | — | 0.035 | 0.020 |
| | $\frac{1}{4}$ | $\frac{7}{64}$ | 0.031 | | | | | | | 0.040 | 0.025 |
| | $\frac{1}{4}$ | $\frac{1}{8}$ | 0.017 | | | | | | | | |
| | $\frac{1}{4}$ | $\frac{1}{8}$ | 0.028 | | | | | | | 0.035 | 0.020 |
| 4 | $\frac{5}{16}$ | $\frac{1}{8}$ | 0.016 | 0.270 | 0.123 | 0.019 | 0.260 | 0.123 | 0.019 | 0.040 | 0.025 |
| | $\frac{5}{16}$ | $\frac{1}{8}$ | 0.025 | | | | | | | | |
| | $\frac{5}{16}$ | $\frac{1}{8}$ | 0.031 | | | | | | | 0.047 | 0.031 |
| | $\frac{3}{8}$ | $\frac{1}{8}$ | 0.036 | | | | | | | | |
| 5 | $\frac{9}{32}$ | $\frac{9}{64}$ | 0.025 | 0.280 | 0.136 | 0.019 | 0.280 | 0.136 | 0.019 | 0.040 | 0.025 |
| | | | | | | | | | | 0.047 | 0.031 |
| | $\frac{5}{16}$ | $\frac{5}{32}$ | 0.027 | | | | | | | 0.040 | 0.025 |
| | $\frac{11}{32}$ | $\frac{11}{64}$ | 0.015 | | | | | | | | |
| 6 | $\frac{3}{8}$ | $\frac{5}{32}$ | 0.016 | 0.295 | 0.150 | 0.021 | 0.320 | 0.150 | 0.022 | 0.047 | 0.031 |
| | $\frac{3}{8}$ | $\frac{5}{32}$ | 0.031 | | | | | | | | |
| | $\frac{3}{8}$ | $\frac{5}{32}$ | 0.036 | | | | | | | 0.055 | 0.040 |
| | $\frac{7}{16}$ | $\frac{5}{32}$ | 0.036 | | | | | | | | |
| | $\frac{3}{8}$ | $\frac{11}{64}$ | 0.016 | | | | | | | 0.047 | 0.031 |
| | $\frac{3}{8}$ | $\frac{11}{64}$ | 0.031 | | | | | | | | |
| 8 | $\frac{3}{8}$ | $\frac{3}{16}$ | 0.036 | 0.340 | 0.176 | 0.023 | 0.381 | 0.176 | 0.023 | 0.055 | 0.040 |
| | $\frac{1}{2}$ | $\frac{3}{16}$ | 0.036 | | | | | | | 0.062 | 0.047 |

| Screw Size | Flat | | | Lock Washers | | | | | | Split | |
| | | | | Internal | | | External | | | | |
| | O.D. | I.D. | Thick. | O.D. | I.D. | Thick. | O.D. | I.D. | Thick. | W | T |
|---|---|---|---|---|---|---|---|---|---|---|---|
| 10 | $\frac{3}{8}$ | $\frac{13}{64}$ | 0.047 | | | | | | | 0.055 | 0.040 |
| | $\frac{13}{32}$ | $\frac{13}{64}$ | 0.036 | | | | | | | | |
| | $\frac{7}{16}$ | $\frac{13}{64}$ | 0.031 | 0.381 | 0.204 | 0.025 | 0.410 | 0.204 | 0.025 | 0.062 | 0.047 |
| | $\frac{7}{16}$ | $\frac{13}{64}$ | 0.062 | | | | | | | | |
| | $\frac{7}{16}$ | $\frac{7}{32}$ | 0.036 | | | | | | | | |
| | $\frac{1}{2}$ | $\frac{7}{32}$ | 0.062 | | | | | | | 0.070 | 0.056 |
| | $\frac{9}{16}$ | $\frac{13}{64}$ | 0.036 | | | | | | | | |
| | $\frac{9}{16}$ | $\frac{1}{4}$ | 0.051 | | | | | | | | |
| 12 | $\frac{1}{2}$ | $\frac{15}{64}$ | 0.040 | 0.410 | 0.231 | 0.025 | 0.475 | 0.231 | 0.025 | 0.062 | 0.047 |
| | | | | | | | | | | 0.070 | 0.056 |
| $\frac{1}{4}$ | $\frac{1}{2}$ | $\frac{17}{64}$ | 0.031 | | | | | | | 0.107 | 0.047 |
| | $\frac{1}{2}$ | $\frac{17}{64}$ | 0.062 | | | | | | | | |
| | $\frac{1}{2}$ | $\frac{9}{32}$ | 0.056 | | | | | | | 0.109 | 0.062 |
| | $\frac{5}{8}$ | $\frac{9}{32}$ | 0.051 | | | | | | | | |
| | $\frac{5}{8}$ | $\frac{9}{32}$ | 0.062 | 0.478 | 0.267 | 0.028 | 0.510 | 0.267 | 0.028 | 0.110 | 0.077 |
| | $\frac{11}{16}$ | $\frac{17}{64}$ | 0.050 | | | | | | | | |
| | $\frac{3}{4}$ | $\frac{9}{32}$ | 0.056 | | | | | | | | |
| | $\frac{3}{4}$ | $\frac{5}{16}$ | 0.051 | | | | | | | | |
| | $\frac{9}{16}$ | $\frac{21}{64}$ | 0.031 | | | | | | | 0.117 | 0.056 |
| | $\frac{9}{16}$ | $\frac{21}{64}$ | 0.062 | | | | | | | | |
| | $\frac{11}{16}$ | $\frac{11}{32}$ | 0.051 | | | | | | | | |
| $\frac{5}{16}$ | $\frac{11}{16}$ | $\frac{11}{32}$ | 0.062 | 0.610 | 0.332 | 0.034 | 0.610 | 0.332 | 0.033 | 0.125 | 0.078 |
| | $\frac{5}{8}$ | $\frac{11}{32}$ | 0.056 | | | | | | | | |
| | $\frac{3}{4}$ | $\frac{11}{32}$ | 0.050 | | | | | | | 0.130 | 0.097 |
| | $\frac{7}{8}$ | $\frac{3}{8}$ | 0.064 | | | | | | | | |
| $\frac{3}{8}$ | $\frac{5}{8}$ | $\frac{25}{64}$ | 0.031 | | | | | | | 0.136 | 0.070 |
| | $\frac{5}{8}$ | $\frac{25}{64}$ | 0.062 | | | | | | | | |
| | $\frac{3}{4}$ | $\frac{13}{32}$ | 0.056 | 0.692 | 0.398 | 0.040 | 0.694 | 0.398 | 0.040 | | |
| | $\frac{13}{16}$ | $\frac{13}{32}$ | 0.062 | | | | | | | | |
| | $\frac{13}{16}$ | $\frac{13}{32}$ | 0.051 | | | | | | | | |
| | 1 | $\frac{7}{16}$ | 0.064 | | | | | | | 0.141 | 0.094 |

All dimensions in inches.

| AWG* Number | Diameter (mils) (approximate) | Area (circular mils) | Maximum Recommended Current Capacity† (amperes) |
|---|---|---|---|
| 00 | 365 | 133,080 | |
| 0 | 325 | 105,560 | |
| 1 | 289 | 83,690 | |
| 2 | 257 | 66,360 | |
| 3 | 229 | 52,620 | |
| 4 | 204 | 41,740 | |
| 5 | 182 | 33,090 | |
| 6 | 162 | 26,240 | |
| 7 | 144 | 20,820 | |
| 8 | 128 | 16,510 | 33.0 |
| 9 | 114 | 13,090 | 26.0 |
| 10 | 102 | 10,380 | 20.5 |
| 11 | 90.7 | 8,226 | 16.5 |
| 12 | 80.8 | 6,529 | 12.0 |
| 13 | 72.0 | 5,184 | 10.0 |
| 14 | 64.1 | 4,109 | 8.0 |
| 15 | 57.1 | 3,260 | 6.5 |
| 16 | 50.8 | 2,580 | 5.0 |
| 17 | 45.3 | 2,052 | 4.0 |
| 18 | 40.3 | 1,624 | 3.0 |
| 19 | 35.9 | 1,289 | 2.5 |
| 20 | 32.0 | 1,024 | 2.0 |
| 21 | 28.5 | 812 | 1.5 |
| 22 | 25.3 | 640 | 1.0 |
| 23 | 22.6 | 511 | 1.0 |
| 24 | 20.1 | 404 | 0.80 |
| 25 | 17.9 | 320 | 0.60 |
| 26 | 15.9 | 253 | 0.50 |
| 27 | 14.2 | 202 | 0.40 |
| 28 | 12.6 | 159 | 0.30 |
| 29 | 11.3 | 128 | 0.25 |
| 30 | 10.0 | 100 | 0.20 |
| 31 | 8.9 | 79 | |
| 32 | 8.0 | 64 | |
| 33 | 7.1 | 50 | |
| 34 | 6.3 | 40 | |
| 35 | 5.6 | 31 | |
| 36 | 5.0 | 25 | |
| 37 | 4.5 | 20 | |
| 38 | 4.0 | 16 | |
| 39 | 3.5 | 12 | |
| 40 | 3.1 | 10 | |
| 41 | 2.8 | 8 | |
| 42 | 2.5 | 6 | |
| 43 | 2.2 | 5 | |
| 44 | 2.0 | 4 | |
| 45 | 1.8 | 3 | |
| 46 | 1.6 | 2.5 | |
| 47 | 1.4 | 2 | |
| 48 | 1.2 | 1.5 | |
| 49 | 1.1 | 1.2 | |
| 50 | 1.0 | 1.0 | |

*American Wire Gauge Standard (AWG).

†Current capacity based on 500 circular mils per ampere.

Proper exposure time for a specific resist is determined empirically for each exposure unit with the use of a *step table* such as the Stouffer 21-step exposure chart shown in Fig. A. This step table consists of a sheet of plastic with 21 areas (windows), each having increasing optical density. Step 1 is essentially clear and totally transparent, whereas step 21 is the most opaque. For any given exposure time and light source, the amount of light striking the resist surface is a function of the density of the window on the step table. Use of this table to determine the correct exposure requires a knowledge of the resist manufacturer's recommendations. Dynachem recommends the following for Laminar AX resist. After exposure and developing, the results of the resist exposed under a 21-step exposure table should be a "solid No. 4" with high-gloss green resist existing totally under that window. Under step 5, some resist will have been attacked during developing and the remaining resist under that window will be a dull-green color. Under step 6 and above to step 21, only clear copper will be visible.

Too long an exposure time will result in "high-gloss" resist occurring under a *higher* step number than 4 as a result of enough light having penetrated the more-opaque windows of the step table. This exposure time will result in resist brittleness. Too short an exposure time will result in high-gloss resist occurring under a *smaller* step number than 4 because of the opaqueness of the smaller step numbers being sufficient to block enough light from striking the resist. Incomplete resist polymerization results. This may appear, after developing, as a dull resist image or simply an incomplete image of the conductor pattern.

The use of a 21-step exposure table to determine the correct exposure time for a particular exposure unit will require that six small pc boards be cleaned, dried, preheated, laminated, and allowed to normalize to room temperature. (These processes are detailed in Chapter 12.) Each board is then exposed with the Stouffer exposure chart as the phototool for a predetermined amount of time with this time designated on the board. The first board should be exposed for 10 seconds, the second for 20 seconds, the third for 30 seconds, and so on, until all boards are exposed. After all boards have been normalized to room temperature for 10 minutes, they should be developed, rinsed, neutralized in a 10% sulfuric acid bath, rinsed, and dried. (Again, refer to Chapter 12.) Six properly processed samples are shown in Fig. B. The time required to obtain a "solid No. 4" on the step table is the proper exposure time for the specific resist and exposure employed.

**Figure A** Stouffer 21-step exposure chart.

**Figure B** Exposure time to obtain a solid No. 4 is approximately 30 seconds.

As discussed in Chapters 11 and 13, plating solutions are extremely susceptible to organic and metallic contaminations. It is therefore essential that all contaminations be removed from all objects that will contact the plating solution. New plating tanks must be thoroughly leached before being used for the first time. The contaminants that must be removed from these tanks are cements, plastic welding residues, mold-release residues, machine oils, airborn particles, and simple dirts and greases. A new tank may look clean but is probably the single largest source of contamination that will result in plating failures.

Leaching involves first exposing the tank to an extremely strong alkaline solution followed by a relatively mild acid solution. The alkaline solution is used to remove the contaminations, and the acid solution neutralizes the alkaline residue.

A 1000-watt quartz heater with temperature control is first positioned in the tank with all hardware removed from the tank. The tank is filled with tap water to within $\frac{1}{2}$ inch of the top rim. The following are then added to the tank:

Trisodium phosphate ($Na_3PO_4$), 6 ounces per gallon

Sodium hydroxide (NaOH)–caustic soda, 3 ounces per gallon

Plastic gloves should be worn when handling these strongly alkaline compounds. The chemicals are stirred into the solution with a plastic paddle and the temperature is raised to 130°F (54°C). The tank is subjected to this temperature for 8 hours, after which the heater is shut off, allowed to cool, and removed from the tank. The alkaline solution is then discarded. The tank is then *twice* filled with cold tap water and emptied.

To neutralize any remaining alkalinity, the tank is then filled with a 10% by volume solution of fluoboric acid. The solution is slowly stirred with a clean plastic paddle. This solution is allowed to remain in the tank for 24 hours at room temperature before being discarded. Again the tank is twice filled with cold tap water and emptied. The use of deionized water as a final rinse is recommended, but cold tap water will suffice if deionized water is unavailable.

The tank is now properly leached and ready for the plating solution.

| °F* → °C | | °F* → °C | | °F* → °C | | °F* → °C | | Weights and Measures |
|---|---|---|---|---|---|---|---|---|
| 0 | −17.8 | 140 | 60.0 | 280 | 137.8 | 420 | 215.6 | 1 pound = 453.6 grams |
| 5 | −15.0 | 145 | 62.7 | 285 | 140.6 | 425 | 218.3 | 1 ounce = 28.35 grams |
| 10 | −12.2 | 150 | 65.6 | 290 | 143.3 | 430 | 221.1 | 1 gallon = 3.785 liters |
| 15 | − 9.4 | 155 | 68.3 | 295 | 146.1 | 435 | 223.9 | 1 quart = 946 milliliters |
| 20 | − 6.7 | 160 | 71.1 | 300 | 148.9 | 440 | 226.7 | 1 ounce = 29.57 milliliters |
| 25 | − 3.9 | 165 | 73.9 | 305 | 151.7 | 445 | 229.4 | 1 inch = 25.4 millimeters |
| 30 | − 1.1 | 170 | 76.7 | 310 | 154.4 | 450 | 232.2 | 1 foot = 304.8 millimeters |
| 35 | 1.7 | 175 | 79.4 | 315 | 157.2 | 455 | 235.0 | 1 square inch = 645.2 square millimeters |
| 40 | 4.4 | 180 | 82.2 | 320 | 160.0 | 460 | 237.8 | |
| 45 | 7.2 | 185 | 85.0 | 325 | 162.8 | 465 | 240.6 | |
| 50 | 10.0 | 190 | 87.8 | 330 | 165.6 | 470 | 243.3 | |
| 55 | 12.8 | 195 | 90.6 | 335 | 168.3 | 475 | 246.1 | |
| 60 | 15.6 | 200 | 93.3 | 340 | 171.1 | 480 | 248.9 | |
| 65 | 18.3 | 205 | 96.1 | 345 | 173.9 | 485 | 251.7 | |
| 70 | 21.1 | 210 | 98.9 | 350 | 176.7 | 490 | 254.4 | |
| 75 | 23.9 | 215 | 101.7 | 355 | 179.4 | 495 | 257.2 | |
| 80 | 26.7 | 220 | 104.4 | 360 | 182.2 | 500 | 260.0 | |
| 85 | 29.4 | 225 | 107.2 | 365 | 185.0 | 505 | 262.8 | |
| 90 | 32.2 | 230 | 110.0 | 370 | 187.8 | 510 | 265.6 | |
| 95 | 35.0 | 235 | 112.8 | 375 | 190.6 | 515 | 268.3 | |
| 100 | 37.8 | 240 | 115.6 | 380 | 193.3 | 520 | 271.1 | |
| 105 | 40.6 | 245 | 118.3 | 385 | 196.1 | 525 | 273.9 | |
| 110 | 43.3 | 250 | 121.1 | 390 | 198.9 | 530 | 276.7 | |
| 115 | 46.1 | 255 | 123.9 | 395 | 201.7 | 535 | 279.4 | |
| 120 | 48.9 | 260 | 126.7 | 400 | 204.4 | 540 | 282.2 | |
| 125 | 51.7 | 265 | 129.4 | 405 | 207.2 | 545 | 285.0 | |
| 130 | 54.4 | 270 | 132.2 | 410 | 210.0 | 550 | 287.8 | |
| 135 | 57.2 | 275 | 135.0 | 415 | 212.8 | 555 | 290.6 | |

*For each °F between those listed add 0.56°C.

$°C = \frac{5}{9}(°F - 32)$       $°F = \frac{9}{5}°C + 32$

Celsius                Fahrenheit

Example: $74°F = 21.1°C + 4(0.56°C)$
$74°F = 23.3°C$

(continued)

| Decimals to Millimeters | | | | Fractions to Decimals to Millimeters | | | | | |
|---|---|---|---|---|---|---|---|---|---|
| *Decimal* | *mm* | *Decimal* | *mm* | *Fraction* | *Decimal* | *mm* | *Fraction* | *Decimal* | *mm* |
| 0.001 | 0.0254 | 0.500 | 12.7000 | $\frac{1}{64}$ | 0.0156 | 0.3969 | $\frac{33}{64}$ | 0.5156 | 13.0969 |
| 0.002 | 0.0508 | 0.510 | 12.9540 | $\frac{1}{32}$ | 0.0312 | 0.7938 | $\frac{17}{32}$ | 0.5312 | 13.4938 |
| 0.003 | 0.0762 | 0.520 | 13.2080 | $\frac{3}{64}$ | 0.0469 | 1.1906 | $\frac{35}{64}$ | 0.5469 | 13.8906 |
| 0.004 | 0.1016 | 0.530 | 13.4620 | | | | | | |
| 0.005 | 0.1270 | 0.540 | 13.7160 | | | | | | |
| 0.006 | 0.1524 | 0.550 | 13.9700 | $\frac{1}{16}$ | 0.0625 | 1.5875 | $\frac{9}{16}$ | 0.5625 | 14.2875 |
| 0.007 | 0.1778 | 0.560 | 14.2240 | | | | | | |
| 0.008 | 0.2032 | 0.570 | 14.4780 | | | | | | |
| 0.009 | 0.2286 | 0.580 | 14.7320 | $\frac{5}{64}$ | 0.0781 | 1.9844 | $\frac{37}{64}$ | 0.5781 | 14.6844 |
| | | 0.590 | 14.9860 | $\frac{3}{32}$ | 0.0938 | 2.3812 | $\frac{19}{32}$ | 0.5938 | 15.0812 |
| 0.010 | 0.2540 | | | | | | | | |
| 0.020 | 0.5080 | | | $\frac{7}{64}$ | 0.1094 | 2.7781 | $\frac{39}{64}$ | 0.6094 | 15.4781 |
| 0.030 | 0.7620 | | | | | | | | |
| 0.040 | 1.0160 | 0.600 | 15.2400 | $\frac{1}{8}$ | 0.1250 | 3.1750 | $\frac{5}{8}$ | 0.6250 | 15.8750 |
| 0.050 | 1.2700 | 0.610 | 15.4940 | | | | | | |
| 0.060 | 1.5240 | 0.620 | 15.7480 | | | | | | |
| 0.070 | 1.7780 | 0.630 | 16.0020 | $\frac{9}{64}$ | 0.1406 | 3.5719 | $\frac{41}{64}$ | 0.6406 | 16.2719 |
| 0.080 | 2.0320 | 0.640 | 16.2560 | $\frac{5}{32}$ | 0.1562 | 3.9688 | $\frac{21}{32}$ | 0.6562 | 16.6688 |
| 0.090 | 2.2860 | 0.650 | 16.5100 | | | | | | |
| | | 0.660 | 16.7640 | $\frac{11}{64}$ | 0.1719 | 4.3656 | $\frac{43}{64}$ | 0.6719 | 17.0656 |
| 0.100 | 2.5400 | 0.670 | 17.0180 | | | | | | |
| 0.110 | 2.7940 | 0.680 | 17.2720 | $\frac{3}{16}$ | 0.1875 | 4.7625 | $\frac{11}{16}$ | 0.6875 | 17.4625 |
| 0.120 | 3.0480 | 0.690 | 17.5260 | | | | | | |
| 0.130 | 3.3020 | | | | | | | | |
| 0.140 | 3.5560 | | | | | | | | |
| 0.150 | 3.8100 | | | $\frac{13}{64}$ | 0.2031 | 5.1594 | $\frac{45}{64}$ | 0.7031 | 17.8594 |
| 0.160 | 4.0640 | 0.700 | 17.7800 | $\frac{7}{32}$ | 0.2188 | 5.5562 | $\frac{23}{32}$ | 0.7188 | 18.2562 |
| 0.170 | 4.3180 | 0.710 | 18.0340 | $\frac{15}{64}$ | 0.2344 | 5.9531 | $\frac{47}{64}$ | 0.7344 | 18.6531 |
| 0.180 | 4.5720 | 0.720 | 18.2880 | | | | | | |
| 0.190 | 4.8260 | 0.730 | 18.5420 | | | | | | |
| | | 0.740 | 18.7960 | $\frac{1}{4}$ | 0.2500 | 6.3500 | $\frac{3}{4}$ | 0.7500 | 19.0500 |
| 0.200 | 5.0800 | 0.750 | 19.0500 | | | | | | |
| 0.210 | 5.3340 | 0.760 | 19.3040 | | | | | | |
| 0.220 | 5.5880 | 0.770 | 19.5580 | $\frac{17}{64}$ | 0.2656 | 6.7469 | $\frac{49}{64}$ | 0.7656 | 19.4469 |
| 0.230 | 5.8420 | 0.780 | 19.8120 | $\frac{9}{32}$ | 0.2812 | 7.1438 | $\frac{25}{32}$ | 0.7812 | 19.8438 |
| 0.240 | 6.0960 | 0.790 | 20.0660 | $\frac{19}{64}$ | 0.2969 | 7.5406 | $\frac{51}{64}$ | 0.7969 | 20.2406 |
| 0.250 | 6.3500 | | | | | | | | |

(*continued*)

| Decimals to Millimeters | | | | Fractions to Decimals to Millimeters | | | | | |
|---|---|---|---|---|---|---|---|---|---|
| Decimal | mm | Decimal | mm | Fraction | Decimal | mm | Fraction | Decimal | mm |
| 0.260 | 6.6040 | | | | | | | | |
| 0.270 | 6.8580 | | | $\frac{5}{16}$ | 0.3125 | 7.9375 | $\frac{13}{16}$ | 0.8125 | 20.6375 |
| 0.280 | 7.1120 | 0.800 | 20.3200 | | | | | | |
| 0.290 | 7.3660 | 0.810 | 20.5740 | | | | | | |
| | | 0.820 | 20.8280 | $\frac{21}{64}$ | 0.3281 | 8.3344 | $\frac{53}{64}$ | 0.8281 | 21.0344 |
| 0.300 | 7.6200 | 0.830 | 21.0820 | $\frac{11}{32}$ | 0.3438 | 8.7312 | $\frac{27}{32}$ | 0.8438 | 21.4312 |
| 0.310 | 7.8740 | 0.840 | 21.3360 | $\frac{23}{64}$ | 0.3594 | 9.1281 | $\frac{55}{64}$ | 0.8594 | 21.8281 |
| 0.320 | 8.1280 | 0.850 | 21.5900 | | | | | | |
| 0.330 | 8.3820 | 0.860 | 21.8440 | | | | | | |
| 0.340 | 8.6360 | 0.870 | 22.0980 | $\frac{3}{8}$ | 0.3750 | 9.5250 | $\frac{7}{8}$ | 0.8750 | 22.2250 |
| 0.350 | 8.8900 | 0.880 | 22.3520 | | | | | | |
| 0.360 | 9.1440 | 0.890 | 22.6060 | | | | | | |
| 0.370 | 9.3980 | | | $\frac{25}{64}$ | 0.3906 | 9.9219 | $\frac{57}{64}$ | 0.8906 | 22.6219 |
| 0.380 | 9.6520 | | | $\frac{13}{32}$ | 0.4062 | 10.3188 | $\frac{29}{32}$ | 0.9062 | 23.0188 |
| 0.390 | 9.9060 | | | $\frac{27}{64}$ | 0.4219 | 10.7156 | $\frac{59}{64}$ | 0.9219 | 23.4156 |
| | | 0.900 | 22.8600 | | | | | | |
| 0.400 | 10.1600 | 0.910 | 23.1140 | | | | | | |
| 0.410 | 10.4140 | 0.920 | 23.3680 | $\frac{7}{16}$ | 0.4375 | 11.1125 | $\frac{15}{16}$ | 0.9375 | 23.8125 |
| 0.420 | 10.6680 | 0.930 | 23.6220 | | | | | | |
| 0.430 | 10.9220 | 0.940 | 23.8760 | | | | | | |
| 0.440 | 11.1760 | 0.950 | 24.1300 | $\frac{29}{64}$ | 0.4531 | 11.5094 | $\frac{61}{64}$ | 0.9531 | 24.2094 |
| 0.450 | 11.4300 | 0.960 | 24.3840 | $\frac{15}{32}$ | 0.4688 | 11.9062 | $\frac{31}{32}$ | 0.9688 | 24.6062 |
| 0.460 | 11.6840 | 0.970 | 24.6380 | $\frac{31}{64}$ | 0.4844 | 12.3031 | $\frac{63}{64}$ | 0.9844 | 25.0031 |
| 0.470 | 11.9380 | 0.980 | 24.8920 | | | | | | |
| 0.480 | 12.1920 | 0.990 | 25.1460 | $\frac{1}{2}$ | 0.5000 | 12.7000 | $1$ | 1.0000 | 25.4000 |
| 0.490 | 12.4460 | 1.000 | 25.4000 | | | | | | |

(continued)

| | | | | | *Millimeters to Decimals* | | | | |
|---|---|---|---|---|---|---|---|---|---|
| *mm* | *Decimal* | *mm* | *Decimal* | *mm* | *Decimal* | *mm* | *Decimal* | *mm* | *Decimal* |
| 0.01 | 0.00039 | 0.41 | 0.01614 | 0.81 | 0.03189 | 21 | 0.82677 | 61 | 2.40157 |
| 0.02 | 0.00079 | 0.42 | 0.01654 | 0.82 | 0.03228 | 22 | 0.86614 | 62 | 2.44094 |
| 0.03 | 0.00118 | 0.43 | 0.01693 | 0.83 | 0.03268 | 23 | 0.90551 | 63 | 2.48031 |
| 0.04 | 0.00157 | 0.44 | 0.01732 | 0.84 | 0.03307 | 24 | 0.94488 | 64 | 2.51969 |
| 0.05 | 0.00197 | 0.45 | 0.01772 | 0.85 | 0.03346 | 25 | 0.98425 | 65 | 2.55906 |
| 0.06 | 0.00236 | 0.46 | 0.01811 | 0.86 | 0.03386 | 26 | 1.02362 | 66 | 2.59843 |
| 0.07 | 0.00276 | 0.47 | 0.01850 | 0.87 | 0.03425 | 27 | 1.06299 | 67 | 2.63780 |
| 0.08 | 0.00315 | 0.48 | 0.01890 | 0.88 | 0.03465 | 28 | 1.10236 | 68 | 2.67717 |
| 0.09 | 0.00354 | 0.49 | 0.01929 | 0.89 | 0.03504 | 29 | 1.14173 | 69 | 2.71654 |
| 0.10 | 0.00394 | 0.50 | 0.01969 | 0.90 | 0.03543 | 30 | 1.18110 | 70 | 2.75591 |
| 0.11 | 0.00433 | 0.51 | 0.02008 | 0.91 | 0.03583 | 31 | 1.22047 | 71 | 2.79528 |
| 0.12 | 0.00472 | 0.52 | 0.02047 | 0.92 | 0.03622 | 32 | 1.25984 | 72 | 2.83465 |
| 0.13 | 0.00512 | 0.53 | 0.02087 | 0.93 | 0.03661 | 33 | 1.29921 | 73 | 2.87402 |
| 0.14 | 0.00551 | 0.54 | 0.02126 | 0.94 | 0.03701 | 34 | 1.33858 | 74 | 2.91339 |
| 0.15 | 0.00591 | 0.55 | 0.02165 | 0.95 | 0.03740 | 35 | 1.37795 | 75 | 2.95276 |
| 0.16 | 0.00630 | 0.56 | 0.02205 | 0.96 | 0.03780 | 36 | 1.41732 | 76 | 2.99213 |
| 0.17 | 0.00669 | 0.57 | 0.02244 | 0.97 | 0.03819 | 37 | 1.45669 | 77 | 3.03150 |
| 0.18 | 0.00709 | 0.58 | 0.02283 | 0.98 | 0.03858 | 38 | 1.49606 | 78 | 3.07087 |
| 0.19 | 0.00748 | 0.59 | 0.02323 | 0.99 | 0.03898 | 39 | 1.53543 | 79 | 3.11024 |
| 0.20 | 0.00787 | 0.60 | 0.02362 | 1.00 | 0.03937 | 40 | 1.57480 | 80 | 3.14961 |
| 0.21 | 0.00827 | 0.61 | 0.02402 | 1 | 0.03937 | 41 | 1.61417 | 81 | 3.18898 |
| 0.22 | 0.00866 | 0.62 | 0.02441 | 2 | 0.07874 | 42 | 1.65354 | 82 | 3.22835 |
| 0.23 | 0.00906 | 0.63 | 0.02480 | 3 | 0.11811 | 43 | 1.69291 | 83 | 3.26772 |
| 0.24 | 0.00945 | 0.64 | 0.02520 | 4 | 0.15748 | 44 | 1.73228 | 84 | 3.30709 |
| 0.25 | 0.00984 | 0.65 | 0.02559 | 5 | 0.19685 | 45 | 1.77165 | 85 | 3.34646 |
| 0.26 | 0.01024 | 0.66 | 0.02598 | 6 | 0.23622 | 46 | 1.81102 | 86 | 3.38583 |
| 0.27 | 0.01063 | 0.67 | 0.02638 | 7 | 0.27559 | 47 | 1.85039 | 87 | 3.42520 |
| 0.28 | 0.01102 | 0.68 | 0.02677 | 8 | 0.31496 | 48 | 1.88976 | 88 | 3.46457 |
| 0.29 | 0.01142 | 0.69 | 0.02717 | 9 | 0.35433 | 49 | 1.92913 | 89 | 3.50394 |
| 0.30 | 0.01181 | 0.70 | 0.02756 | 10 | 0.39370 | 50 | 1.96850 | 90 | 3.54331 |
| 0.31 | 0.01220 | 0.71 | 0.02795 | 11 | 0.43307 | 51 | 2.00787 | 91 | 3.58268 |
| 0.32 | 0.01260 | 0.72 | 0.02835 | 12 | 0.47244 | 52 | 2.04724 | 92 | 3.62205 |
| 0.33 | 0.01299 | 0.73 | 0.02874 | 13 | 0.51181 | 53 | 2.08661 | 93 | 3.66142 |
| 0.34 | 0.01339 | 0.74 | 0.02913 | 14 | 0.55118 | 54 | 2.12598 | 94 | 3.70079 |
| 0.35 | 0.01378 | 0.75 | 0.02953 | 15 | 0.59055 | 55 | 2.16535 | 95 | 3.74016 |
| 0.36 | 0.01417 | 0.76 | 0.02992 | 16 | 0.62992 | 56 | 2.20472 | 96 | 3.77953 |
| 0.37 | 0.01457 | 0.77 | 0.03032 | 17 | 0.66929 | 57 | 2.24409 | 97 | 3.81890 |
| 0.38 | 0.01496 | 0.78 | 0.03071 | 18 | 0.70866 | 58 | 2.28346 | 98 | 3.85827 |
| 0.39 | 0.01535 | 0.79 | 0.03110 | 19 | 0.74803 | 59 | 2.32283 | 99 | 3.89764 |
| 0.40 | 0.01575 | 0.80 | 0.03150 | 20 | 0.78740 | 60 | 2.36220 | 100 | 3.93701 |

| Band Colors | 1st Band | 2nd Band | 3rd Band | 4th Band | |
|---|---|---|---|---|---|
| Black ➡ | 0 | 0 | 0 | | |
| Brown ➡ | 1 | 1 | 10 | | |
| Red ➡ | 2 | 2 | 100 | | |
| Orange ➡ | 3 | 3 | 1,000 | | |
| Yellow ➡ | 4 | 4 | 10,000 | | |
| Green ➡ | 5 | 5 | 100,000 | | |
| Blue ➡ | 6 | 6 | 1,000,000 | | |
| Violet ➡ | 7 | 7 | — | | |
| Gray ➡ | 8 | 8 | — | | |
| White ➡ | 9 | 9 | — | | |
| Gold ➡ | — | — | 0.1 | ± 5% | |
| Silver ➡ | — | — | 0.01 | ± 10% | |
| No color ➡ | — | — | — | ± 20% | |
| | (A) | (B) | (C) | (D) | |

Arrows point to possible numerical values for each BAND COLOR depending upon position

| Example: | R | = | (A) | before | (B) | × | (C) | ± | (D)% |
|---|---|---|---|---|---|---|---|---|---|
| | | | Yellow | | Violet | | Orange | | Silver |
| | R | = | 4 | | 7 | × | 1,000 | ± | 10% |

Engineering notation
1,000 = K
10,000 = 10 K
100,000 = 100 K
1,000,000 = M

R = 47,000 Ω ± 10% or 47 KΩ ± 10%

## Material Safety Data Sheet
May be used to comply with
OSHA's Hazard Communication Standard,
29 CFR 1910.1200. Standard must be
consulted for specific requirements.

## U.S. Department of Labor
Occupational Safety and Health Administration
(Non-Mandatory Form)
Form Approved
OMB No. 1218-0072

IDENTITY (*As Used on Label and List*)

Note: Blank spaces are not permitted. If any item is not applicable, or no
information is available, the space must be marked to indicate that.

### Section I

| Manufacturer's Name | Emergency Telephone Number |
|---|---|
| Address (*Number, Street, City, State, and Zip Code*) | Telephone Number for Information |
| | Date Prepared |
| | Signature of Preparer (*optional*) |

### Section II – Hazardous Ingredients/Identity Information

| Hazardous Components (Specific Chemical Identity; Common Name(s)) | OSHA PEL | ACGIH TLV | Other Limits Recommended | % (*optional*) |
|---|---|---|---|---|
| | | | | |
| | | | | |
| | | | | |
| | | | | |
| | | | | |
| | | | | |
| | | | | |
| | | | | |
| | | | | |
| | | | | |

### Section III – Physical/Chemical Characteristics

| Boiling Point | | Specific Gravity (H$_2$O = 1) | |
|---|---|---|---|
| Vapor Pressure (mm Hg.) | | Melting Point | |
| Vapor Density (AIR = 1) | | Evaporation Rate (Butyl Acetate = 1) | |

Solubility in Water

Appearance and Odor

### Section IV – Fire and Explosion Hazard Data

| Flash Point (Method Used) | Flammable Limits | LEL | UEL |
|---|---|---|---|

Extinguishing Media

Special Fire Fighting Procedures

Unusual Fire and Explosion Hazards

(Reproduce locally)

OSHA 174, Sept. 1985

## Section V – Reactivity Data

| Stability | Unstable | | Conditions to Avoid |
|---|---|---|---|
| | Stable | | |

Incompatibility (*Materials to Avoid*)

Hazardous Decomposition or By-products

| Hazardous Polymerization | May Occur | | Conditions to Avoid |
|---|---|---|---|
| | Will Not Occur | | |

## Section VI – Health Hazard Data

Route(s) of Entry:     Inhalation?          Skin?          Ingestion?

Health Hazards (*Acute and Chronic*)

Carcinogenicity          NTP?          IARC Monographs?          OSHA Regulated?

Signs and Symptoms of Exposure

Medical Conditions
Generally Aggravated by Exposure

Emergency and First Aid Procedures

## Section VII – Precautions for Safe Handling and Use

Steps to Be Taken in Case Material is Released or Spilled

Waste Disposal Method

Precautions to Be Taken in Handling and Storing

Other Precautions

## Section VIII – Control Measures

Respiratory Protection (*Specify Type*)

| Ventilation | Local Exhaust | | Special |
|---|---|---|---|
| | Mechanical (*General*) | | Other |

| Protective Gloves | | Eye Protection |
|---|---|---|

Other Protective Clothing or Equipment

Work/Hygienic Practices

American National Standards Institute (ANSI)
11 W. 42nd Street, New York, NY 10018
Phone (212) 642-4900

Chemical Abstracts Service (CAS)
Division of American Chemical Society
Box 3012, Columbus, OH 43210
Phone (614) 421-3600

Environmental Protection Agency (EPA)
401 M Street, SW, Washington, DC 20460
Phone (202) 382-2080

National Fire Protection Association (NFPA)
1 Battery Park, Quincy, MA 02269
Phone (617) 770-3000

National Institute of Occupational Safety and Health (NIOSH)
4676 Columbia Parkway, Cincinnati, OH 45226
Phone (513) 533-8236
Telephone Hot Line 1-800-356-4674

Occupational Safety and Health Administration
(OSHA), Health Standards
200 Constitution Avenue, Washington, DC 20210
Phone (202) 523-7075

Substance Identification
(Identify chemical substance by CAS number and name)
Phone 1-800-848-6538

Superintendent of Documents
Government Printing Office
Washington, DC 20402
Phone (202) 783-3238

Distributors

Lab Safety Supply, Inc.
PO Box 1368, Janesville, WI 53547
Phone (608) 754-2345
Safety TechLine 1-800-356-2501

Global Occupational Safety
22 Harbor Park Drive, Dept. 7147
Port Washington, NY 11050
Phone 1-800-433-4848

# Bibliography

Allied Chemical Corporation Technical Data. Allied Chemical, Morristown, NJ.

*AutoCAD—Release 12 Handbook.* Autodesk, Inc., Sausalito, CA.

**Baer, C. J., and J. R. Ottaway.** *Electrical and Electronics Drafting,* 4th ed. McGraw-Hill Book Company, New York, 1980.

*Chemical Hazard Communication,* OSHA #3084. U.S. Department of Labor, Occupational Safety and Health Administration.

*Chemical Labels: Your Signposts to Safety,* #34G, 8/91 Edition. Business and Legal Reports Inc. Madison, CT.

**Coombs, C. F.** *Printed Circuit Handbook,* 2d ed. McGraw-Hill Book Company, New York, 1979.

**Coughlin, R. F., and F. F. Driscoll.** *Operational Amplifiers and Linear Integrated Circuits,* 3d ed. Prentice-Hall, Inc., Englewood Cliffs, NJ, 1987.

**Driscoll, F. F.** *Introduction to 6800/68000 Microprocessors.* Breton Publishers, North Scituate, MA, 1987.

**Driscoll, F. F.** *Microprocessor—Microcomputer Technology.* Breton Publishers, North Scituate, MA, 1983.

**Driscoll, F. F., R. F. Coughlin, and R. S. Villanucci.** *Data Acquisition and Process Control With the M68HC11 Microcontroller.* Macmillan Publishing Company, New York, 1994.

Dynachem Technical Data. Dynachem Corporation, Santa Fe Springs, CA.

*EAGLE 2.6 Manual.* CadSoft Computer, Inc., Delray Beach, FL, 1992.

*EIA Standards.* Electronics Industries Association, Washington, DC.
    EIA-PDP-100 *Mechanical Outlines for Registered Standard Electronic Parts.*
    EIA-JEP-95 *JEDEC Registered and Standard Outlines for Semiconductor Devices.*
    EIA-46 *Test Procedure for Resistance to Soldering for Surface Mount Devices.*
    EIA-49 *Solderability Test Procedures.*

**Fuller, J. E.** *Using AutoCAD® Release 12.* Delmar Publishers, Inc., Albany, NY, 1993.

*Hazard Communication Guidelines for Compliance,* OSHA #3111. U.S. Department of Labor, Occupational Safety and Health Administration.

*IPC Standards.* The Institute for Interconnecting and Packaging Electronic Circuits, Lincolnwood, IL.
    IPC-D-310 *Artwork Generation and Measurement Techniques.*
    IPC-A-600 *Acceptability of Printed Wiring Boards.*

IPC-R-700 *Modification and Repair for Printed Boards and Assemblies.*

IPC-M-675 *Microsectioning Handbook.*

IPC-CM-770 *Guidelines for Printed Circuit Component Mounting.*

IPC-SM-780 *Electronic Component Packaging and Interconnection With Emphasis on Surface Mounting.*

IPC-SM-782 *Surface Mount Land Patterns.*

IPC-S-804 *Solderability Test Methods for Printed Wiring Boards.*

IPC-S-805 *Solderability Test for Component Leads and Terminations.*

IPC-S-815 *General Requirements for Soldering Electronic Interconnections.*

IPC-SM-817 *General Requirements for Surface Mount Adhesives.*

IPC-SF-818 *General Requirements for Electronic Soldering Fluxes.*

IPC-SM-819 *General Requirements and Test Methods for Electronic-Grade Solder Paste.*

*Kahn Companies Technical Data.* Kahn Companies, Wethersfield, CT.

*LabTech Acquire Handbook.* Laboratory Technologies Corporation, Wilmington, MA, 1989.

*Lea Ronal Technical Data.* Lea Ronal, Freeport, NY.

**Lindsey, D.** *The Design and Drafting of Printed Circuits,* 2d ed. Bishop Graphics, Inc., Westlake Village, CA, 1982.

**Manko, H. H.** *Soldering Handbook for Printed Circuits and Surface Mounting.* Van Nostrand Reinhold, New York, 1986.

**Military Standard MIL-STD-275.** *Printed Wiring for Electronic Equipment.*

*MSDS Pocket Dictionary,* Edited by Joseph O. Accrocco. Genium Publishing Corporation, Schenectady, NY, 1991.

*MSDS - Your Guide to Chemical Safety,* #200 039 00 9/91 Edition. Business and Legal Reports Inc., Madison, CT.

*nuArc Technical Data.* nuArc Company, Inc., Chicago.

**Prasad, R. P.** *Surface Mount Technology Principles and Practice.* Van Nostrand Reinhold, New York, 1989.

PSpice Circuit Analysis Manual. MicroSim Corporation, Irvine, CA.

*SCR Manual,* 4th ed. General Electric Company, Auburn, NY.

*Shipley Technical Data.* Shipley Company, Inc., Newton, MA.

*6B Series User's Manual.* Analog Devices Corporation, Norwood, MA, 1989.

**Villanucci, R. S.** Adding one breakpoint linearizes thermistor, *EDN,* 7 January 1993.

**Villanucci, R. S.** Using computer-assisted techniques to design and analyze nonlinear analog circuits, *ASEE Journal of Engineering and Technology,* Spring 1993.

# Index